Saas-Fee Advanced Course 25
Lecture Notes 1995

Springer
*Berlin
Heidelberg
New York
Barcelona
Budapest
Hong Kong
London
Milan
Paris
Santa Clara
Singapore
Tokyo*

S. D. Kawaler   I. Novikov   G. Srinivasan

Stellar Remnants

Saas-Fee Advanced Course 25
Lecture Notes 1995
Swiss Society for Astrophysics and Astronomy

Edited by Georges Meynet and Daniel Schaerer

With 120 Figures

 Springer

Professor S. D. Kawaler
Department of Physics and Astronomy, Iowa State University,
Ames, IA 50011, USA

Professor I. Novikov
Theoretical Astrophysics Center
Blegdamsvej 17, DK-2100 Copenhagen 0, Denmark

Professor G. Srinivasan
Raman Research Institute (RRI), Sadashivanagar,
C.V. Raman Avenue, Bangalore 560 080, India

Volume Editors:

Dr. Georges Meynet
Dr. Daniel Schaerer
Observatoire de Genève, 51 chemin des Maillettes,
CH-1290 Sauverny, Switzerland

This series is edited on behalf of the Swiss Society for Astrophysics and Astronomy:

Société Suisse d'Astrophysique et d'Astronomie
Observatoire de Genève, ch. des Maillettes 51, CH-1290 Sauverny, Switzerland

Cover photograph: Ground-based image of the supernova remnant of the Crab Nebula *(left)* and HST image of the central Crab Pulsar *(right)*. Courtesy of Jeff Hester and Paul Scowen (Arizona State University), and NASA.

Library of Congress Cataloging-in-Publication Data
Kawaler, Steven D.
Stellar remnants / S. D. Kawaler, I. Novikov, G. Srinivasan; edited by Georges Meynet and Daniel Schaerer.
p. cm. – (Saas-Fee advanced course 25 lecture notes; 1995) Includes bibliographical references and index.
ISBN 3-540-61520-2 (hc: alk. paper)
1. Neutron stars – Congresses. 2. White dwarfs – Congresses. 3. Black holes (Astronomy) – Congresses.
4. Stars – Evolution–Congresses. I. Novikov, I. D. (Igor'Dmitrievich) II. Srinivasan, G. (Ganesan), 1942– .
III. Meynet, G. IV. Schaerer, Daniel. V. Title. VI. Series: Saas-Fee advanced course ...lecture notes; 1995.
QB843.N4K39 1996 523.8'87–dc20 96-45985 CIP

ISBN 3-540-61520-2 Springer-Verlag Berlin Heidelberg New York

This work is subject to copyright. All rights are reserved, whether the whole or part of the material is concerned, specifically the rights of translation, reprinting, reuse of illustrations, recitation, broadcasting, reproduction on microfilm or in any other way, and storage in data banks. Duplication of this publication or parts thereof is permitted only under the provisions of the German Copyright Law of September 9, 1965, in its current version, and permission for use must always be obtained from Springer-Verlag. Violations are liable for prosecution under the German Copyright Law.

© Springer-Verlag Berlin Heidelberg 1997
Printed in Germany

The use of general descriptive names, registered names, trademarks, etc. in this publication does not imply, even in the absence of a specific statement, that such names are exempt from the relevant protective laws and regulations and therefore free for general use.

Data conversion by Springer-Verlag
SPIN 10519556   55/3144 - 5 4 3 2 1 0 - Printed on acid-free paper

Preface

The 25th "Saas-Fee" Advanced Course organized by the Swiss Society for Astronomy and Astrophysics (SSAA), which was held in Les Diablerets from April 3rd to 8th, 1995, had for subject "Stellar Remnants".

End points of stellar evolution, white dwarfs, neutron stars and black holes, give researchers the unique opportunity to explore the consequences of extreme physical conditions never met in the laboratory. At the crossroads of quantum and relativistic effects, the study of these objects offers the possibility to achieve a better understanding of these two theories. From the astrophysical point of view, which is the one emphasized in this book, condensed objects are at the center of numerous observed processes, as for instance novæ, pulsars, X-ray binaries, γ-ray bursts, active galactic nuclei ... An understanding of these fascinating phenomena requires a very good knowledge of the main physical processes occurring inside and/or in the vicinity of these compact stellar remnants.

The three lecturers, S. Kawaler from Ames, I. Novikov from Copenhagen and G. Srinivasan from Bangalore, made a wonderful effort to present in the clearest way the developments in their field of research and succeeded to maintain, despite the concurrence of snow and sun, a high level of interest throughout the whole course. We hope that the reader will also perceive the pleasure we had listening to the lecturers.

76 participants of 16 nationalities attended the course. 31 of them represented Swiss institutes. Five participants from Eastern European countries benefited from a fellowship of the Swiss National Fund for Scientific Research.

The Eurotel provided its well-known and very much appreciated hospitality. Sun and snow were also present and contributed to the success of the course. A trumpet and organ concert sponsored by the Swiss Society for Astronomy and Astrophysics was organized by Paul Bartholdi in the church of Vers-l'Eglise. The performance of the two musicians, Patrick Chappuis and Pierre Pilloud, was highly appreciated and provided the highlight to celebrate the quarter of a century of the SSAA Advanced Courses.

We would like to extend our thanks to all those who helped in the organization of this course, in particular to Mrs. I. Scheffre and Mrs. F. Buchschacher, whose administrative support was extremely helpful.

Geneva
May 1996

Georges Meynet
Daniel Schaerer

Contents

White Dwarf Stars
By *Steven D. Kawaler* (With 37 figures)

1 **Introduction** 1
   1.1 White Dwarfs as Useful Stars 2
   1.2 Origins: the Clue of White Dwarf Masses 2
   1.3 The Main Channel 3
   1.4 Why Such a Narrow Mass Distribution? 4
2 **Observed Properties of White Dwarfs** 6
   2.1 Discovery of White Dwarfs 6
   2.2 Finding White Dwarfs 7
   2.3 White Dwarf Colors and the White Dwarf Luminosity Function 8
   2.4 White Dwarf Optical Spectra 10
   2.5 Distribution of Spectral Types with Effective Temperatures 16
   2.6 Magnetic White Dwarfs 19
   2.7 Pulsating White Dwarfs 20
3 **Physics of White Dwarf Interiors** 21
   3.1 Equation of State 21
   3.2 Heat Transport in Degenerate Matter 28
   3.3 Nonideal Effects 30
   3.4 Specific Heat 31
4 **White Dwarf Formation and Early Cooling** 32
   4.1 Thermal Pulses on the AGB 32
   4.2 Departure from the AGB 36
   4.3 The PNN Phase 39
   4.4 Nuclear Shutdown and Neutrino Cooling 41
5 **Chemical Evolution of White Dwarfs** 43
   5.1 Diffusive Processes 44
   5.2 Accretion of "Fresh" ISM vs. Mass Loss 47
   5.3 Convection 48
   5.4 Chemical Evolution Scenarios 50
6 **White Dwarf Cooling and the White Dwarf Luminosity Function** 52
   6.1 A Simplified Cooling Model 52
   6.2 Complications: Neutrinos and Crystallization 54
   6.3 Realistic Cooling Calculations 55
   6.4 Construction of Theoretical Luminosity Functions 58
   6.5 The Age of the Galactic Disk 59

7	**Nonradial Oscillations of White Dwarfs: Theory**	61
	7.1 Review of Observations	62
	7.2 Hydrodynamic Equations	62
	7.3 Local Analysis and the Dispersion Relation	66
	7.4 g-mode Period Spacings	66
	7.5 Mode Trapping	67
	7.6 Rotational and Magnetic Splitting	69
	7.7 The Seismological Toolbox	70
8	**Pulsating White Dwarfs**	71
	8.1 The Whole Earth Telescope	71
	8.2 PG 1159 Stars and Pulsating PNNs	72
	8.3 GD 358: A Pulsating DB White Dwarf	76
	8.4 The ZZ Ceti Stars	78
9	**Astrophysical Applications of White Dwarfs**	78
	9.1 Stellar Evolution as a Spectator Sport	78
	9.2 The White Dwarf Luminosity Function and Our Galaxy	82
	9.3 White Dwarfs and Cluster Ages	83
	9.4 The Planetary Nebula Luminosity Function and Galaxy Distances	84
	9.5 Driving and Damping of Pulsations and Convective Efficiency in White Dwarfs Ceti Stars	87
	9.6 Final Thoughts	90
	References	91

Neutron Stars

By *G. Srinivasan* (With 67 figures)

1	**A Historical Introduction**	97
	1.1 What Are the Stars?	97
	1.2 Why Are the Stars as They Are?	99
	1.3 White Dwarfs	101
	1.4 Can All Stars Find Peace?	104
	1.5 Neutron Stars and Supernovae	107
	1.6 Black Holes	109
2	**The Nature of Pulsars**	110
	2.1 Phenomenology	110
	2.2 The Pulsar Population	111
	2.3 The Emission Mechanism	111
	2.4 Pulsars as a Dynamo	117
	2.5 The Near and Wind Zone	118
	2.6 The Dipole Radiator	121
	2.7 Gaps, Sparks and Pairs	121
	2.8 Coherence of the Radio Radiation	125
	2.9 Period Evolution	126
	2.10 High-Frequency Radiation	128

2.11 The Crab Nebula	133
3 The Physics of Neutron Stars	139
3.1 Internal Structure	140
3.2 The Equation of State	142
3.3 Stability of Stars	147
3.4 Neutron Star Models	149
3.5 The Observed Masses of Neutron Stars	155
3.6 Exotic States of Matter	157
3.7 Cooling of Neutron Stars	163
4 The Progenitors of Neutron Stars	167
4.1 Pulsar Birth Rate	171
4.2 Pulsar Current	171
4.3 Do Pulsars Trace Spiral Arms?	177
5 Binary Pulsars	181
5.1 Spinning Up a Star!	185
5.2 The Population of Binary Pulsars	189
5.3 Are Many Pulsars Processed in Binaries?	189
5.4 Pulsar Velocities	191
5.5 Migration from the Plane	193
5.6 The Origin of Pulsar Velocities	194
6 Millisecond Pulsars	196
6.1 The Moral of Millisecond Pulsars	200
6.2 A New Population of Gamma-Ray Sources?	204
7 Magnetic Field Evolution	205
7.1 The Nature of the Field	206
7.2 Field Decay	209
7.3 Mechanism of Field Decay	213
8 Glitches	218
9 Plate Tectonics	224
9.1 Stresses on the Crustal Lattice	225
9.2 Plate Tectonics	228
9.3 Some Astrophysical Consequences	228
Acknowledgements	232
References	232

Black Holes

By *Igor Novikov* (With 16 figures)

1 Astrophysics of Black Holes	237
1.1 Introduction	237
1.2 The Origin of Stellar Black Holes	237
2 A Nonrotating Black Hole	239
2.1 Introduction	239
2.2 Schwarzschild Gravitational Field	240
2.3 Motion of Photons Along the Radial Direction	243

	2.4	Radial Motion of Nonrelativistic Particles	244
	2.5	The Puzzle of the Gravitational Radius	245
	2.6	R and T Regions	248
	2.7	Two Types of T-Regions	250
	2.8	Gravitational Collapse and White Holes	251
	2.9	Eternal Black Hole?	254
	2.10	Black Hole Celestial Mechanics	258
	2.11	Circular Motion Around a Black Hole	261
	2.12	Gravitational Capture of Particles by a Black Hole	263
	2.13	Corrections for Gravitational Radiation	264
3	**A Rotating Black Hole**		266
	3.1	Introduction	266
	3.2	Gravitational Field of a Rotating Black Hole	267
	3.3	Specific Reference Frames	269
	3.4	General Properties of the Spacetime of a Rotating Black Hole; Spacetime Inside the Horizon	275
	3.5	Celestial Mechanics of a Rotating Black Hole	278
	3.6	Motion of Particle in the Equatorial Plane	278
	3.7	Motion of Particles off the Equatorial Plane	281
	3.8	Peculiarities of the Gravitational Capture of Bodies by a Rotating Black Hole	281
4	**Electromagnetic Fields Near a Black Hole**		284
	4.1	Introduction	284
	4.2	Maxwell's Equations in the Neighborhood of a Rotating Black Hole	285
	4.3	Stationary Electrodynamics	287
	4.4	Boundary Conditions at the Event Horizon	296
	4.5	Electromagnetic Fields in Vacuum	301
	4.6	Magnetosphere of a Black Hole	304
5	**Some Aspects of Physics of Black Holes, Wormholes, and Time Machines**		311
6	**Observational Appearance of the Black Holes in the Universe**		312
	6.1	Black Holes in the Interstellar Medium	312
	6.2	Disk Accretion	315
	6.3	Black Holes in Stellar Binary Systems	317
	6.4	Black Holes in Galactic Centers	320
	6.5	Dynamical Evidence for Black Holes in Galaxy Nuclei	322
7	**Primordial Black Holes**		324
	Acknowledgements		327
	References		327
	Index		335

List of Previous Saas–Fee Advanced Courses:

* 1971 Theory of the Stellar Atmospheres
    D. Mihalas, B. Pagel, P. Souffrin

* 1972 Interstellar Matter
    N.C. Wickramasinghe, F.D. Kahn, P.G. Mezger

* 1973 Dynamical Structure and Evolution of Stellar Systems
    G. Contopoulos, M. Hénon, D. Lynden–Bell

* 1974 Magnetohydrodynamics
    L. Mestel, N.O. Weiss

* 1975 Atomic and Molecular Processes in Astrophysics
    A. Dalgarno, F. Masnou–Seeuws, R.V.P. McWhirter

* 1976 Galaxies
    K. Freeman, R.C. Larson, B. Tinsley

* 1977 Advanced Stages in Stellar Evolution
    I. Iben jr., A. Renzini, D.N. Schramm

* 1978 Observational Cosmology
    J.E. Gunn, M.S. Longair, M.J. Rees

* 1979 Extragalactic High Energy Astrophysics
    F. Pacini, C. Ryter, P.A. Strittmatter

* 1980 Star Formation
    J. Appenzeller, J. Lequeux, J. Silk

! 1981 Activity and Outer Atmospheres of the Sun and Stars
    F. Praderie, D.S. Spicer, G.L. Withbroe

* 1982 Morphology and Dynamics of Galaxies
    J. Binney, J. Kormendy, S.D.M. White

! 1983 Astrophysical Processes in Upper Main Sequence Stars
    A.N. Cox, S. Vauclair, J.P. Zahn

! 1984 Planets, Their Origin, Interior and Atmosphere
    D. Gautier, W.B. Hubbard, H. Reeves

! 1985 High Resolution in Astronomy
    R.S. Booth, J.W. Brault, A. Labeyrie

! 1986 Nucleosynthesis and Chemical Evolution
    J. Audouze, C. Chiosi, S.E. Woosley

!	1987	Large Scale Structures in the Universe *A.C. Fabian, M. Geller, A. Szalay*
!	1988	Radiation in Moving Gaseous Media *H. Frisch, R.P. Kudritzki, H.W. Yorke*
!	1989	The Milky Way as a Galaxy *G. Gilmore, I. King, P. van der Kruit*
*	1990	Active Galactic Nuclei *R. Blandford, H. Netzer, L. Woltjer*
!!	1991	The Galactic Interstellar Medium *W.B. Burton, B.G. Elmegreen, R. Genzel*
*	1992	Interacting Binaries *S.N. Shore, M. Livio, E.J.P. van den Heuvel*
!!	1993	The Deep Universe *A.R. Sandage, R.G. Kron, M.S. Longair*
!!	1994	Plasma Astrophysics *J.G. Kirk, D.B. Melrose, E.R. Priest*
**	1996	Galaxies Interactions and Induced Star Formation *Joshua Barnes, Robert Kennicutt, François Schweizer*

---

* Out of print

! May be ordered from **Geneva Observatory**

 Saas–Fee Courses
 Geneva Observatory
 CH–1290 SAUVERNY / Switzerland

!! May be ordered from **Springer-Verlag**

** In preparation

White Dwarf Stars

Steven D. Kawaler

Department of Physics and Astronomy, Iowa State University, Ames, IA 50011 USA

1 Introduction

White dwarf stars represent the final stage of evolution of stars like our sun. They represent the ultimate fate of all stars with masses less than about 8 $M_\odot$, and are the natural consequence of the finite fuel supply of these stars. Upon exhaustion of their nuclear fuels of hydrogen and helium, these stars lack sufficient mass to take advantage of the limited return of carbon fusion, and are doomed to gravitational collapse. Upon reaching a radius comparable to that of the Earth, the inner cores of these stars support themselves almost entirely by electron degeneracy pressure. Their final transformation is through the gradual release of heat that has been stored within them during the prior stages of nuclear burning.

The basic theory of white dwarf stars was stimulated by the discovery, early in this century, of a class of stars with observable luminosity but very small radius. The solution to this mystery, by Chandrasekhar and Fowler, neatly integrated the then–new sciences of quantum mechanics and relativity with astrophysics. Their work was essentially a complete description of the inner workings of cool white dwarf stars, and the Nobel Prize in physics justly went to S. Chandrasekhar in recognition of his contribution to stellar structure theory. While Chandrasekhar's theory is indeed elegant, in the later half of this century we have come to recognize that these stars represent a rich storehouse of information on the evolution of all stars. The devil, as they say, is in the details.

There is a simple and compelling reason to care about white dwarf stars. The initial mass function (IMF) of stars in the galaxy, $\Psi_s(M)dM$ is the number of stars formed per year per cubic parsec within an interval of masses between M and $M + dM$. Salpeter (1955) found that

$$\Psi_s\, dM = 2 \times 10^{-12} M^{-2.35}\, dM \ \ \mathrm{stars/yr/pc^3}, \tag{1}$$

which is the famous (and well–tested) "Salpeter mass function". White dwarf stars are the remnants of stars with initial main sequence masses between $0.5 M_\odot$ and $8 M_\odot$. Thus

$$\frac{\#\ \mathrm{of\ proto-WD\ stars}}{\#\ \mathrm{of\ proto-NS/BH\ stars}} = \frac{\int_{0.5}^{8} \Psi_s\, dM}{\int_{8}^{\infty} \Psi_s\, dM}. \tag{2}$$

I leave it as an exercise to the reader to show that, using the Salpeter mass function, the above equation shows that *98% of all stars are or will be white dwarfs !* To be fair, one might wish to consider not the number of stars, but the mass of the stars. In that case, 94% of all matter in stars is either already locked-up in white dwarfs, or is in stars that will eventually become white dwarfs; a mere 6% of matter is destined to be, or already incorporated in neutron stars, and an insignificant amount is on the black hole path. Therefore, the study of white dwarf stars is really the study of stars, to a high level of precision...

1.1 White Dwarfs as Useful Stars

The relative numbers of white dwarfs per unit luminosity interval provides important information on the rate of formation of white dwarfs as well as the rate at which they cool. Since the cooling rate decreases with decreasing luminosity, if we assume a constant rate of white dwarf formation there would be an ever-increasing number of white dwarfs that lie within fixed ranges of luminosity. However, the observed luminosity function turns downwards at luminosities below about $10^{-4}L_\odot$ (Liebert et al. 1988). This deficit is caused by the finite age of our galaxy. Simply put, the oldest white dwarfs in our galaxy have had a long but finite time to cool. Comparison of the observed luminosity function with theoretical cooling rates therefore provides a determination of the age of the oldest white dwarfs and therefore a lower limit for the age of the Universe (Winget et al. 1987). The current best estimate for this number is between 8 and 11 Gyr (Wood 1992).

At the high–luminosity part of the white dwarf luminosity function, the number of stars per luminosity interval is also a reflection of the rate of cooling (or fading). At the hot end, though, this rate is determined by the nuclear and/or gravitational luminosity sources present in the stars. Shutdown of nuclear burning, the species of nuclear burning, the rate of neutrino emission, and the gravitational luminosity all contribute to the observed luminosity function. The principal motivation for exploring the hot end of the luminosity function, however, comes from studies of extragalactic planetary nebulae (PN). George Jacoby and coworkers (1992 and references therein) have shown that the PNs in the Milky Way and external galaxies have very similar luminosity functions, and in particular show a sharp drop at a fixed luminosity. Thus by observing the planetary nebulae nebulosity function (PNLF) in external galaxies and determining the apparent magnitude of the cutoff, they can estimate the distance to that galaxy. Efforts to calibrate the PNLF with galactic and Magellanic Cloud PNs have provided PNLFs that can also be used to constrain the early stages of white dwarf evolution (Dopita et al. 1992).

1.2 Origins: the Clue of White Dwarf Masses

An excellent review of the ways of determining the masses of white dwarf stars, and the results of mass determinations, can be found in Weidemann (1990). Briefly, the mass distribution of white dwarfs is much narrower than that of the progenitor stars. The mean mass for single white dwarfs in the solar neighborhood is about $0.56 M_\odot$ with a width of about $0.14 M_\odot$ (Bergeron et al. 1992). Mass determinations for single white dwarfs rely on spectroscopy. One technique employs the modest gravitational redshift of the hydrogen lines to determine the mass, while the other uses the spectroscopic determination of the surface gravity. Both assume as known the mass–radius relation for white dwarfs. Since the gravitational redshift and gravities are difficult quantities to measure, and the mass-radius relation is in fact temperature dependent, individual mass determinations for white dwarf stars are usually of low precision. However, with so many objects in the data set, the mean mass and the form of the distribution can still be determined with confidence (Bergeron et al. 1992).

An interesting (and still open) question is whether the mass distributions are different for DB (helium–rich) and DA (hydrogen–rich) white dwarfs. Unfortunately the lack of precision in determining masses prevents any firm conclusions from being drawn.

1.3 The Main Channel

Following exhaustion of hydrogen in the core of main sequence stars with masses between $0.8 M_\odot$ and about $8 M_\odot$, helium eventually ignites in the center. With exhaustion of helium, these stars climb the Asymptotic Giant Branch (AGB) with increasing luminosities and decreasing temperatures. The source of the stellar luminosity is the burning of hydrogen and helium in thin shells surrounding a degenerate core composed of carbon and oxygen (and perhaps neon and magnesium near the upper mass limit). The degenerate core contains between 0.5 and 1.4 $M_\odot$ of material in a region with a radius similar to that of the Earth. The remainder of the stellar mass lies in an enormously extended cool envelope, reaching radii of several hundred $R_\odot$. At this time the gravitational binding of the envelope is extremely weak, so that radiative processes along with energy input through pulsations causes them to lose mass at a rate that increases with luminosity. Thus the core of an AGB star is a proto-white dwarf which is slowly shedding the surrounding cocoon of the stellar envelope. Evolution up the AGB features shell flashes: somewhat periodic instabilities which cause mild thermonuclear runaways in the helium burning shells For an excellent overview of evolution leading up to the white dwarf phase, and indeed most aspects of stellar evolution, see Iben (1991).

Along the AGB the luminosity of the star depends primarily on the mass of the degenerate core, as discovered by Paczynski (1970) and Uus (1970):

$$\frac{L}{L_\odot} \approx 60,000 \left(\frac{M_c}{M_\odot} - 0.5 \right) \tag{3}$$

Eventually, with the rate of mass loss approaching $10^{-4} M_\odot/\text{yr}$, the remaining outer envelope is lost to the star. Once most of the envelope has been lost, and the envelope contains less than about 1% of the mass of the core, the effective temperature increases very rapidly with decreasing envelope mass. Evolution to high $T_{\rm eff}$ occurs at almost constant luminosity, resulting in exposure of the white–hot core along a horizontal evolutionary path in the H–R diagram.

The rate at which the newly born white dwarf crosses the H–R diagram is a sensitive function of its mass. For those with masses near $0.6M_\odot$, evolution to $T_{\rm eff} > 10^5$ K occurs in about 10^4 years. Within that time, the remnant AGB envelope material is close enough to the hot star that it can be largely ionized by the UV flux from the star, and it becomes a planetary nebula. Thus certain newly formed white dwarfs are identified with "neon" signs, simplifying identification of these rapidly–evolving objects. Other very young and rapidly evolving white dwarfs without associated nebulae are usually discovered in surveys for anomalously blue stars, such as quasar surveys like the Palomar–Green survey (Green et al. 1986).

From theoretical evolution studies as well as space density arguments, AGB stars are the most likely progenitors for most white dwarf stars. However, other channels into the white dwarf cooling track must exist. In particular, normal AGB evolution can only produce white dwarfs with masses greater than about $0.5M_\odot$, but the observed mass distributions all show significant numbers of white dwarfs with smaller masses. Binary star evolution can produce low–mass white dwarfs, even with degenerate helium cores (Iben 1991). In addition, low mass post–horizontal branch evolution can produce white dwarfs without evolution proceeding all the way to the AGB. Such failed AGB stars are those which lose significant amounts of mass up to and through the core helium flash, and are left with thin envelopes that cannot expand to large radii (Dorman et al. 1993). For the remainder of this review, however, I will concentrate on those white dwarfs which result from AGB evolution.

1.4 Why Such a Narrow Mass Distribution?

The narrow distribution in mass of isolated white dwarfs is remarkable. Given the many things that happen to stars between the main sequence and white dwarf formation, why should the final distribution of masses be so narrow? And why should they cluster around a mass near $0.6M_\odot$? The fact that shell flashes, pulsation, and mass loss on the AGB can indeed conspire to funnel this primary white dwarf formation channel give considerable credibility to the entire scenario. Here, I would like to outline the arguments for such "convergent" evolution.

The current idea is that mass loss on the AGB is driven by large–amplitude radial pulsation as Mira variables. The models of Bowen & Willson (1991), discussed later in this paper, show that mass loss increases with increasing luminosity on the AGB until mass loss is sufficient to bring the hydrogen–burning shell to within a few percent of the stellar mass of the surface. During this stage, a *crude* fit of the Bowen models gives a mass loss rate of

$$\dot{M} \approx -2 \times 10^{-8} L_{1000}^{3.7} \left(\frac{M}{M_\odot}\right)^{-3.1} R_{100} \frac{M_\odot}{\rm yr}, \qquad (4)$$

where L_{1000} is the luminosity in units of $1000L_\odot$, and R_{100} is the stellar radius in units of $100R_\odot$ (adapted from Blöcker 1993). From the mass loss rate, we can derive a time scale for mass decrease of the stellar envelope by simply taking the inverse of the mass loss rate divided by the mass of the envelope:

$$\tau_{env} = \left(\frac{\dot{M}}{M_{env}}\right)^{-1}. \tag{5}$$

While on the AGB, the proto–white–dwarf (the AGB core) is growing as the result of processing of hydrogen (and helium) through the nuclear burning shells. If we assume that the stellar luminosity is provided by hydrogen burning, then the rate of growth of the core is proportional to the luminosity and inversely proportional to the efficiency of energy production:

$$\dot{M}_c = \frac{L}{0.007\, c^2} \tag{6}$$

or, in convenient units

$$\dot{M}_c = 10^{-8} L_{1000}\, \frac{M_\odot}{\text{yr}} \tag{7}$$

Now we can use these two relations to approximate the conditions at the birth of a white dwarf. A white dwarf will be born when the time scale for envelope shrinkage becomes shorter than the time scale for growth of the core. After that time, the core is effectively frozen at its current size, and envelope mass loss will expose the core. The mass of the core will be at least $0.5 M_\odot$; this is the minimum mass for core helium ignition during earlier evolution. This value provides a lower limit for the time scale of core growth of

$$\tau_c \geq 5 \times 10^7 L_{1000}^{-1}\ \text{yr}. \tag{8}$$

The luminosity (and radius) at which this time scale becomes, say, ten times longer than the time scale for envelope shrinkage gives us the luminosity at which the white dwarf forms, as a function of the mass of the star.

$$L_{1000} \approx 2.35 M^{1.5} R_{100}^{-0.4}. \tag{9}$$

Now, to obtain the core mass at this luminosity, we use the Paczynski mass–luminosity relation (3), obtaining

$$\frac{M_c}{M_\odot} \approx 0.5 + \left[0.04 \left(\frac{M}{M_\odot}\right)^{1.5} R_{100}^{-0.37}\right]. \tag{10}$$

In deriving the above equation, I cheated in assuming that the stellar mass is constant when, in fact, it is clearly decreasing through AGB evolution. As the mass on the AGB decreases, the mass loss rate increases and τ_{env} decreases. Accounting for this would effectively lower the luminosity threshold for white dwarf formation. This equation does illustrate the convergent behavior of the combined process of mass loss and core growth. A factor of 4 change in the "initial" mass results in a maximum change in the final core mass of only 0.3 solar masses. With more careful treatment of the mass in the envelope time scale, the process becomes even more convergent on a single mass with a small dispersion.

2 Observed Properties of White Dwarfs

2.1 Discovery of White Dwarfs

Perhaps fittingly, the first human being to actually see a white dwarf star was probably William Herschel. One of Herschel's many achievements included his catalog of double stars, which was published in two parts in 1782 and 1785 (Herschel 1782, 1785); he later published a paper proving that the motions of some multiple stars must result from orbital motions about a common center of gravity. In his second catalog, he describes observations he made on the evening of 31 January 1783 of a new double star in the constellation Eridanus. This double star was about $1\frac{1}{3}'$ southeast of the star 40 Eri. The components were very unequal in brightness, and difficult to separate; the fainter star lay at a position angle $56°42'$ north and west of the brighter star. We now know this pair as 40 Eri B/C, and the brighter component is a DA white dwarf star. Of course, Herschel did not recognize 40 Eri B as an unusual star; the identification of it and other similar stars as peculiarly dense required more advanced technology and physics than available to him. The story of the discovery of the special nature of white dwarfs parallels the development of astronomy from a taxonomical pursuit into a mature physical science that rides on the cutting edge of theoretical physics.

The story continues with the work of F.W. Bessel. In the years from 1834 to 1843 he detected a departure from linear proper motion in Sirius and Procyon. Upon investigation of older measurements, he stated in a letter to the Royal Astronomical Society, communicated by Sir John Herschel, that "it follows ... that a period, not very different from that of a half century, would serve ... for a sufficient explanation of the observations." The true period of Sirius A/B is now known as 50.09 years. As the first stars to exhibit peculiar proper motions, Bessel considered various possibilities to explain the motions, and concluded that "If we were to regard *Sirius* and *Procyon* as double stars, the change of their motions would not surprise us... The existence of numberless visible stars can prove nothing against the existence of numberless invisible ones" (Bessel 1844).

The "invisible" companion to Sirius was, of course, eventually detected. Precisely 81 years after Herschel's discovery of 40 Eri B/C, in the process of testing a newly completed $18''$ refractor Alvan Clark Jr. discovered a faint companion to Sirius. This discovery proved Bessel's remarkable assertion. Between 1844 and 1862 it was found that the mass of the companion had to be about the same as the Sun; yet the companion shined with merely 10^{-4} times the brightness of Sirius itself.

Henry Norris Russell relates the next step in a remarkable prelude to a paper on white dwarfs that appeared much later (Russell 1944). In 1910, he and E.C. Pickering were discussing the spectra of stars with large parallax; all were turning out to be G stars or cooler. Upon considering the companion of 40 Eri, Pickering "telephoned to Mrs. (Willamina Paton) Fleming who reported within an hour or so, that it was of Class A. I (Russell) saw enough of the physical implications of this to be puzzled, and expressed some concern. Pickering smiled and said, 'It is just such discrepancies which lead to the increase of our knowledge.' Never has the soundness of his judgment better illustrated." The first correspondence of the spectral type of 40 Eri B was in 1915, in a footnote to a paper by Hertzsprung (1915).

Thus the first white dwarf to be identified as such was 40 Eri B, in 1910. Within four years the spectral type of Sirius B was ascertained by Adams (1914, 1915) and that of the dwarf van Maanen 2 by van Maanen (1913), certifying them as additional members of the new class of stars known as white dwarfs.

2.2 Finding White Dwarfs

The relatively recent discovery of white dwarfs is evidence that they are relatively hard to find. In fact, identification of white dwarf stars can be made in a number of ways. Since they are all so under–luminous, almost all white dwarfs that we know of are in the immediate solar neighborhood; 50% of known white dwarfs lie within 24 parsecs of the Sun. White dwarf discoveries have come from surveys of stars with large proper motions, searches for faint blue objects, and examination of faint members of common proper motion pairs. All three produce candidate objects that require spectroscopic follow-up observations for confirmation.

The basic catalog of white dwarf stars, which is indispensable for all workers in the field, is that of McCook & Sion (1987). Rumor has it that this is to be updated shortly, but the 1987 catalog (available electronically as well as in print) is essential reference material in this field.

Surveys of stars with large proper motions include Luyten's *Half-Second Catalog* (Luyten 1979). Spectroscopic follow-up has identified a large number of white dwarf stars. However, the selection effects involved in such surveys are thorny to account for, and make statistical studies suspect.

Common proper motion binaries provide another discovery channel. In this procedure, proper motion catalogs are scanned for pairs of stars with similar proper motions, but possibly large angular separation (i.e. Luyten 1979, 1969 and Giclas et al. 1978, 1971). Such surveys can be claimed to be complete; for example, the survey analyzed by Oswalt & Smith (1995) is complete for pairs with proper motions greater than about 0.2″/yr and photographic magnitudes brighter than +18.

The most successful recent surveys for producing new white dwarf identifications have been surveys for faint but excessively blue stars. A prime example is the Palomar–Green (PG) survey (Green et al. 1986) for blue objects out of the galactic plane. While the PG survey (and others like it) are designed for discovery of quasars, they are almost optimally designed to pick out white dwarf stars. Once again, spectroscopic follow-up is required, but such surveys have an advantage in that they produce magnitude–limited samples, and therefore are without the peculiar selection effects associated with proper–motion identifications. With the success of the the PG survey, others are recently underway, such as the Montreal–Cambridge survey (Demers et al. 1986) and the Edinburgh–Cape survey (Stobie et al. 1987, Kilkenny et al. 1991). These all sample a large area of the sky. Another approach is that of Claver (1995), who concentrates on smaller fields, but goes extremely deep using a wide–field CCD.

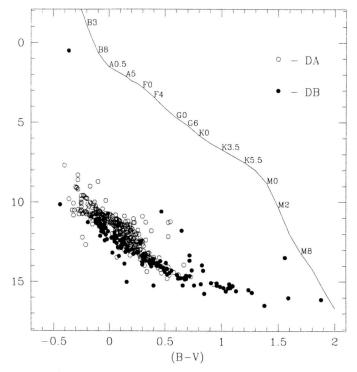

Fig. 1. A color–magnitude diagram for white dwarfs in the McCook & Sion (1987) catalog. Open symbols represent DA (hydrogen atmosphere) white dwarfs while DB white dwarfs are represented by solid symbols. The solid line shows the position of the main-sequence. From Hansen & Kawaler (1994).

2.3 White Dwarf Colors and the White Dwarf Luminosity Function

White dwarf have spectra that are characterized by very broad lines. As a result, broadband colors are affected by the lines to a much larger degree than stars of similar $T_{\rm eff}$ on the main sequence. For placement in an H–R diagram, for example, photometric systems designed for high–gravity stars differ from the usual Johnson or Strömgren colors. Greenstein (1976) derived a system using multichannel color indices to determine white dwarf temperatures and to estimate their luminosities; the system was later refined in a number of papers (i.e. Greenstein 1984, 1986, and Greenstein & Liebert 1990).

Because of the relatively narrow range in masses of white dwarfs (and their confinement to a line of constant radius in the H–R diagram) the Greenstein colors allow a fair determination of the absolute visual magnitude M_v. In this system, the (U-V) color provides a useful indicator for the temperature of hot white dwarfs, while the (G-R) color index works best for cooler white dwarfs. Fleming et al. (1986) give the following empirical relations for the absolute visual

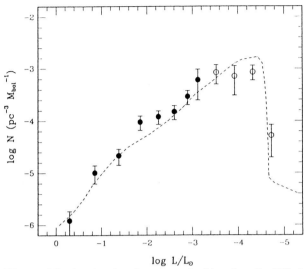

Fig. 2. The luminosity function of white dwarfs. Filled circles represent data from Fleming et al. (1986) for DA white dwarf stars; open circles are the stars analyzed by Liebert et al. (1988). The dotted line is the representative theoretical luminosity function of Winget et al. (1987).

magnitude using the Greenstein colors:

$$M_v = 11.857 + 2.882(U - V) - 0.519(U - V)^2 \qquad (11)$$

for warm white dwarfs, and

$$M_v = 13.033 + 3.114(G - R) - 0.799(G - R)^2 \qquad (12)$$

when (G-R) is smaller than −0.5. When Strömgren photometry is available, one may use

$$M_v = 11.50 + 7.56(b - y) \qquad (13)$$

(Green 1980), while Sion & Liebert (1977) give the following approximation that is useful for cases where Johnson UBV photometry is the only available color:

$$M_v = 11.246(B - V + 1)^{0.60} - 0.045. \qquad (14)$$

Conversion of M_v to bolometric magnitude $M_{\rm bol}$ requires model atmosphere bolometric corrections, such as given in Greenstein (1976) for cool hydrogen–atmosphere white dwarfs. Of course, when parallax measurements are available, more accurate estimates of individual values of M_v and $M_{\rm bol}$ are possible. Figure 1 shows the locations of those white dwarfs from McCook & Sion (1987) for which $(B - V)$ and M_v are available, along with a representative main sequence.

Using estimates of M_v, one can construct a luminosity function for white dwarfs. The luminosity function is a representation of the relative numbers of white dwarfs with different absolute magnitudes, usually plotted in terms of number of stars per cubic parsec per unit bolometric magnitude, versus luminosity. The still–definitive white dwarf luminosity function of "warm" white dwarfs with luminosities greater than $10^{-3.3} L_\odot$ and T_eff greater than 10,000 K is that of Fleming et al. (1986). They used data from the PG survey, supplemented with spectroscopic follow–up and the above color–magnitude relations. This luminosity function was extended to lower luminosities by Liebert et al. (1988), who used data on stars in the Luyten (1979) survey. An example of the combined luminosity function from these two studies is shown in Figure 2, from Liebert et al. (1988). More will be said about the white dwarf luminosity function in Section 6; for now, note simply that the luminosity function increases steadily with decreasing luminosity. As could be expected, the cooler a white dwarf is, the more slowly it cools and fades, and so the number increases with lower temperatures. However, there is a steep turn–down in the luminosity function below $10^{-4.4} L_\odot$ that requires explanation. This turn–down is the result of the finite time that white dwarfs in the solar neighborhood, and by extension in our galaxy, have had to cool. The time it takes for white dwarfs to fade to below this luminosity must be longer than the age of the galaxy itself. Thus the luminosity of this cut–off is a direct measure of the age of our galaxy.

2.4 White Dwarf Optical Spectra

Ground–based studies of white dwarf spectra have been going on for many decades; a still–useful review is by Liebert (1980). The principal conclusion of these studies was the demonstration that white dwarfs have very pure surfaces of either hydrogen or helium resulting from diffusive separation (Schatzman 1950). The purity of these zones is not 100%. Pollution may come from above, in the case of accretion from the interstellar medium, or from below in the form of radiative levitation or simple thermal diffusion. The metallic component includes calcium, magnesium, and iron. These are thought to be the result of accretion of interstellar material, since they are usually seen in cooler white dwarfs that have had time to encounter interstellar clouds (Aannestad and Sion 1985). Other polluting elements (such as carbon and silicon, to name a few) are found in white dwarfs in relative abundances that cannot be the result of accretion because of their abundance ratios. These elements are thought to be the diffusion "tail" of deeper layers. For instance, trace amounts of carbon are frequently seen in DB (see below) white dwarf stars. To be seen in cool DB white dwarfs, the surface helium layer must be between several times 10^{-3} and $10^{-4} M_\odot$, according to models by Pelltier et al. (1986).

To handle the peculiar spectral properties of white dwarfs, a systematic classification scheme has been developed for white dwarfs. The system in use now is detailed in a paper by Sion et al. (1983). Spectral type designations are of the form DαN, where the "D" prefix signals that this is a degenerate dwarf, α is assigned via the criteria listed in Table 1, and N is determined by the effective temperature: $N = 50400/T_\mathrm{eff}$. In common usage, the "N" is frequently omitted.

Table 1. White dwarf spectroscopic classification scheme

Spectral Type	Characteristics
DA	Balmer Lines only: no He I or metals present
DB	He I lines (4026Å, 4471Å, 4713Å) : no H or metals present
DO	He II lines (4686Å)
DZ	Metal lines only (CaII, Fe, O): no H or He
DQ	Carbon features, C_2
DC	Continuous spectrum; no lines

DA, DB, and DO stars are easy to understand. As in garden–variety A stars, the DA stars have spectra that are dominated by the Balmer lines of hydrogen. DB and DO stars show only lines of helium; DB stars show strong HeI lines, and DO stars show lines of singly ionized helium. Beyond these all similarities to traditional spectral types end. DA stars have atmospheres that are almost pure hydrogen; the dominance of the Balmer lines is not simply a temperature effect. Such pure spectra can result only if the helium abundance is less than approximately 6×10^{-4} by number. Similarly, the DB and DO stars have spectra dominated by helium lines primarily because of their nearly pure helium surfaces. In the remainder of this review, I will refer to white dwarfs with hydrogen–dominated surfaces as DA stars, and with helium–dominated surfaces as DB or DO stars. The surface abundances of the DZ, DQ, and DC stars will be discussed later.

Not all white dwarfs fit neatly into a single spectral classification; when we talk about a composite classification for a white dwarf this does not mean that it is a blended spectrum of two stars, but means that the stars shows some combination of spectral features from two (or more) categories. For example, DAB stars are DA stars with some HeI lines, DOZ stars are DO stars with lines of heavier elements, DOZQ stars are DO stars with heavier elements and carbon features, etc. This system can lead to some very unusual–looking spectral designations.

The rich variety of white dwarf spectra are displayed in an important (and remarkable) review paper by Wesemael et al. (1993, hereafter "W93"). This paper shows, on as uniform a scale as possible, examples of all of the known white dwarf spectral varieties. I will rely heavily on this paper in the remainder of this section, and recommend it highly to anyone interested in white dwarf stars. The spectra shown in W93 are all real spectra of real stars from a variety of sources.

Figure 3, from W93, shows representative spectra for DA stars ranging from the hot DA star G191-B2B (at $T_{\rm eff} \approx 80,000\,{\rm K}$) to the coolest. Above about 80,000 K, the Balmer lines effectively disappear, though they have been detected in the spectra of hydrogen–rich planetary nebula central stars that may be significantly hotter. Note that G191-B2B is actually classified as a DAZ star because it shows some photospheric metal lines in its ultraviolet spectrum. As is clearly shown in Figure 3, the Balmer lines are extremely broadened. Hot DA stars show broad but shallow Balmer lines; as the temperature decreases, the lines become stronger. They reach a maximum strength near the DA4-DA4.2 type, with $T_{\rm eff} \approx 12,000\,{\rm K}$ (not far from normal stars). For DA stars cooler than DA9, ($T_{\rm eff} < 5600\,{\rm K}$), only $H\alpha$ belies their identity as DA stars. Note also that the signal–to–noise ratio decreases for the coolest white dwarfs in Figure 3; this is simply because the coolest white dwarf stars are extremely faint.

The DB stars show a similar progression. Figure 4, again from W93, shows spectra of DB stars ranging from the hottest star to show HeI lines, PG 0112 (at $T_{\rm eff} \approx 25,000\,{\rm K}$) down to the coolest. The spectrum of HeI is complex; this hinders our ability to accurately assign effective temperatures and gravities to these stars. As $T_{\rm eff}$ decreases, the strength of the helium lines decreases, until the dominant HeI line at 4471Å is barely visible in the DB5 stars.

Note that the disappearance of HeI lines at approximately 10,000 K is not necessarily caused by the disappearance of helium. Stars cooler than DB5 can appear as DZ or DQ stars as trace elements still contribute enough opacity to produce lines that are stronger than the HeI lines. At the cool end of the DB sequence (around DB5), the existence of several DBZ and DBQ stars link the DB stars and the cooler DZ and DQ stars. In the DZ stars, examples of which are shown in Figure 5, the metal species usually seen is CaII, via the strong CaII H and K doublet at 3933Å and 3968Å; other species seen in the DZ stars include MgI, FeI, and CaI (see W93 for a more complete inventory). With temperatures exceeding 10,000 K, the DZ and DQ stars must be helium rich because no trace of the Balmer lines are seen.

There are no DB white dwarfs that are hotter than about 25,000 K. Is this simply a spectroscopic effect; i.e. helium–rich stars hotter than that temperature show only HeII lines, and are therefore classified as DO stars? The hottest DB star, PG 0112, is the uppermost spectrum in Figure 4. It shows a rich variety of strong HeI lines. Stars slightly hotter would indeed show HeI lines. But no such star has ever been found! The next hottest helium–atmosphere white dwarf is a DO1 star with $T_{\rm eff} \approx 45,000\,{\rm K}$. This lack of helium–rich white dwarfs between 30,000 K and 45,000 K is a perplexing problem; it is perhaps the most important issue facing white dwarf studies today.

Spectra of DO stars are shown in Figure 6. These stars are hot, with temperatures ranging from 45,000 K up to about 100,000 K. Over the wavelength range presented by W93, the continuum slopes continuously upwards towards shorter wavelength. The dominant spectral feature in these stars is the HeII line at 4686Å. The somewhat cooler DO stars, with $T_{\rm eff} < 70,000\,{\rm K}$ can also show weak HeI lines. Stars that show strong HeI features, along with HeII 4686Å should in principle exist, but none has been found. Again, the lack of such helium–atmosphere stars, which would populate the effective temperature range between 30,000 K and 45,000 K, presents a challenging and as yet unsolved mystery.

While most white dwarfs show only hydrogen or helium in their spectra. Figure 3 shows the star G191-B2B, a hot DA with metallic features in its ultraviolet spectrum. The hottest DA stars frequently show a 4686Å line of HeII in their spectra, as shown in the first spectrum in Figure 7. These are the DAO stars.

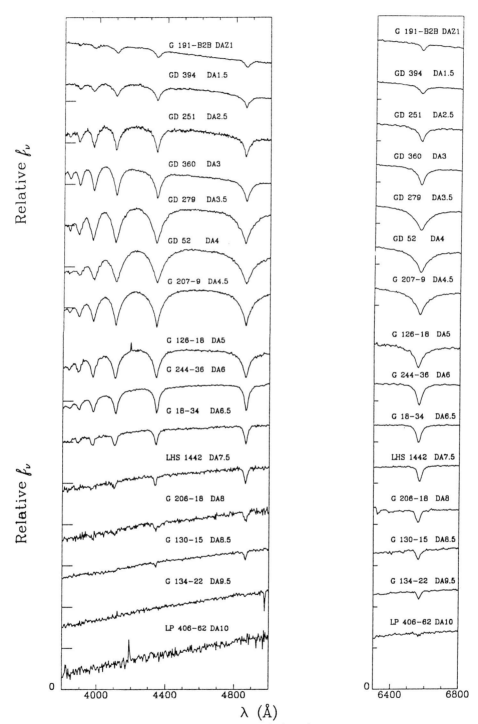

Fig. 3. Spectra of DA stars, from Wesemael et al. (1993).

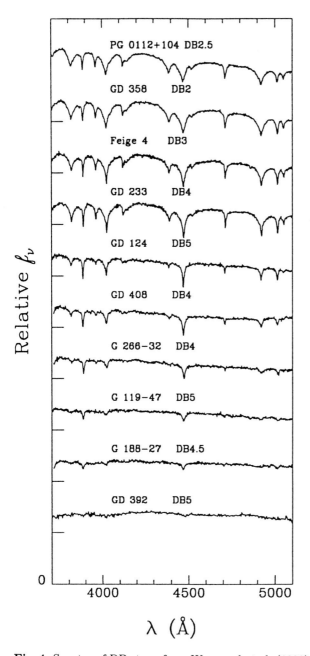

Fig. 4. Spectra of DB stars, from Wesemael et al. (1993).

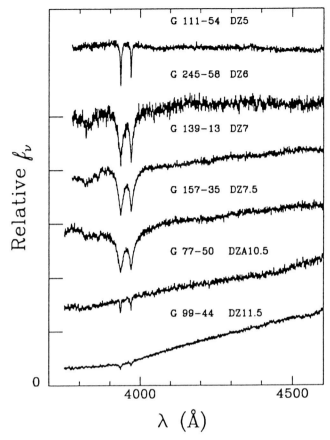

Fig. 5. Spectra of DZ stars, from W93.

Cooler stars that show predominantly hydrogen spectra, but that also show some helium features are, logically, the DAB stars, such as the second spectrum in Figure 7. Nearly all DAB stars are DAB2 stars, with temperatures near 25,000 K. Some of the cooler DB stars also show hints of Balmer lines of hydrogen. A representative of this DBA class is shown as the third spectrum in Figure 7. The incidence of trace hydrogen in the DB stars is possibly dependent on signal–to–noise; many DB stars show very weak hydrogen lines. At least three DBA stars also show lines of ionized calcium blended with a HeI line. These stars make up the DBAZ subclass; all lie near $T_{\rm eff} = 14,000$ K.

Perhaps the most interesting composite subclass of white dwarf stars are the very hot hydrogen–deficient stars known as "PG 1159" stars. These are the hottest white dwarfs; perhaps more accurately, they appear to be *pre*–white dwarf stars. These stars have very unusual spectra. They show no evidence of hydrogen or HeI lines. They are dominated by broad and shallow absorption lines of CIV and OVI. The CIV lines near 4670Å blend with the HeII 4686Å. In

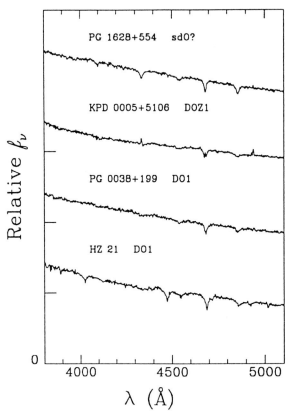

Fig. 6. Spectra of DO stars, from W93.

the hotter PG 1159 stars, the CIV–HeII feature shows strong emission features as well. The spectra of a variety of PG 1159 stars are shown in Figure 8. The spectrum of the prototype also shows nitrogen lines (NV). The spectra of the PG 1159 stars have been investigated in great detail by Klaus Werner and his colleagues (i.e. Werner 1993 and references therein). Because of the presence of HeII, CIV, and OVI, the PG 1159 stars have the official spectroscopic designation DOQZ, but are most commonly called PG 1159 stars. Several central stars of planetary nebula show PG 1159–type spectra. As transition objects between the AGB and the white dwarf cooling track, the PG 1159 stars are currently being very actively studied by a variety of researchers; more on these stars will come shortly.

2.5 Distribution of Spectral Types with Effective Temperatures

In the previous section, we saw that there is no clear continuum between the helium–atmosphere spectral classes DO and DB; in other words, there appear to be no helium–atmosphere white dwarfs with $T_{\rm eff}$ between 45,000 K and 30,000 K.

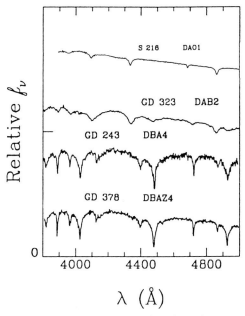

Fig. 7. Examples of white dwarfs with composite spectra. The top spectrum shows a DAO1 star (S 216). The next spectrum downwards shows a DAB star, the prototype GD 323, a DAB2 star that shows Balmer lines of hydrogen along with weak HeI lines. The next spectrum is an example of a DBA star, which is dominated by HeI lines but also shows weak Balmer lines, and the bottom spectrum shows the DBAZ4 star GD 378. From W93.

Such stars would show strong HeI lines, and so it is not simply because they are spectroscopically selected against. This fact is one feature of the distribution of white dwarf surface abundances with effective temperature. Figure 9, adapted from Fontaine & Wesemael (1987), shows the spectral type distribution in graphical form. As expected, the numbers increase (in general) with decreasing effective temperatures for both flavors of white dwarfs. On a finer scale, though, these functions show interesting detail.

Above about 80,000 K, there are no DA white dwarf stars. According to the theory of white dwarf formation by cooling of planetary nebula central stars, the maximum temperature achieved by hydrogen–rich planetary nebula central stars is well in excess of 100,000 K for masses near $0.60 M_\odot$, but the hottest pre–white dwarfs are cooler than this. At the very hot end of the white dwarf sequence, then, nearly all stars appear to be helium rich, and the ratio of non-DA to total white dwarfs is almost unity. From 80,000 K down to 45,000 K, the DA white dwarfs increase in number rapidly compared to the helium–dominated DO white dwarf stars, and the ratio of non-DA to total goes down. We then come to the region between 45,000 K and 30,000 K where no helium–surface white dwarfs have been seen. Below 30,000 K, the DB stars appear, and the ratio of non-DA to total rises again, to reach about 25 percent. Below 10,000 K, few if any white

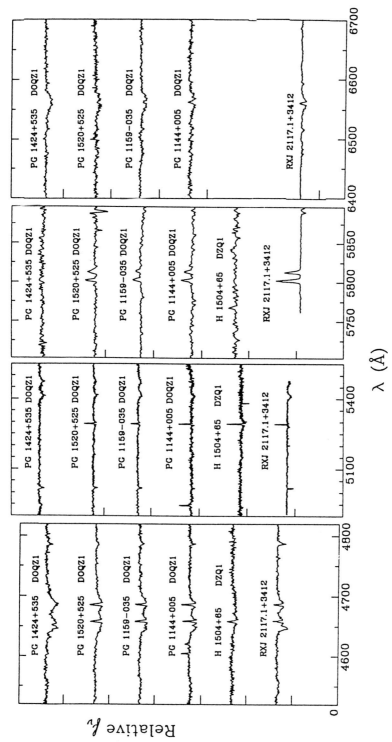

Fig. 8. Spectra of several PG 1159 stars, adapted from W93.

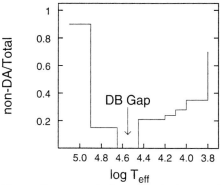

Fig. 9. Fraction of white dwarfs that are helium–rich as a function of decreasing $T_{\rm eff}$. Adapted from Fontaine & Wesemael (1987)

dwarfs show spectral signature of hydrogen, and again the non-DA to total ratio rises to nearly 1, though determination of surface abundances in the coolest white dwarfs is a difficult job.

The "DB gap", and the changes in the ratio of the numbers of non-DA white dwarfs to the total number, are statistically significant observations. The hottest, and therefore newest white dwarfs are all hydrogen–poor. Working down the white dwarf cooling track, one finds no DAs at 100,000 K to all DAs at 45,000 K. This would be easy to understand if all helium–dominated stars disappear never to return; simply, the helium that dominates the spectra of the hotter white dwarf sinks below the visible surface, and whatever hydrogen existed in the white dwarf interior has floated to the surface by the time the stars cool to 45,000 K. However, below 30,000 K white dwarfs with helium–rich surfaces reappear! The observation of chemical evolution of white dwarf stars indicates that something strange is happening in the outer layers of white dwarfs that causes helium to disappear from the surface and then, in a significant fraction of white dwarfs, to reappear at lower effective temperatures. Understanding the chemical evolution of white dwarfs is a primary goal of much of current white dwarf research; we will come back to the processes involved in white dwarf chemical evolution in a later section.

2.6 Magnetic White Dwarfs

There are two primary ways of detecting magnetic fields in white dwarfs: the fields manifest themselves by imparting polarization of the spectrum, and or by linear or quadratic Zeeman effects in the spectral lines themselves. This spectroscopic technique requires strong lines, which are not always to be found. The polarization method depends on the detection of circular polarization and is especially useful when magnetic field strengths are high. At the present time, there are 42 known magnetic white dwarfs (Schmidt & Smith 1995); the lower limit for detectable fields is about 10^4 Gauss (and that small only under unusually favorable circumstances), and that is still 10,000 times bigger than the Sun's average magnetic field!

Most magnetic white dwarf have hydrogen atmospheres, and are between 30,000 K and 5000 K. Angel et al. (1981), and Schmidt (1987) describe three broad categories of magnetic white dwarfs; a currently complete list is presented in Schmidt & Smith (1995). At the "small" field end are those stars for which field strengths of 10 MG or less; approximately 15 stars fall into this category. In objects with fields less than 1MG ($= 10^6$G), fields are measured through weak circular polarization across the Balmer line profile (Schmidt & Smith 1994, 1995). Fields in excess of 1 MG produce polarization across the continuum. Approximately 14 stars fall within the "intermediate" category, with fields of between 10 and 100 MG. In this range, the Balmer lines split into multiple Zeeman components. The third category, comprising approximately 12 stars with fields ranging from 100 MG to nearly 1000 MG. At these field strengths, Zeeman effects cause extreme distortion of the spectrum (lines and continuum). Approximately 4% of all white dwarfs show magnetic fields greater than several thousand MG (Schmidt & Smith 1995).

These stars are peculiar in many ways. Since the magnetic field is not necessarily aligned with the rotation axis, changes in projected field strength with rotation result in periodic variations in magnetic splittings and/or polarization. Deduced rotation periods range from several hours to several days, with no clear correlation between field strength and rotation period. However, the high–field white dwarfs show little or no change in spectra over time; some show apparently secular changes that, if caused by rotation, suggest rotation periods of several hundred years!

An immediate question is the origin of magnetic white dwarfs. Why should some white dwarfs show huge magnetic fields, while others show little and most show none? One possible answer is that the magnetic fields are somehow primordial, and that magnetic white dwarfs are the descendants of the magnetic main–sequence Ap stars. Given the warm temperatures of most white dwarf stars, they had to have formed from relatively massive progenitors, in the range from 1 to 8 $M_\odot$. The surface fields of Ap stars are about 1000G; such a field could easily be amplified to 1 MG or more if it is conserved to the white dwarf phase. With a space density similar to the magnetic white dwarfs with significant magnetic field, the Ap stars remain as possible ancestors to the magnetic white dwarfs.

2.7 Pulsating White Dwarfs

Pulsating white dwarfs are found in three regions in the H-R diagram, corresponding to three different ranges of effective temperature. At the hot end, some of the PG 1159 (DOQZ) stars are observed to vary in luminosity by small amounts (usually less than 10%) on time scales of minutes. These are the result of multiperiodic pulsation with periods of 300–1000 seconds. The pulsating PG 1159 stars comprise the GW Vir stars. Some of the planetary nebulae with PG 1159 type spectra also show variations, but with longer periods of 1000–2500 seconds (Bond et al. 1993). The DB stars also show pulsations, with periods comparable to the GW Vir stars; the DBV stars are found at effective temperatures near 27,000 K. The coolest known pulsating white dwarf stars are the ZZ Ceti stars. The ZZ Ceti stars are DA stars that are confined to a narrow instability strip near $T_{\rm eff} \approx 12,000$ K.

In later sections of this review, I will discuss how models of the evolution of white dwarfs are constrained by analysis of the pulsations of white dwarfs during several stages in their evolution. The observed pulsations will be described in much more detail in these sections.

3 Physics of White Dwarf Interiors

This section reviews some results from statistical mechanics, which will eventually be used to derive equations of state for stellar material appropriate for the conditions found in white dwarf stars. We then treat the thermal properties of the material in white dwarfs, and the transport of radiation in their interiors.

3.1 Equation of State

Several excellent texts on statistical mechanics, in addition to specialized texts in stellar interiors, are available for students of white dwarf stars. For the statistical mechanics background, the text by Landau & Lifshitz (1958) and the monograph by Pippard (1957) are classic and unparalleled sources. For the astrophysical applications, see for example Cox (1968), and Kippenhahn & Weigert (1990), and Hansen & Kawaler (1994). Material in this subsection has been adapted from the latter reference.

The thermodynamic variables needed in structure and evolution calculations are P, T, ρ, S, E, Q, and various isotopic number densities, n_i. N_i is the number density of an ith species in the units of number per gram of material with $N_i = n_i/\rho$. It is the Lagrangian version of n_i and it will prove useful because it remains constant even if volume changes.

Another useful thermodynamic quantity is the *chemical potential*, μ_i, defined by

$$\mu_i = \left(\frac{\partial E}{\partial N_i}\right)_{S,V}$$

as associated with an ith species in the material (not to be confused with μ_I, the ion molecular weight). If there exist "chemical" reactions in the stellar mixture involving some subset of species (ions, electrons, photons, molecules, etc.) whose concentrations could, in principle, change by dN_i as a result of those reactions, then thermodynamic (and chemical) equilibrium requires that

$$\sum_i \mu_i \, dN_i = 0. \qquad (15)$$

Changing N_i by dN_i in a real mixture usually means that other components in the mixture must change by an amount related to dN_i so that not all the dN_i are independent.

The above quantities are not independent; given a subset of the above quantities, one can calculate the others. The equation of state (hereafter "EOS") relates the thermodynamic quantities, allowing determination of one quantity given several others. We concentrate here on expressing the pressure P in terms of temperature T and composition (parameterized by the mean molecular weight μ). A definition of pressure is that it is the momentum flux perpendicular to a differential surface element $d\sigma$. Equivalently, it is the product of the instantaneous particle flux and the mean momentum per particle. The particle flux is

$$\text{particle flux} = n(p,\theta)\,dp\,v(p)\,\cos\theta d\theta \tag{16}$$

where p is the magnitude of the momentum for a particle propagating at an angle θ to the outward normal of the surface element. Integrating the product of the above particle flux and the component of the particle momentum perpendicular to $d\sigma$ gives the following general expression for the pressure:

$$P = \int_0^\infty \int_0^\pi d\theta\, p\, v(p)\, n(p,\theta)\, \cos^2\theta\, d\theta \tag{17}$$

Now, assuming an isotropic distribution of particle momenta, the above reduces to

$$P = \frac{1}{3}\int_0^\infty p\,v(p)\,n(p)\,dp. \tag{18}$$

Now the problem is reduced to knowing the velocity and number densities of the particles as a function of particle momentum p.

For elementary particles at their lowest energy state E_o (corresponding to their rest-mass energy mc^2), the number density of a given species is given by

$$n(p) = \frac{1}{h^3}\frac{g}{\exp\left[\frac{-\mu+E_o+E(p)}{kT}\right]\pm 1}. \tag{19}$$

where g is the degeneracy of the reference level state, and μ is the chemical potential of the species. The total number density follows directly as

$$n = \int_0^\infty 4\pi p^2 dp\, n(p). \tag{20}$$

For general Fermions such as electrons, protons, and neutrons, we have $g=2$, $E_o = mc^2$,

$$E(p) = mc^2\left[\sqrt{1+\left(\frac{p}{mc}\right)^2}-1\right], \tag{21}$$

and

$$v(p) = \frac{\partial E}{\partial p} = \frac{p}{m}\left[\sqrt{1+\left(\frac{p}{mc}\right)^2}\right]^{-1}. \tag{22}$$

Then

$$n = \frac{8\pi}{h^3}\int_0^\infty p^2\,dE\,f(E), \tag{23}$$

which defines a function of energy

$$f(E) \equiv \frac{1}{\exp\left[\frac{E-(\mu-mc^2)}{kT}\right]+1}. \qquad (24)$$

The function $f(E)$ contains the information on the distribution of particle energies in a Fermion gas. Inspection of the denominator shows that at very low temperatures, a characteristic energy appears at a value of $\mu - mc^2$. Define E_f as $\mu - mc^2$ and the momentum corresponding to this energy as p_f; the f subscript denotes the special energy and momentum of Fermions, in honor of Enrico Fermi. It will also be useful to define a parameter x_f as p_f/mc.

Complete Degeneracy The distribution function $f(E)$ is shown in Figure 10 for two representative temperatures, including $T = 0$. As we approach $T = 0$, if E is greater than E_f, $f(E)$ approaches zero, while if E is less than E_f, then $f(E)$ approaches unity. In the low–temperature limit, the equation of state simplifies conveniently as follows. For the number density n we have

$$n = \frac{8\pi}{h^3} \int_0^{p_f} p^2 \, dp = 8\pi h^3 \left(\frac{h}{mc}\right)^{-3} x_f^3 \qquad (25)$$

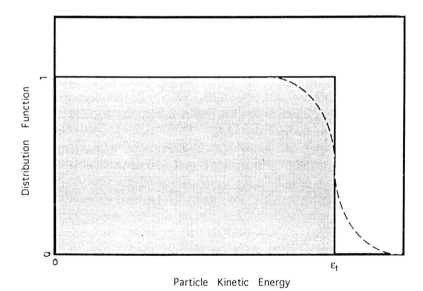

Fig. 10. The distribution function for particle kinetic energy in the limit of zero temperature. Fermions are confined to energies less than E_f, where the distribution function has a value of unity. The dashed line shows how the distribution is eroded for finite (but still low) temperature.

For electrons, then, the mass density (ρ) becomes

$$\frac{\rho}{\mu_e} = \frac{8\pi}{3N_A} \left(\frac{h}{mc}\right)^{-3} x_f^3 = 9.7 \times 10^5 \times x^3 \qquad (26)$$

where we have dropped the f subscript on x_f. The corresponding pressure of the electrons is

$$P_e = \frac{8\pi}{h^3} \int_0^{p_f} v(p) \, p^3 \, dp = 6.0 \times 10^{22} F(x) \qquad (27)$$

which includes the function $F(x)$ defined by Chandrasekhar (1939) as

$$F(x) \equiv x \left(2x^2 - 3\right) \left(1 + x^2\right)^{1/2} + 3\sinh^{-1}(x). \qquad (28)$$

The energy density of electrons is given in a straightforward way as

$$E_e = 6.0 \times 10^{22} G(x) \qquad (29)$$

with $G(x)$ given by

$$G(x) = 8x^3 \left[\left(1 + x^2\right)^{1/2} - 1\right] - F(x). \qquad (30)$$

When $p_f \ll m_e c$ (i.e. $x \ll 1$), then the velocity corresponding to the momentum of the electrons is small compared to c, and we call the gas nonrelativistic. In the nonrelativistic case,

$$\lim_{x \to 0} F(x) = \frac{8}{5} x^5 \qquad (31)$$

and therefore

$$P_e \to 10^{13} \left(\frac{\rho}{\mu_e}\right)^{5/3} \quad \text{nonrelativistic.} \qquad (32)$$

On the other hand, when $x \gg 1$ we find ourselves in the relativistic regime, and

$$\lim_{x \to \infty} F(x) = 2x^4 \qquad (33)$$

and then

$$P_e \to 1.2 \times 10^{15} \left(\frac{\rho}{\mu_e}\right)^{4/3} \quad \text{relativistic.} \qquad (34)$$

Note the famous result that in both cases, the electron pressure depends only on the density of the material and not on the temperature. This is not a redundant statement; the above results are obtained in the limit of $T \to 0$ but is a very close approximation at temperatures significantly above. For example, when the density is approximately 10^6 g/cm^3, the pressure of the electrons in a zero-temperature gas with $\mu_e = 2$ is 3×10^{22} dyncm^{-2}. This same pressure would be exerted by the electrons in a nondegenerate electron gas with the same density and a temperature of 2×10^5 K. In the last part of this section, we show the conditions under which the degenerate pressure dominates, and you will see the regions of the (ρ, T) plane where the zero-temperature approximation is valid.

We can apply the above expressions to the conditions in white dwarfs which allows us to understand their existence, first argued by Chandrasekhar and Fowler in the 1930s. Integrating the equation of hydrostatic equilibrium of a zero-temperature gas with these equations allows us to derive a mass-radius relation in the nonrelativistic case:

$$\frac{R}{R_\odot} = 0.012 \left(\frac{M}{M_\odot}\right)^{-1/3} \left(\frac{\mu_e}{2}\right)^{-5/3} \quad \text{nonrelativistic.} \tag{35}$$

Miraculously, in the relativistic limit, the 4/3 exponent of the density results in a limiting mass for a stable star supported by electron degeneracy pressure:

$$M_\infty = 1.46 \left(\frac{\mu_e}{2}\right)^{-2} \quad \text{relativistic} \tag{36}$$

which is the famous "Chandrasekhar" mass. At higher masses, the increased densities and pressure cannot halt the collapse because the relativistic limit has already been reached. In the nonrelativistic limit, on the other hand, a new configuration may be reached by decreasing the radius. Extreme relativistic equations of state, including that for photons, are too "soft" compared to the effects of self-gravity. (You can't make the particles exceed the speed of light to try to increase pressures!) This conclusion might have been anticipated because extreme relativistic effects imply $\gamma \to 4/3$.

The astrophysical significance of the Chandrasekhar limiting mass is as follows. If electron degenerate configurations are good representations of white dwarfs, and if those objects are the final product of evolution for most stars, then the late stages of evolution are severely constrained. That is, if a star does not finally rid itself of enough mass to eventually leave a white dwarf with $M \approx 1.46 M_\odot$ (assuming $\mu_e = 2$), then something catastrophic will happen at some time in its life. Since there are so many white dwarfs in the sky, a large fraction of stars either start off with sufficiently low masses, or they manage to rid themselves of the excess mass.

For cases of intermediate degeneracy, the integrations are a bit more difficult; Eggleton (as quoted by Truran & Livio 1986) has parameterized the mass radius relationship conveniently as

$$\frac{R}{R_\odot} = 0.026 \left[1 - \left(\frac{M}{M_\infty}\right)^{4/3}\right]^{1/2} \left(\frac{M}{M_\infty}\right)^{-1/3}. \tag{37}$$

The boundary between the regions in the $(\rho - T)$ plane where relativistic degeneracy holds is the locus of points for which $p_f \approx mc^2$. It is a simple exercise to show that this condition corresponds to a value of ρ/μ_e of 10^6 for electron degenerate material.

Figure 11 shows the mass–radius relation for several different cases. This figure, taken from the important paper by Hamada & Salpeter (1961) shows the original Chandrasekhar solution as dashed lines. Hamada & Salpeter investigated the effects of different ionic species, and found that the mass-radius relation does indeed depend on the internal composition; the relations for various species are

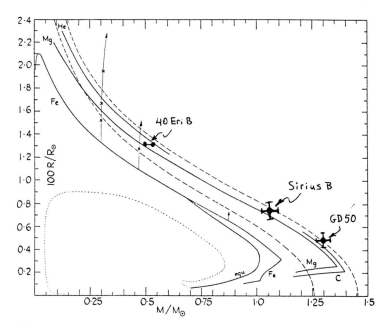

Fig. 11. The mass–radius relation for fully degenerate zero–temperature configurations. Curves are labeled with the composition; the dashed curves are the Chandrasekhar curves for $\mu_e = 2.0$ and $\mu = 2.15$ (from *left* to *right*). Note that the heavier the core composition, the smaller the star at a given mass. Also shown are the positions of three white dwarfs with secure mass and radius determinations.

shown in this figure. The positions of three white dwarfs of known mass and radius are also shown here. It is interesting to note that 40 Eri B, the first white dwarf recognized, by itself does not distinguish between the relativistic and nonrelativistic case. Sirius B and GD 50, on the other hand, require the relativistic equation of state to explain their dimensions.

Partial Degeneracy The simplifications in the above arguments came from assuming that the temperature was zero. In this way, we could use a step function for the energy distribution of the electrons, being 1 for energies below the Fermi energy and 0 above. If now we allow for the temperature to be non-zero, then the energy of a given electron can be boosted by energies of order kT. Electrons with energies at well below the Fermi energy would receive a boost in energy that is insufficient to boost their energy above E_f, and so they will be excluded from such a boost. That is, at small temperatures, the distribution function $f(E)$ will be relatively unaffected at low E. However, for electrons nearer the top of the Fermi sea, a boost in energy by kT can boost their energy above E_f. Thus the number of electrons at energies near but below E_f will be eroded somewhat, and will cause $f(E)$ to begin to drop from 1 for energies a bit below E_f, and to rise a bit above 0 for energies above E_f. As illustrated in Figure 10, the distribution function $f(E)$ looks like a worn step at E_f, with the erosion working towards lower energies as kT increases.

In the limiting case of sufficiently high temperatures, the Fermi–Dirac statistics give way to the Maxwell–Boltzmann distribution, and we call the gas nondegenerate. This suggests a parameter to estimate the importance of degeneracy effects as E_f/kT. When this number exceeds unity, we call the gas degenerate and the equation of state approaches that of the previous section. When it is small, the gas is nondegenerate, and the ideal perfect gas equation of state applies. In between, when $E_f/kT \approx 1$, things get complicated. This is the regime of partial degeneracy, and in the nonrelativistic case, it occurs when

$$kT = E_f = \frac{mc^2}{2}\left(\frac{3N_A}{8\pi}\right)^{2/3}\left(\frac{h}{m_e c}\right)^2 \left(\frac{\rho}{\mu_e}\right)^{2/3} \tag{38}$$

or

$$\frac{\rho}{\mu_e} \approx 6.0 \times 10^{-9}\, T^{3/2} \quad \text{nonrelativistic.} \tag{39}$$

I leave it to you to show that the corresponding equation in the relativistic case is

$$\frac{\rho}{\mu_e} \approx 4.6 \times 10^{-24}\, T^3 \quad \text{relativistic.} \tag{40}$$

If ρ/μ_e exceeds these values at a given temperature, then the gas will show degenerate effects. Please do not consider this as an abrupt transition! The transition from degenerate to nondegenerate conditions is a gentle one in real stellar systems.

So given the complexity of the equation of state of matter that is partially degenerate, it should be no surprise that expressions for the pressure as a function of density and temperature require effort to derive and solve. Hansen & Kawaler (1994) and other sources give expansions for n_e and P_e in the case of *almost complete degeneracy*. For the purposes of numerical modeling of white dwarf interiors (which are highly degenerate) and envelopes (which are frequently nondegenerate), the electron densities and pressures can be computed numerically.

Allow $\eta \equiv \mu/kT$ to define a new degeneracy parameter. With this definition, the integrals over the distribution $f(E)$ in the nonrelativistic limit become

$$n_e = \frac{4\pi}{h^3}(2mkT)^{3/2} \int_0^\infty \frac{u^{1/2}}{e^{u-\eta}+1}\, du \tag{41}$$

$$= \frac{4\pi}{h^3}(2mkT)^{3/2}\, F_{1/2}(\eta) \tag{42}$$

and

$$P_e = \frac{8\pi kT}{3h^3}(2mkT)^{3/2} \int_0^\infty \frac{u^{3/2}}{e^{u-\eta}+1}\, du \tag{43}$$

$$= \frac{8\pi kT}{3h^3}(2mkT)^{3/2}\, F_{3/2}(\eta). \tag{44}$$

The parameter η is another indicator of degeneracy. When $\eta < -4$, the material is essentially nondegenerate, while strong degeneracy occurs for $\eta > 10$ or so. The integrals above are known as the Fermi-Dirac integrals, and have been tabulated in a number of places. For numerical computations, Eggleton et al. (1973) provide a convenient parameterization that is widely used.

3.2 Heat Transport in Degenerate Matter

The transport of energy through degenerate matter is quantitatively different than through the bulk of normal stars. In normal stars, heat is transported by convection and/or radiation, depending on the state of the gas through which the heat is flowing. In the case of radiative transport, we usually assume that the energy leaks out by diffusion of photons through a mostly opaque medium. In the case of radiation, then, the flux (integrated over all wavelengths) is given by a diffusion equation of the form

$$F_r = -D_r \frac{dT}{dr} \qquad (45)$$

where dT/dr is the total temperature gradient. The diffusion coefficient D_r is given in terms of the specific radiative opacity averaged in some way over all wavelengths, $\kappa_r [\text{cm}^2/\text{g}]$:

$$D_r = \frac{4acT^3}{3\kappa_r \rho}. \qquad (46)$$

In this definition, a is the radiation density constant and c is the speed of light. Frequently, the total radiative opacity is the sum of several terms, including opacity from electron scattering, bound–free and free–free transitions.

In degenerate matter, electrons near the top of the Fermi sea are essentially "free to roam"; that is, they can conduct heat in a manner analogous to conduction in terrestrial metals. We can write the energy flux by conduction in a form analogous to that of the radiative flux, with F_r replaced by F_{cond} and D_r replaced by D_{cond}. The diffusion coefficient for conduction is of the general form

$$D_{cond} \sim \frac{c_v v_e \lambda}{3} \qquad (47)$$

where c_v is the specific heat of the electrons, v_e is the relevant electron velocity, and λ is the electron mean-free-path. When material is mostly degenerate, the specific heat of the electrons scales with the temperature T and with the momentum parameter x. Under degenerate conditions, both v_e and λ are constrained in that only electrons near the Fermi energy E_f are free. This allows us to specify the value of v_e from the condition that $m_e v_e \approx p_f$. Recalling

$$p_f \propto x_f \propto \left(\frac{\rho}{\mu_e}\right)^{1/3}, \qquad (48)$$

v_e therefore scales with the 1/3 power of ρ/μ_e, and c_v for the electrons scales as $T(\rho/\mu_e)^{1/3}$

The electron mean-free-path λ is determined by the probability of Coulomb scattering of the ions. Thus

$$\lambda \sim \frac{1}{\sigma_c n_{ion}}. \tag{49}$$

The Coulomb scattering cross section σ_c depends on the square of the mean impact parameter, which will be large for small electron kinetic energies. Equating the Coulomb potential energy with the kinetic energy of the electrons results in the impact parameter being inversely proportional to v_e^2, and so $\sigma_c \propto v_e^{-4}$, and so we have

$$\lambda \propto \left(\frac{\rho}{\mu_e}\right)^{4/3} \left(\frac{\rho}{\mu_I}\right)^{-1} \tag{50}$$

and

$$D_{cond} \propto \left(\frac{\rho}{\mu_e}\right)^2 T \left(\frac{\rho}{\mu_I}\right)^{-1} \tag{51}$$

where μ_I is the mean atomic weight per ion.

With more care with the constants, we can now estimate the conductive opacity κ_{cond} using the above diffusion coefficient, finding

$$\kappa_{cond} \approx 4 \times 10^{-8} \left(\frac{\mu_e^2}{\mu_I}\right) Z^2 \left(\frac{T}{\rho}\right)^2 \tag{52}$$

For typical white dwarf interior conditions, where $T = 10^7$ K and $\rho = 10^6$ g/cm^3, the above equation gives a value of $\kappa_{cond} \approx 5 \times 10^{-5}$ cm^2/g. With this definition of κ_r and κ_{cond}, the total flux is the sum of the radiative flux and the conductive flux:

$$F_{tot} = F_r + F_{cond} = -\frac{4ac}{3\rho} T^3 \frac{dT}{dr} \left(\frac{1}{\kappa_r} + \frac{1}{\kappa_{cond}}\right) \tag{53}$$

Thus the radiative and conductive opacities add in parallel. For the above conditions, the dominant radiative opacity is electron scattering, with a value of roughly 0.2 cm^2/g. Therefore, the conductive flux is about 4000 times bigger than the radiative flux. This has very important implications for the thermal structure of the interiors of white dwarf stars. With such efficient heat transport, only a small temperature gradient is needed to drive a large heat flux. This means that the interiors of white dwarf stars can be nearly isothermal. Along with the lack of temperature dependence of the pressure, that means that the mechanical support of the star by electron degeneracy pressure can occur with an isothermal core that can still support a significant photon luminosity.

Of course, the above estimate of the conductive opacity is crude. More accurate expressions are needed for accurate models of white dwarf stars and stars with degenerate cores. Conductive opacity computations need to incorporate the interionic forces with some precision, allowing for nonideal effects of the equation of state, and for various compositions. The pioneering work in this area is the calculation by Hubbard & Lampe (1969), with more modern calculations published by Itoh et al. (1983, 1984).

3.3 Nonideal Effects

The sections above that deal with the equation of state of degenerate matter considered the fictional cases of ideal gases, in which there are no electromagnetic interactions between the particles. In a fully ionized medium, the ionic charge is essentially assumed to be surrounded by a cloud of electrons that exactly cancel the positively charged ion. As long as the ions do not get too close, this screening effectively neutralizes the electromagnetic forces, and the nuclei can be considered as non-interacting.

This is, after all, the material in the core of a white dwarf star, where densities are enormous by terrestrial standards. Above some density (and at finite temperature) the ions are indeed sufficiently crowded that their electrostatic interaction can affect the equation of state. The computation of nonideal effects for stellar plasmas is extremely complicated, and I only outline the background here. More details relevant to white dwarf interiors are available from numerous sources; see for example Shapiro & Teukolsky (1983) and references therein.

Lets consider a simplification of the problem, and assume that each ion is surrounded by a sphere of electrons that occupies a volume equal to $1/n_{ion}$. The Coulomb energy of such a sphere is

$$E_c = \frac{Z^2 e^2}{a} = Z^2 e^2 \left(\frac{4\pi \rho N_A}{3A}\right)^{1/3} \tag{54}$$

If E_c is smaller than kT, then the gas acts effectively as an ideal gas. However, if $E_c > kT$ then nonideal effects can occur. As a criterion, then, define

$$\Gamma_c \equiv \frac{E_c}{kT} = 2.27 \frac{Z^2}{A^{1/3}} \frac{\rho_6^{1/3}}{T_7} \tag{55}$$

where ρ_6 is the density in units of 10^6g/cm^3 and T_7 is the temperature in units of 10^7 K. When Γ_c exceeds 1, the material is considered as a quantum liquid state, where long–range forces between ions play a role in the thermodynamics of the material. Note that at a given density, the value of Γ_c increases with decreasing temperature. Thus as white dwarfs cool, the effects of Coulomb interactions become more important. Also, since the interiors of white dwarfs are roughly isothermal, the Coulomb effects are largest at the center.

If Γ_c becomes large enough, then Coulomb effects overwhelm those of thermal agitation, long–range forces organize the small-scale structure of the material, and the gas settles down into a crystal. The best estimates as to how this takes place yield a Γ_c of around 170 for the transition. With this value of Γ_c, the condition for crystallization becomes

$$T_{\text{xtal}} = 1.33 \times 10^5 \frac{Z^2}{A^{1/3}} \rho_6^{1/3} \, \text{K}. \tag{56}$$

For conditions relevant to white dwarf stars, with central densities of order $10^6 [\text{g/cm}^3]$, carbon crystallizes at about 3.4×10^6 K, while oxygen crystallizes at 2.1×10^6 K. These central temperatures are reached when the white dwarf has been cooling for about 10^9 years. Thus crystallization can be an important process in the evolution of white dwarf stars.

The crystallization condition is dependent on the value of Γ_c at which crystallization takes place. Shapiro & Teukolsky (1983) outline how this parameter can be estimated. Note however that this constant assumes a one component plasma, while material within most white dwarfs is probably a mixture of at least carbon and oxygen. Thus the actual process of crystallization must be extremely complex. Still we use the above criterion as a guide to when it must be important.

3.4 Specific Heat

As a white dwarf cools, the specific heat for the degenerate interior changes in ways most familiar to condensed matter physicists. Here, I only sketchily summarize the behavior of the specific heat; further details (in language appropriate for astrophysicists) are clearly described by Shapiro & Teukolsky (1983). Equations (32) and (34), which present the equation of state in the limit of complete degeneracy, show that the pressure is a function only of the density. Therefore, the internal energy under such extreme conditions is independent of the temperature. For such a system, the specific heat is identically zero. At the other extreme, the ideal perfect gas, the specific heat at constant density is independent of temperature, and proportional only to the mean atomic weight μ. Therefore, as a material becomes degenerate, the specific heat of the electrons becomes very small. Since the ions are nondegenerate, however, the specific heat of the ions remains unchanged. Therefore the specific heat of degenerate material is determined solely by the specific heat of the ions; that is

$$c_V = \left(\frac{\partial Q}{\partial T}\right)_\rho \tag{57}$$

$$\approx c_{V,\text{ions}} = \frac{3}{2}\frac{1}{\mu_I}kN_A \;\; \text{erg/g/K}. \tag{58}$$

Note that the specific heat of the ions includes the three translational components, each of which contributes $kT/2$ per ion.

From equation (55), we see that the assumption that the ions behave as an ideal gas begins to break down when $\Gamma_c \geq 1$; at that point interactions between incompletely shielded ions affect the internal energy via the Coulomb interaction. At that time, the matter is in a quantum liquid state, and a decrease in temperature allows significant long-range ordering of the ionic structure. The corresponding lattice vibrations (in three dimensions) ultimately result in a doubling of the specific heat as the material cools towards crystallization.

When the temperature of the material cools further, the internal energy of the lattice is determined by quantum mechanical effects. Upon cooling below the Debye temperature, the further adventures of the specific heat produce Debye cooling, where the specific heat becomes

$$c_V \approx \frac{16\pi^4}{5}\left(\frac{T}{\theta_D}\right)kN_A \;\; \text{erg/g/K} \tag{59}$$

when the temperature is below θ_D, the Debye temperature:

$$\theta_D = 4 \times 10^3 \rho^{1/2} \text{ K}. \tag{60}$$

4 White Dwarf Formation and Early Cooling

As discussed in the first section of this paper, there is overwhelming evidence that most white dwarfs represent the final stage of evolution of stars that have spent some time on the Asymptotic Giant Branch. For that reason, I will concentrate on the formation of white dwarfs from AGB stars. Other channels, no doubt, produce stars that are now classified as white dwarfs, and will produce stars with (initially) different thermal profiles in their interiors, and different elemental composition in the outer layers. But by the time a white dwarf cools sufficiently, it has forgotten most of its prior history. Thus the story of later white dwarf cooling, which I defer to the next section, depends little on the earlier stages described (or ignored) in the current section.

Figure 12 shows, in a schematic way, the evolution of a "typical" white dwarf, beginning on the main sequence. Though this is a schematic figure (a copy of one shown in the lectures), the scales are correct. The evolution of a star as a white dwarf does indeed cover a much larger portion of the H–R diagram than during previous phases (from the pre–main sequence through the AGB). This chapter describes the early evolution of white dwarfs, from the AGB to the white dwarf cooling track. During the course of this section, we will discuss some of the largest (in radius) stars in the Universe, with radii of several hundred solar radii, and some of the smallest ordinary stars, the white dwarfs.

4.1 Thermal Pulses on the AGB

We begin with a description of evolution along the AGB. Figure 13 shows a schematic cross–section of the central parts of an AGB star, emphasizing the AGB core, which will become a white dwarf. The core consists of a central region of degenerate carbon and oxygen, at a temperature exceeding 10^8 K. Overlaying this is a thin layer of nearly pure helium. Between the helium shell and the C/O core is a region where nuclear fusion is converting helium into carbon and oxygen. Separating the helium layer and the AGB envelope is where the occasionally active hydrogen burning shell. Nearly the entire radius of the star consists of the convective AGB envelope, which extends from the core (at less than 10% of a solar radius) out to over 100 $R_\odot$. Still, the tiny volume of the AGB core determines the stellar luminosity.

The eventual evolution will result in the ejection of the AGB envelope and the exposure of the AGB core as a proto-white dwarf. The final phases of evolution on the AGB determine the outer layer structure of the remnant white dwarf. Those late phases feature large and quasi–periodic excursions of luminosity as the principal energy source oscillates between helium and hydrogen shell burning. Depending on the phase of these oscillations at which the star leaves the AGB, it can become either a DA or a DB white dwarf star. Therefore, understanding

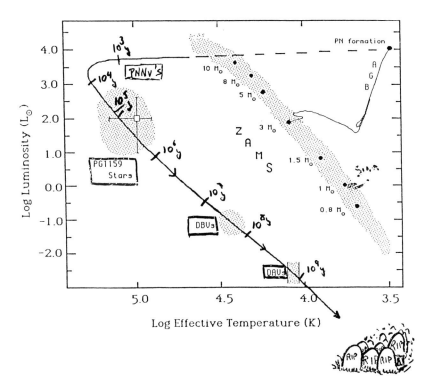

Fig. 12. Schematic evolutionary track of a star from the main sequence through the white dwarf cooling track, adapted from Kawaler (1986). The shaded band represents the Main Sequence. Three regions where white dwarfs are observed to pulsate are also shaded. The evolutionary track eventually leads to the stellar graveyard.

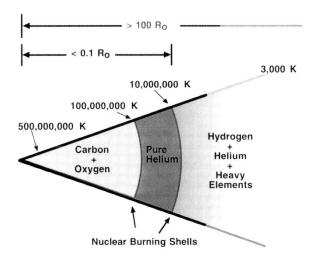

Fig. 13. Schematic diagram of the central portion of an AGB star.

these so-called thermal pulses on the AGB is important in understanding the distribution of white dwarfs of various spectral types.

One can think of the helium–burning shell as a carbon/oxygen (C/O) factory, which takes the raw material (helium) from above, passing it through the nuclear furnace, and sending the product (C/O) out the bottom. Similarly, the hydrogen–burning shell is a helium "factory" which uses the raw material in the stellar envelope to make helium. This helium, in turn, is processed by the helium–burning shell to produce C/O. As in any manufacturing chain, one wants the chain to work in synchrony; that is one wants the output rate from one factory to match to the input rate for the next down the line. If the first factory works too fast, then its product piles up. Such a pile–up, in industry, results in lost profits; in an AGB star its results are nearly catastrophic. Indeed, the hydrogen–burning shell produces helium faster than the underlying helium–burning shell can further process it into C/O. So the helium piles up, compressing to nearly degenerate densities. Still, the helium shell is not hot enough to process helium quickly enough. With mildly degenerate helium, finally, the rate of helium burning increases dramatically as the helium can heat up without a correspondingly large increase in pressure (and subsequent expansion). The acceleration of helium burning can be quite rapid, though not nearly as devastating as the much more degenerate helium core flash. The shell sources undergo a "thermal pulse", or "helium shell flash".

Many investigators have studied the shell flash phenomenon; Iben (1991) succinctly reviews the anatomy of these pulses. During most of the time on the AGB, the stellar luminosity is generated by the hydrogen–burning shell. Eventually, enough helium accumulates above the helium-burning region that vigorous helium burning begins. The luminosity of the helium–burning shell increases on a very short time scale (of order 30 years), and helium burning becomes convective. This very rapid input of energy into that part of the star causes that region of the star to expand (not all of the energy can escape as photons on such a short time). This pushes out the region where hydrogen is burning, cooling the region and extinguishing the hydrogen burning shell.

Since the hydrogen–burning shell was providing the bulk of the observed stellar luminosity, when it goes out, the luminosity of the star decreases somewhat. Eventually, the energy from the helium burning shell diffuses outward, and the luminosity of the star recovers, only to drop again as helium burning quiets down. During the pulse, helium burning is sufficiently vigorous that it processes most of the accumulated helium. As the hydrogen shell recovers and hydrogen reignites, the luminosity of the star continues to rise until it reaches approximately the luminosity before the flash. With hydrogen burning reestablished, and the helium burning shell quiet, the process of accumulation of helium above the helium–burning shell returns, setting the stage for the next thermal pulse. The pulse itself takes about 400 years from the time of the ignition of the helium shell to the time when hydrogen burning again becomes the dominant source of luminosity for the star. It takes about 25% of the time between pulses for the luminosity of the star to reach its pre–flash luminosity, while thermal pulses themselves repeat every 10^5 years or so in AGB stars with masses near

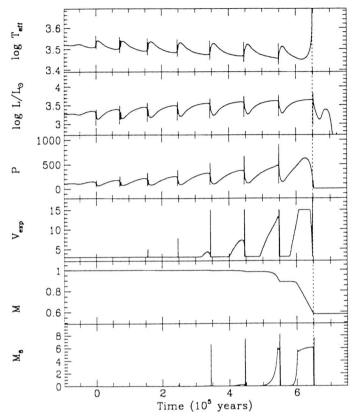

Fig. 14. Evolution of the properties of a 1 solar mass AGB model through several thermal pulses. From Vassiliadis & Wood (1993).

about 1–2 solar masses. Figure 14 shows how several properties of a thermally pulsing AGB star change with time.

The interior structure undergoes rapid transformations through a shell flash because of the transient convective helium–burning shell. Figure 15 is a rough sketch, shown in lecture, of the helium abundance as a function of fractional mass. The helium abundance is zero in the C/O core, rises quickly to nearly 1.0 through the helium–burning shell, and drops down to the primordial abundance as one moves further outward through the hydrogen–burning shell. The abundance of helium is constant in the outer layers because of convective mixing in the AGB envelope. Upon ignition of the helium–burning shell, convection within the shell reduces the helium abundance and increases the C/O abundance, as does the nuclear processing of helium into C/O. Following the shell flash, helium burning continues at a low level, but the helium abundance retains the multistep profile sketched in the figure. Hydrogen burning begins again, keeping the convective envelope at bay. Successive helium shell flashes can result in a very complex helium abundance profile. These steps have important consequences

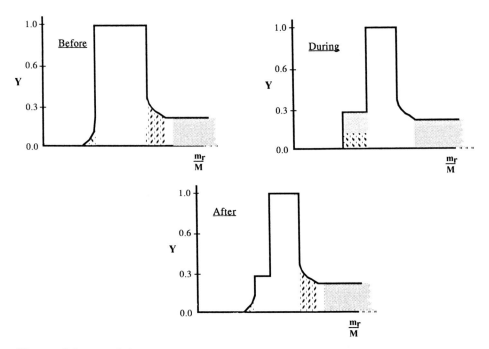

Fig. 15. Schematic helium mass fraction before, during, and after an early helium shell flash. Dotted areas indicate convective areas, and striped areas have active nuclear burning.

in terms of the chemical evolution of white dwarfs later on, and are a direct reflection of the thermally–pulsing AGB phase.

4.2 Departure from the AGB

AGB stars are observed to be mass–losing stars, with mass loss rates approaching $10^{-4} M_\odot/\text{yr}$. They also show large–amplitude radial pulsations as Mira and/or semi–regular variable stars. With mass loss along the AGB, the envelope mass is decreasing with time. As the core mass increases, the luminosity of the AGB star increases, and the mass loss rate increases accordingly.

The envelopes are extremely extended and tenuous; they are held only weakly by the gravitational attraction of the core. As a result, radiation pressure on molecules and dust grains can contribute to mass loss. In addition, pulsational energy is being deposited in the envelope, causing it to be extended even further than static atmospheric calculations indicate. The connection between pulsation and mass loss is well established in AGB stars both observationally and theoretically. The observational correlation between mass loss rate and pulsation properties has been reviewed recently by J. Chapman (1995); while the theoretical results are compactly described by Bowen & Willson (1991). Figure 16, from the paper by Bowen & Willson, illustrates the evolution of AGB stars with

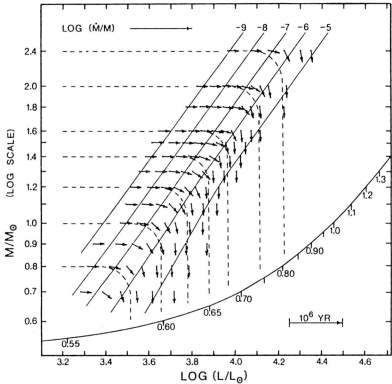

Fig. 16. The evolution of AGB stars including mass loss from pulsation. The mass loss rates are represented by diagonal lines, labeled with the log of the mass loss rate in solar masses per year. Arrows signify the direction of evolution of models at various positions on the plane. Mass loss rates are those computed using hydrodynamical models of pulsating Mira envelopes. From Bowen & Willson (1991).

pulsation–driven mass loss. A given AGB model evolves from left to right as its luminosity increases, and downwards as its mass decreases. As its luminosity increases, the mass loss rate slowly rises, until the model reaches a luminosity where pulsation–driven mass loss brings the mass loss rate up to nearly $10^{-5} M_\odot/yr$. At that point, mass loss occurs on a time scale much shorter than the growth of the core (and increase in luminosity) and the model loses its entire envelope. In Figure 16, this is indicated by the track heads mostly downwards, towards the upward–curving line that represents the core mass for the given luminosity. Bowen & Willson describe the evolution in this plane as a model reaching, and falling over, the cliff.

With mass loss, the envelope mass (M_{env}, defined as the mass of material that lies outside of the hydrogen burning shell) decreases. Since the evolution of the core and envelope are largely decoupled in AGB stars, the decrease in envelope mass has little effect on the effective temperature of the AGB star. In fact, an intermediate mass AGB star can lose a large fraction of its total mass

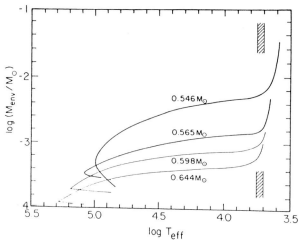

Fig. 17. Dependence of T_{eff} on envelope mass, from Schönberner (1983), for several different core masses in AGB and PN central star models with hydrogen–burning shells.

and still remain on the AGB. The effective temperature is a very weak function of M_{env} until M_{env} falls to about $10^{-3} M_\odot$, at which point T_{eff} becomes a very strong function of M_{env}, as first shown by Paczynski (1970). Figure 17, from Schönberner (1983), illustrates this effect for several different core masses. As mass loss continues on the AGB, the effective temperature changes only slowly until the envelope mass thins to $0.001 M_\odot$. Further mass loss causes a rapid increase in the effective temperature. This is a fancy way of saying that the white dwarf core is being exposed as the last part of the envelope is removed.

Since the luminosity of the star is determined by the core mass, this increase in effective temperature is manifest as a rapid evolution to high T_{eff} at nearly constant luminosity as the envelope is removed. How quickly does the star cross the H–R diagram ? The envelope mass decreases because of mass loss, but hydrogen shell burning also acts to reduce the mass of the envelope from below. Thus the speed at which the star crosses the H–R diagram is determined by both the mass loss rate and the rate of nuclear burning. An approximation to the curves in Figure 17 is that $d(\log T_{\text{eff}})/d(\log M_{env})$ is about -3.3. The rate of envelope shrinkage is given by the rate at which hydrogen burning consumes hydrogen and the rate of mass loss. In general, the time scale for increase in T_{eff} is

$$\tau^{-1} = \left(\frac{d \log T_{\text{eff}}}{d \log M_{env}} \right) \frac{1}{M_{env}} \frac{dM_{env}}{dt}. \tag{61}$$

With nuclear burning alone, (see equation 6), it is easy to show that

$$\tau_{\text{nuc}} \approx 60,000 \,\text{yr} \left(\frac{L}{1000 L_\odot} \right)^{-1} \left(\frac{M_{env}}{10^{-3} M_\odot} \right) \tag{62}$$

when hydrogen burning is occurring. Thus even in the absence of mass loss, the nuclear evolution of the core can result in a rapid rise in the effective temperature with an e-folding time of 60,000 years. Mass loss accelerates this evolution because it assists the reduction of the mass of the envelope. Acting alone, the time scale for $T_{\rm eff}$ increase is

$$\tau_{\dot M} \approx 30{,}000\,{\rm yr}\left(\frac{\dot M}{10^{-8}M_\odot/{\rm yr}}\right)^{-1}\left(\frac{M_{env}}{10^{-3}M_\odot}\right). \qquad (63)$$

The representative value of the mass loss rate above is typical of PN central stars for which mass loss rates have been measured. Thus mass loss contributes significantly to the rate of evolution of post–AGB stars. The resulting time scale is the combination of the two effects above:

$$\tau^{-1} = \tau_{\rm nuc}^{-1} + \tau_{\dot M}^{-1}. \qquad (64)$$

At the terminal stages of AGB evolution, the extremely high mass loss rate that leads to exposure of the core has been termed the "superwind" by Renzini (1981). This superwind has been identified by Bowen & Willson (1991) as mass loss induced by pulsation in long–period variables near the "cliff" in Figure 16. Once the star becomes hot enough that the large-amplitude pulsation ends, the mass loss rate again drops down to smaller levels, but now the core has been exposed and detached from the envelope. The material that remains near the star, if dense enough, can be ionized by the increasingly hard spectrum from the hot, exposed core, producing a planetary nebula.

4.3 The PNN Phase

Depending on the power source for stars that are entering the white dwarf cooling track from the AGB, the evolution across the H–R diagram and at the top end of the white dwarf cooling track can be relatively fast or slow. The previous subsection described the factors governing the rate of evolution across the H–R diagram in general terms. For more realistic model calculations with hydrogen–burning shells, several workers have shown (i.e. Schönberner 1983, Iben 1984) that the time for crossing the H–R diagram is inversely proportional to the mass of the star to the 10th power! That is, massive remnants cross the H–R diagram much faster than less massive remnants.

The temperature of the central star of a planetary must be at least 30,000 K to ionize the surrounding material; that is, to cause the PN to fluoresce. The remnant must reach this temperature before the surrounding material (the AGB envelope remnant) expands away and dissipates. Similarly, if the remnant is of sufficiently high mass, it will heat up, and then fade in luminosity before the expanding material becomes optically thin. For a planetary nebula to become visible, the central star (or planetary nebula nucleus, PNN) must evolve to 30,000 K before the nebula expands too much, but not so fast that the envelope remains optically thick. Therefore, true planetary nebula central stars are a subset of the class of pre–white dwarfs that have come from the AGB. The severe selection effect caused by the blazing neon sign aids the discovery of pre–white dwarfs that have surrounding ionized nebulae, and since a small range of remnant masses evolve at the right rate to produce visible nebulae, one would expect the mass distribution of planetary nebula central stars to be even narrower than that of cooler white dwarfs.

Indeed, Weidemann (1990) indicates that the dispersion in planetary nebula central star masses is indeed small, but with a mean that is almost the same as cooler white dwarf stars. The narrower mass dispersion in PN central stars indicates that PNNs are not the only feeding channel to the white dwarf cooling track. It appears to be entirely coincidental that the matching of the time scale for nebular expansion to the heating (and fading) rates of the central star that produce observable planetary nebulae occurs at a mean mass of $0.58 M_\odot$, and that in turn is very close to the mean mass of white dwarf stars. There is no reason why this is so; it is just a wonderful coincidence that allows us to connect directly AGB stars and white dwarfs.

The surface structure of PNNs depends on exactly how it left the AGB. Iben (1984) shows that low–mass AGB stars spend about 80% of their time with their luminosity provided by the hydrogen–burning shell, and only 20% of the time (during and soon after a shell flash) dominated by the helium–burning shell. If the probability of leaving the AGB is uniformly random for any luminosity on the AGB, then 80% of stars should leave the AGB while burning hydrogen, and become hydrogen–burning PNNs. If that is the case, then mass loss will stop above the hydrogen–burning shell, and result in a remnant with a hydrogen shell of about $10^{-4} M_\odot$. This hydrogen shell thickness is set by the minimum mass required to support hydrogen burning; if the mass of the shell is less than this number, the temperature at the base of the hydrogen shell is insufficient for hydrogen fusion to occur. Below this layer lies the nearly pure helium layer, with a mass of approximately $10^{-2} M_\odot$; at the center lies the carbon/oxygen core, with the bulk of the mass of the remnant. If the star leaves the AGB during the 20% of the time that it is burning helium (i.e. during or soon after a thermal pulse), then it will evolve into a helium–burning PNN. The helium layer is approximately $10^{-2} M_\odot$, and it surrounds the C/O core.

The two possible compositional structures that follow departure from the AGB are illustrated schematically in Figure 18. A third possibility has been explored extensively by Iben (see Iben 1991). Some fraction of AGB stars may leave the AGB just prior to a thermal pulse. If the time taken for the star to cross the H–R diagram is longer than the time it would have taken to reach the next shell flash, a final thermal pulse may occur while the star is still at high luminosity but near the white dwarf cooling track. This final thermal pulse can cause the star to return (briefly) to the AGB, where mass loss can again be significant. In this case, most of the helium layer can be ejected as well as any residual hydrogen, leaving a pre–white dwarf with a surface depleted not only of hydrogen, but of helium as well. Since helium burning will occur at the base of the helium–rich shell, the material removed can reach down to one of the "steps" in the composition profile that were developed during prior shell flashes. This third possibility is illustrated in Figure 18 as well; Figure 19 (from Iben 1991) shows a possible evolutionary track for such a "born–again" AGB star.

a) if departs AGB burning hydrogen:

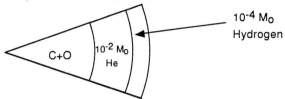

b) if departs AGB burning helium:

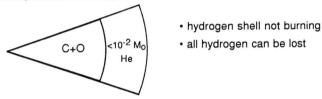

c) if "Born Again" -> late shell flash (Iben 1983):

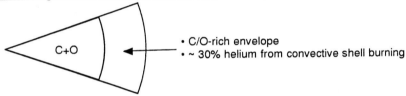

Fig. 18. Schematic cross–sections of pre–white dwarf stars that follow departure from the AGB a) during hydrogen burning, b) during helium burning and c) after a late shell flash.

4.4 Nuclear Shutdown and Neutrino Cooling

Eventually, either from mass loss or from nuclear burning from below, the mass of the outer hydrogen (or helium) layer is reduced to the point where there is insufficient mass above for the temperature to be high enough to sustain nuclear fusion. When this occurs, nuclear energy ceases to be its source of luminosity. However, the layers where nuclear burning was active retain the thermal structure associated with nuclear burning for a time; the reduction of their heat content cannot occur faster than the local thermal time scale. As the nuclear shell source quits, contraction occurs to prevent the star from cooling too quickly where nuclear burning cannot be sustained. Energy generation by nuclear burning is gradually replaced by energy provided by gravitational contraction.

In terms of the evolution of the star in the H–R diagram, this phase is represented by the "knee" in the evolutionary track where the model leaves the constant–luminosity track and begins to cool and to fade. The rate at which its luminosity drops (and at which it "moves" in the H–R diagram) accelerates when nuclear burning dies out. For hydrogen–burning central stars, this may result in a very rapid acceleration of the evolution and fading of the star, as the outer layers

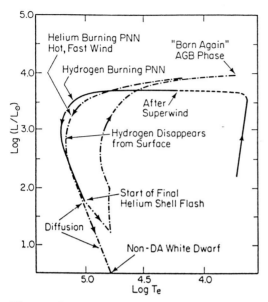

Fig. 19. Evolutionary track of a post–AGB star that experiences a final helium shell flash while on the hot end of the white dwarf cooling track. From Iben (1991).

contract (Schönberner 1983). For helium–burning central stars, the acceleration is more modest. Once the energy of the (now fossil) nuclear burning shell has diffused away, the evolution is steadily slowed by the decreasing luminosity (i.e. smaller rate of energy loss).

The different rates at which hydrogen–burning and helium–burning central stars fade give us (in principle) a way to determine which luminosity source is the preferred one for PNNs (Schönberner 1983). If an accurate luminosity function for PNNs could be constructed, for example, then a gap in the planetary nebula luminosity function (PNLF) corresponding to rapid evolution following nuclear shell shutdown would indicate that most PNNs are hydrogen burning. However, due to the very peculiar selection effects involved in galactic PNN observations, such an effect is not possible to demonstrate with current data (Kawaler 1990b). Other studies of the PNLF in external galaxies (i.e. Jacoby 1989, and Ciardullo et al. 1991, Dopita et al. 1992) show that the PNLF is somewhat more consistent with helium–burning central stars than with hydrogen burners. Jacoby (1989), for example, shows that the best fit with hydrogen–burning models requires a mean mass ($0.65 M_\odot$) significantly larger than that for galactic PNNs, while helium–burning models fit the data nicely with a mean mass of $0.60 M_\odot$. Still, the question of what powers central stars of planetary nebulae remains an open one.

In models of PNNs with masses of about $0.60 M_\odot$, at about the time that nuclear burning ceases to be an important luminosity source, another nuclear process begins to dominate. Figure 20 shows the luminosity as a function of time for simple white dwarf models from Kawaler (1986). Energy loss by neutrino

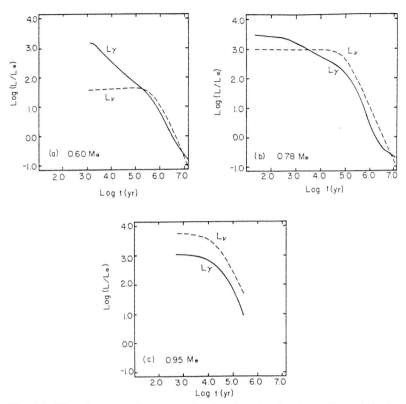

Fig. 20. The photon and neutrino luminosities in simple cooling white dwarf models with masses of $0.60 M_\odot$, $0.78 M_\odot$, and $0.95 M_\odot$. From Kawaler (1986).

emission exceeds the photon luminosity during the time the white dwarf models cool from $30 L_\odot$ down to about $0.3 L_\odot$. In more massive white dwarfs, the neutrino luminosity can dominate the photon luminosity for most of the cooling of the star from higher luminosities to lower. The processes responsible for production of neutrinos plasmon and bremsstrahlung processes; see Beaudet et al. (1967) for a general treatment of neutrino production in dense plasmas. In lower mass white dwarfs, plasmon neutrinos are the dominant species, while in more massive white dwarfs bremsstrahlung neutrinos carry away most of the energy. Thus neutrinos are the dominant coolant (even over photon emission) for about 10^7 years in ordinary white dwarf stars.

5 Chemical Evolution of White Dwarfs

The fundamental observation that white dwarfs come in two flavors (hydrogen–rich DAs and helium–rich DB/DOs) represents a modest challenge to understanding white dwarf formation and evolution. This primary observation, that about 80% of white dwarfs are DAs and nearly all of the rest are DOs, can be explained as reflecting the phase at which the white dwarf left the AGB. As pointed out by Iben (1984), hydrogen burning is the dominant luminosity source during roughly 80% of the time on the AGB (between thermal pulses); if a star leaves the AGB while burning hydrogen, then it will be left with a dominantly hydrogen surface layer. The remaining 20% of the time, during and immediately following a thermal pulse, helium–burning dominates and the hydrogen–burning shell is not active. Thus the star can lose the entire hydrogen–rich envelope and expose the nearly pure helium layer.

However, the ratio of non–DA to DA white dwarfs changes dramatically when considering white dwarfs in restricted temperature ranges. At high temperatures, very few white dwarfs show measurable hydrogen. The ratio of non–DA to DA white dwarfs increases with decreasing temperature; by 45,000 K there are no helium dominate white dwarfs. Below 30,000 K, however, the DB stars (magically, so far) reappear, reaching the 20% level until still lower temperatures where they again become the dominant species. These changes, as well as the purification of the surface layers, must result from a combination of history and internal processes; that is, the surface abundances and their changes with time reflect the conditions at the formation of the white dwarf, the microscopic process of diffusion, and the macroscopic processes of accretion, mass loss, and convective mixing. This section discusses these processes.

5.1 Diffusive Processes

The purity of white dwarf atmospheres was an early puzzle in astrophysics that was solved by Schatzman nearly 50 years ago (Schatzman 1950, 1958). He demonstrated that in stars with such high gravity, heavier elements will settle down and lighter elements will float to the top on time scales that were relatively short compared to the stellar ages. Diffusive processes was later included in realistic white dwarf models when the computational tools were developed in the 1970s by Michaud, Fontaine, and others.

At birth, white dwarfs are naturally stratified by shell burning, but the outer hydrogen–rich layer contains the primordial mix of hydrogen, helium, and heavier elements; similarly, the helium–rich layer contains the primordial metal abundance. These elements can experience accelerations simply because of gradients in their number densities. In most general terms, the number densities of various species change with radius in a white dwarf, and therefore are subject to diffusive motion through the action of several processes. The basic equation governing diffusion is

$$\frac{\partial n_j}{\partial t} = -\frac{1}{r^2}\frac{\partial \left(r^2 w_j n_j\right)}{\partial r} \tag{65}$$

where w_j contains all of the magic; it is the diffusive velocity of species j with respect to the center of mass of the system. The diffusion velocities for each species are given in terms of diffusion coefficients K_{ij} which, in turn, are related to physical processes. For example,

$$\frac{1}{n_i}\sum_j K_{ij}\left(w_i - w_j\right) - Z_i eE = -A_i m_H g - kT\frac{d\ln n_i}{dr} - kT\frac{d\ln T}{dr} \tag{66}$$

provides w_i in terms of the diffusion velocities of the other species and the local gradients in pressure, temperature, and number density (Iben & MacDonald 1985).

One sees that at high gravity, heavier elements have higher differential velocities inwards; this (the first term on the right–hand side of the above equation) leads to gravitational settling. Acting alone, it would eventually result in discontinuous composition profiles, with lighter elements floating above deeper layers with successively increasing atomic weight. However, the second term on the right–hand side, which represents ordinary diffusion, works against the formation of excessively steep composition gradients. In the vicinity of a discontinuity, the gradient becomes strongly negative, resulting in a diffusive force pushing heavier elements outward and lighter elements inward. The third term above represents thermal diffusion which, for white dwarfs, is usually a much smaller effect than either of the other two. Though gravitational settling, and ordinary diffusion are usually the dominant processes acting in white dwarfs, at certain times other processes, such as radiative levitation, may be important. These other diffusive processes can be incorporated into the right–hand side of the above equation.

With gravitational settling competing with ordinary diffusion, it is clear that an equilibrium condition will eventually develop when the composition profile results in zero diffusive velocity for all components. That is, the time derivative in equation (65) can vanish when gravitational settling balances ordinary diffusion. The local time scale for composition change (and therefore of the approach to diffusive equilibrium) can be very short near the stellar surface, but increases rapidly with depth. Since other physical processes, such as stellar winds and mass loss, can prevent the operation of diffusion during early evolution, the asymptotic approach to equilibrium, in practice, means that true diffusive equilibrium is not obtained during the life time of most white dwarf stars. For the oldest white dwarf stars, however, diffusive equilibrium is a reasonable approximation.

Gravitational settling and ordinary diffusion are the two dominant diffusive processes that affect the outer layer abundances of hydrogen and helium in white dwarfs. As a simple illustration of the action of diffusion, Figure 21 shows an evolutionary calculation of the evolution of a helium–rich white dwarf model from Dehner & Kawaler (1995). This model started with a surface layer thickness of several $\times 10^{-3} M_\odot$ of uniform composition (30% helium, 35% carbon, and 35% oxygen). The helium abundance is shown as a function of the log of the surface mass fraction; the surface of the model is to the left on this scale. This figure shows how helium "floats" to the surface as the carbon and oxygen sinks. Once diffusion was "turned on", the helium rapidly concentrated near the surface. Since there was no hydrogen in this model, this shows how the lightest element present in a mixture rapidly rises to the surface. Still, the abundance deeper within the star continues to change with time, and the composition transition zone moves inwards into the model. The equilibrium profile would be centered at the approximate position of the transition to pure C/O on this scale, with a shape similar to that of the coolest model.

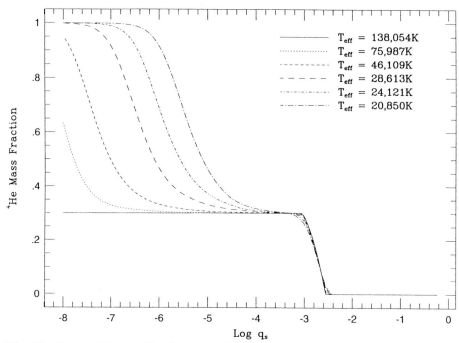

Fig. 21. Composition profiles for an evolutionary sequence of $0.58 M_\odot$ helium–rich white dwarf models models. The abscissa is the log of the surface mass fraction, given by $q_s = (M_* - M_r)/M_\odot$. From Dehner & Kawaler (1995).

While gravitational settling causes heavier elements to sink, another process tends towards pushing higher Z elements outwards. This process is known as "radiative levitation". Heavier elements have large photon absorption cross sections when partially ionized. Hence they are good absorbers of photons from the radiation field of hot white dwarf stars. Photon absorption can result in some fraction of the absorbed photon's momentum being converted into outward translational momentum in the absorbing ion. Thus heavy (or opaque) ions can be floated outward by the radiation field. Ions such as iron and calcium are particularly susceptible to radiative levitation when the radiation field is intense enough (i.e. in the hot white dwarf stars).

Clearly, computation of radiative levitation is extremely complex, as it requires knowledge of the frequency–dependent cross sections of all elements of interest, computed in a radiation field that is affected by dynamical motions. Though radiative levitation can be treated formally using terms similar to those in the equations above, the complexity of the associated diffusion coefficients makes it difficult even to perform equilibrium (i.e. stationary) calculations. Still, calculations have been performed by a number of groups for specialized problems, which show that the effect is real, and (in general) produces spectral features and behavior that is consistent with observations. A recent equilibrium calculation is presented by Chayer et al. (1995a, 1995b). One primary reason to suspect that radiative levitation can be important comes from the surface abundances of metals in the PG 1159 DOZQ and DAZ stars. Another is that the number of hot white dwarfs detected in the EUV and X-ray bands is smaller than would be expected if they had pure helium (or hydrogen) surfaces (see, for example, Barstow et al. 1993). Some source of opacity in the EUV must be present in these stars; heavy elements appear to be the most likely suspect.

5.2 Accretion of "Fresh" ISM vs. Mass Loss

Diffusive processes can affect the surface abundances of white dwarfs. But in the absence of other effects, once a surface abundance pattern has been established, diffusion will no longer change it; the time scale to reach equilibrium concentrations at the photosphere is extremely short. Other processes, however, can change this surface abundance actively or by selectively suppressing diffusive transport.

One process is accretion of the interstellar medium. White dwarfs live a long time; most that we see are located in the Milky Way disk, where the interstellar medium can be relatively dense. Since the ISM is usually hydrogen and metal rich, recent accretion by a white dwarf could show it to be either hydrogen rich or metal rich independent of its prior (self-contained) history. Another process that could, for example, turn a hydrogen–surface white dwarf into a DB or DO white dwarf is mass loss. Even moderate mass loss can strip a DA star of its hydrogen in a time less than its cooling time. Mass loss competes directly with accretion in this sense.

Accretion rates for white dwarfs in an ISM can be estimated using Bondi/Hoyle accretion as

$$\dot{M}_{\rm accretion} \approx 10^{-15} \frac{M_\odot}{\rm yr} \left(\frac{n}{\rm cm^{-3}}\right) \left(\frac{v}{\rm 10km/s}\right)^{-3} \left(\frac{M}{M_\odot}\right)^2 \quad (67)$$

where n is the number density of the ISM, and v is the white dwarf velocity with respect to the ISM. Thus in a very short time, a white dwarf can sweep up a photosphere's worth of hydrogen if accretion is allowed to proceed. However, we see cool DB stars that have cooling ages of 10^8 years and more. In that time, they would have accreted more than $10^{-7} M_\odot$ of hydrogen. To prevent accretion, then, DB white dwarfs must have some very modest wind associated with them... in the absence of magnetic fields or other anti–accretion measures. To first order, a mass loss rate equaling the accretion rate, at a level of $10^{-15} M_\odot/{\rm yr}$, could then stop accretion. Mass loss at such a low level would only result in a total mass loss of $10^{-5} M_\odot$ over the lifetime of a white dwarf. Thus a white dwarf that began as a DA with $10^{-4} M_\odot$ of hydrogen would remain a DA white dwarf, and a DB white dwarf would remain a DB white dwarf without accreting a hydrogen envelope.

A more careful calculation of accretion and mass loss in white dwarfs by Jim MacDonald (1992) that treats interaction between the white dwarf wind and the ISM finds that mass loss rates of 3×10^{-18} to $10^{-21} M_\odot/{\rm yr}$ are sufficient to prevent accretion. As quoted by MacDonald (1992) radiative wind mass loss is given approximately by

$$\dot{M} \approx 2 \times 10^{-15} \left(\frac{L}{L_\odot}\right)^2 \frac{Z}{0.02} \frac{M_\odot}{\text{yr}} \qquad (68)$$

where Z is the metallicity. Thus even for cool white dwarfs, with luminosities significantly less than solar, the mass loss rate can exceed the accretion rate.

The corresponding flow velocity for a given mass loss rate is

$$v \approx 10^{-8} \text{cm/s} \left(\frac{\dot{M}}{10^{-15} M_\odot/\text{yr}}\right) \left(\frac{R}{0.01 R_\odot}\right)^{-2} \left(\frac{\rho}{1 \text{ g/cm}^3}\right)^{-1} \qquad (69)$$

At these very low mass loss rates, the resulting mass flux in the outer layers of white dwarfs is tiny enough that it will not affect gravitational settling and diffusion; the diffusive velocities of 10^{-7} cm/s are much larger than the velocities associated with the mass flux. Mass loss from white dwarfs can indeed prevent accretion of normal ISM material and still allow diffusive separation of light and heavy elements in the interior.

However, earlier in their evolution (i.e. at higher luminosities) the mass loss rate is significantly larger. For example, at the luminosity of PG 1159 (about 200 $L_\odot$) the radiative wind should result in a mass loss rate of several times $10^{-12} M_\odot/\text{yr}$ or more. The flow velocity corresponding to this high a mass loss rate is several times 10^{-5} cm/s, which is much larger than the diffusive velocities. Thus radiative mass loss will suspend diffusive processes such as gravitational settling and ordinary diffusion until the mass loss rate drops below about $10^{-14} M_\odot/\text{yr}$ or so (i.e. at luminosities of a few times solar and $T_\text{eff} \approx 60,000$ K). This may explain why we see carbon and oxygen at large abundances in the PG 1159 stars down to about $T_\text{eff} \approx 75,000$ K without having to resort to radiative levitation.

5.3 Convection

Another process that can change the surface abundance of white dwarfs is surface and subsurface convection. Convection is a rapid and active mixing process; as a white dwarf cools and surface convection deepens, convection mixes material up from below the photosphere. If the surface convection zone penetrates through a pure surface layer and into a deeper layer of different composition, the surface abundance can change very rapidly with further cooling. Figure 22 shows the depth of the surface convection zones of DA and DB white dwarf models, from Tassoul et al. (1990. In DB models, convection first appears in the very outermost layers at fairly high temperatures. However, the zone deepens below the photosphere at T_eff of about 18,000 to 20,000 K in models that employ standard mixing–length theory (the "1" curves). For models simulating more efficient convection, the surface convection zone becomes deeper at higher T_eff, as shown in the figure by lines labeled "2" and "3". In the DA models, convection begins at

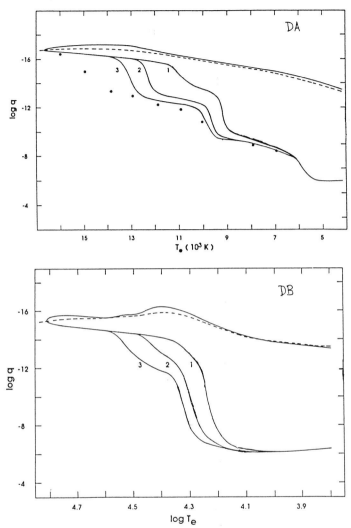

Fig. 22. Development of surface convective layers in DA and DB white dwarf models. Solid lines enclose the convectively mixed layers, with lower boundaries labeled by the mixing length formalism used. The ordinate of both panels is the logarithm of the surface mass fraction. Lines are labeled by the convective prescription used. ML1 is standard mixing length theory; ML2 and ML3 use progressively more efficient transport of energy in convection zones. Adapted from Tassoul et al. (1990).

T_{eff} of approximately 15,000 K. The zone deepens significantly at temperatures between 13,000 K and 11,000 K, depending on the efficiency of convection.

The figures show the range over which convective mixing occurs. Note that in DA models, the convective zone reaches down close to the bottom of the hydrogen–rich layer even in those white dwarfs that have the maximum permissible hydrogen layer. Below about 6000 K — that is, in the coolest white dwarf stars — convection may mix the surface hydrogen with the deeper (and thicker) helium, converting DA white dwarfs into helium–rich white dwarf stars. No such dramatic conversion of DB white dwarf stars into DQ or DZ stars is expected unless the surface layer of helium is less than about $10^{-5} M_\odot$.

5.4 Chemical Evolution Scenarios

As outlined previously, the relative numbers of white dwarfs with hydrogen or helium–dominated atmospheres changes several times as one considers cooler and cooler stars. This distribution is shown schematically in Figure 9. Such apparent evolution in the surface abundances of white dwarfs is widely interpreted as changes in surface abundances of individual white dwarfs as they cool. The above sections outlined the various processes that can change the surface (and subsurface) abundances of white dwarf stars. Presumably, some combination of these processes results in the observed distribution of white dwarf surface abundances. Any description of white dwarf chemical evolution has to address these relative distributions, and in particular, must be able to explain the deficit of helium–rich white dwarfs, the "DB Gap", between 45,000 K and 30,000 K.

The future development of a white dwarf must be influenced by its particular formation history. Depending on precisely when a star leaves the AGB, it could be left with a hydrogen–rich or helium–rich envelope (see for example Schönberner1983, Iben 1984, or Wood and Faulkner 1986,). Those stars which leave the AGB between helium shell flashes do so while hydrogen burning is the luminosity source. These stars presumably continue to burn hydrogen through the PN central star phase. Eventually, the hydrogen shell source is extinguished and, barring further mass loss, these stars will retain a surface hydrogen layer of about $10^{-4} M_\odot$, overlaying a helium layer of $\approx 10^{-2} M_\odot$. Stars which leave the AGB during a helium shell flash can lose most or all of their entire hydrogen–rich outer envelope, and retain a helium–rich envelope of less than $10^{-2} M_\odot$. To understand the relative numbers of white dwarfs of different surface abundances at different temperatures, consideration of the initial surface abundance is critical. Attempts at understanding the later changes in relative numbers fall into two categories that are described below.

Multichannel In the multichannel scenarios (see, for example, Sion 1986), DB white dwarfs are formed by stars which leave the AGB during a thermal pulse. These stars begin with helium rich (or at least hydrogen–poor) surfaces and evolve to very low temperatures with permanent helium surfaces. Similarly, DA white dwarfs are formed when stars leave the AGB between thermal pulses, during times of hydrogen shell burning. These schemes allow for subtle abundance variations to occur as a result of accretion of hydrogen (and metals) from the interstellar medium, mass, loss, diffusion and radiative levitation of carbon, oxygen, and other elements within the stellar photosphere.

The multichannel scenarios allow for natural explanation of the relative numbers above and below this gap in terms of the precursor phase. That is, the relative number of DBs to DAs reflects the relative fraction of the time an AGB star spends in and between shell flashes. Thus if the probability of a star leaving the AGB is independent of the shell burning phase, these models produce the overall ratio of DA stars to non–DA stars.

Unfortunately, these scenarios all face the major difficulty of explaining why we see no DB white dwarfs between 30,000 K and 45,000 K. If cooler DB white dwarfs started out without hydrogen, then they must always have been DB white dwarfs, contrary to the existence of the gap.

Single–Channel Single channel schemes for the origin of white dwarfs from AGB progenitors have grown in popularity in recent years. In all of these various models, a variety of physical processes are responsible for converting a given white dwarf from a DB to a DA (and back) as it cools. A general scenario (see Fontaine & Wesemael, 1987) involves a white dwarf starting off with a predominantly helium surface, but with trace amounts of hydrogen, upon leaving the AGB during a shell flash. This star will initially appear as a DO white dwarf. As it cools, the hydrogen floats to the surface (and heavier elements sink), eventually producing a hot DA white dwarf when the surface hydrogen–rich layer becomes deeper than a few $\times 10^{-16} M_\odot$. Eventually, subsurface convection in the helium–rich interior mixes this hydrogen back down, and the star becomes a DB once again. In a model such as this, the DB gap can be explained as the effective temperatures at which the hydrogen has reached the surface but has not yet been mixed down. Such a scenario places strong (and sometimes contradictory) constraints on the composition of the outer layers of white dwarfs at their formation, and on the amount of mass lost over their lifetime.

Advantages of the single–channel models include explanations for the DB gap, the lack of very hot DA stars, and the increasing ratio of DA to non–DA white dwarfs above the gap. Additionally, it allows for a range of progenitor hydrogen and helium abundances within narrowly specified and observationally testable ranges. However, the details of such a model depend on physical processes described above (convection, accretion, diffusion, and mass loss) that are extremely difficult to model. More significantly, a possible flaw of the single–channel idea is that it requires a distribution of hydrogen layer thickness in cool white dwarfs. The results of pulsation observations and modeling suggests that the ZZ Ceti stars all have thick hydrogen layers (near $10^{-4} M_\odot$), and that few (if any) DA white dwarfs that lie within the ZZ Ceti instability strip for their mass do not pulsate (Clemens 1993a). Therefore, since all DA stars must at one time or another cool through the ZZ Ceti instability strip, thick surface hydrogen appears to be a rule.

It is safe to say that despite sophisticated models and detailed observations, understanding the chemical evolution of white dwarf stars in terms of a simple and coherent evolutionary scheme is still not possible.

6 White Dwarf Cooling and the White Dwarf Luminosity Function

As Don Winget has stated many times, the history of star formation in the disk of our galaxy is written in the coolest white dwarf stars. These cool white dwarfs are the oldest stars in the disk of our galaxy; by understanding their statistics, we can recreate the star formation rate and its changes over the entire history of the galaxy. In this section, we discuss the cooling of white dwarfs, with the goal of understanding the white dwarf luminosity function in terms of the cooling history of white dwarfs. White dwarf cooling, along with the initial mass function and star formation history, combine to determine the white dwarf luminosity function. If we can claim an understanding of white dwarf cooling, then the observed luminosity function can constrain the other inputs.

6.1 A Simplified Cooling Model

In a remarkable 1952 paper, Leon Mestel laid the groundwork for the study of white dwarf cooling (Mestel 1952). Following the discovery of the peculiar nature of white dwarf stars, the question arose as to what their power source might be. Nuclear burning, identified as the energy source in ordinary stars, was an early suspect, but Ledoux & Sauvenier–Goffen (1950) showed that if white dwarfs were powered by nuclear burning, then they would be vibrationally unstable. Since at that time white dwarfs were known to be non–variable, their conclusion was that another mechanism was needed. Mestel (1952) identified the luminosity source as release of stored thermal energy.

The reservoir of thermal energy stored in the core of a pre–white dwarf star is slowly depleted by leakage of this thermal energy into space. The cooling process is not unlike that undergone by a heated brick placed in a cool environment. Such a brick would cool rapidly when exposed to the cool outside air unless the brick is surrounded by some sort of insulating blanket. With insulation, the cooling of the brick is slowed as the temperature gradient between it and the cool outside is reduced by the poor thermal transport properties of the blanket. Recall that in a white dwarf star, heat is transported through the degenerate core by the efficient process of electron conduction, while in the nondegenerate envelope energy is transported by the (much less efficient) form of photon diffusion. Thus in a white dwarf star, the degenerate core corresponds to the hot brick, and the nondegenerate outer layers play the role of an insulating blanket.

Mestel (1952) began with the equations of stellar structure, and then made several simplifying assumptions for the case of white dwarf stars. Nuclear energy generation is assumed to play no role. In that case, the energy equation becomes

$$\frac{dL_r}{dm_r} = -T\frac{\partial s}{\partial t} \qquad (70)$$

where s is the specific entropy. The right hand side of the above equation can be rewritten, using thermodynamic identities, as

$$-T\frac{\partial s}{\partial t} = -c_v\frac{\partial T}{\partial t} + \frac{T}{\rho^2}\left.\frac{\partial P}{\partial T}\right|_\rho\frac{\partial \rho}{\partial t}. \tag{71}$$

This simplifies further when we note that for white dwarfs, the rate of contraction is very slow; when sufficiently cool, white dwarfs cool at constant radius. Therefore, the second term on the right–hand side above is zero. Also, as discussed in the section on the equation–of–state for white dwarf interiors, the specific heat in the core is set by the nondegenerate ions, so $c_v = -3/2N_Ak/A$ where A is the atomic weight of the ions within the degenerate core. The next step is to assume that the degenerate core is isothermal; this too is a reasonable assumption because of the efficient energy transport in the degenerate interior through electron conduction. Combining these assumptions, and integrating through the degenerate core, the luminosity of a white dwarf star can be expressed as

$$L_c = -\int_M \frac{3}{2}\frac{N_Ak}{A}\frac{\partial T_c}{\partial t}\,dm_r \tag{72}$$

or, in more practical terms

$$\frac{L}{L_\odot} \approx 6.42 \times 10^7 \frac{1}{A}\frac{M}{M_\odot}\frac{\partial T_c}{\partial t}. \tag{73}$$

Our goal is to obtain an expression for the luminosity of the star as a function of time; the time dependence of the above equation is isolated in the time derivative of the core temperature. Our next step, then, is to relate the core temperature to the luminosity in an independent way; then we will integrate with respect to time to obtain luminosity as a function of age.

The above expression was derived for the degenerate core. Atop this core is the nondegenerate envelope, which contains (by assumption) only a small fraction of the mass of the star. If the envelope is radiative, then the rate of energy transfer is governed by the radiative opacity of the material. We adopt a Kramers opacity law of the form

$$\kappa = \kappa_o \rho T^{-3.5} \tag{74}$$

and also assume a zero boundary condition (i.e. $P = T = 0$ at $R = R_*$). Together, these assumptions lead to the so–called radiative zero solution for T as a function of P (see for example Hansen & Kawaler 1994):

$$\frac{1}{8.5}T^{8.5} = \frac{3}{4ac}\kappa_o\frac{\mu}{N_Ak}\frac{L}{4\pi GM}\frac{P^2}{2} \tag{75}$$

where a and c are the usual physical constants. We now match this solution to the degenerate core, with a matching point defined as where the material begins to become significantly degenerate. At this point, the Fermi energy E_f for the electrons is approximately equal to kT, or the ideal gas pressure is roughly equal to the pressure of the degenerate electrons. At this point, the outer limit of the core, the pressure P_c is

$$P_c = \frac{2}{3}\pi k \left(\frac{20mk}{h^2}\right)^{3/2} \frac{\mu_e}{\mu} T_c^{5/2} \tag{76}$$

where μ_e is the mean molecular weight per electron in the core, and μ is the mean molecular weight of the envelope. Combining this equation with the radiative zero solution provides the desired expression relating the stellar luminosity to the core temperature:

$$\frac{L}{L_\odot} = 1.7 \times 10^{-3} \left(\frac{M}{M_\odot}\right) \frac{\mu}{\mu_e^2} T_{c,7}^{3.5} \tag{77}$$

where the core temperature is expressed in units of 10^7 K.

We now substitute for $L/L_\odot$ using this expression into that for the luminosity as a function of the core cooling rate, integrate with respect to time, and then recast the result to obtain the cooling time to reach a given luminosity:

$$t_{\text{cool}} = 9.41 \times 10^6 \text{ yr} \left(\frac{A}{12}\right)^{-1} \left(\frac{\mu_e}{2}\right)^{4/3} \mu^{-2/7} \left(\frac{M}{M_\odot}\right)^{5/7} \left(\frac{L}{L_\odot}\right)^{-5/7} \tag{78}$$

Some features to note about this remarkable result include the dependence of t_{cool} on A and M. Cores with larger A (cores composed of heavier elements) cool faster than "lighter" cores. In other words, cores composed of, say, carbon, will take longer to fade to a given luminosity than those composed of heavier elements. Also, more massive white dwarfs cool and fade more slowly. This comes about because of two reasons built into this analysis: they have more mass and therefore a correspondingly larger store of thermal energy, and they have a smaller radius at a given temperature, which means they have a smaller surface area with which to radiate away the core energy. If we assume representative values for mass ($0.60 M_\odot$), atomic weight ($A = 14$, a 50/50 mix of carbon and oxygen), $\mu = 1.4$, $\mu_e = 2$, and a luminosity corresponding to the faintest white dwarfs ($L = 10^{-4.5} L_\odot$), then this analysis results in a cooling time of 7×10^9 years.

6.2 Complications: Neutrinos and Crystallization

This estimate of the cooling time is remarkably close (within 30%) of the most modern computation of the cooling age of these coolest white dwarf stars. This despite the fact that it leaves out several effects that are now known to be very important: neutrino cooling, prior evolutionary history, crystallization effects, and nuclear burning. The relative contributions of these effects to the Mestel assumptions can be significant, but they largely cancel each other out, leaving white dwarfs to cool as the simple Mestel model suggests. Van Horn (1971) reviews the Mestel theory and evaluates the precision of the assumptions; Iben & Tutukov (1984) discuss the Mestel cooling law in light of their more detailed evolutionary models. Both these papers provide an excellent overview of the relevant physics of white dwarf cooling.

Most of the complications are minor at best; the two largest effects which cause departures of cooling from the Mestel law are neutrino emission and thermal effects involved in crystallization. Neutrinos are an additional sink of energy for the core, and accelerate core cooling when the neutrino luminosity is significant. As we saw in the previous section, neutrino energy losses can exceed the photon luminosity in average–mass white dwarfs when the luminosity lies between 30 and $0.3L_\odot$. Thus a white dwarf at a given luminosity will be younger than predicted by the Mestel law at the high–luminosity end. By the time the white dwarf luminosity drops below about $0.1L_\odot$, the Mestel law provides an accurate cooling age; the accelerated cooling occurs at young ages and so does not influence the age estimate for white dwarfs of sufficient age that neutrinos have ceased to be an important coolant. For white dwarfs of $0.60M_\odot$, this time corresponds to an age of approximately 3×10^7 to 10^8 years.

At the cooler end of the white dwarf cooling function, thermal effects associated with crystallization become important. The Mestel law assumes that thermal energy is stored in the form of the specific heat of the ions. However, when crystallization begins, the latent heat release associated with the growing long–range ordering provides an additional luminosity source. As crystallization proceeds (and the core temperature drops) the specific heat rises (as described in Section 3). With a higher specific heat, the rate of change of the luminosity decreases with time as the latent heat of crystallization provides an additional source of energy. In average white dwarfs, this occurs at $\log L/L_\odot \approx -3.6$ to -4.2. In this luminosity range, a white dwarf at a given luminosity is probably older than the Mestel law would indicate.

As a white dwarf becomes fully crystallized, and the core cools below the Debye temperature discussed in Section 3, the specific heat drops quickly with further cooling. With c_v dropping, the rate of change of luminosity increases dramatically; a low specific heat means that the star cannot hold in thermal energy easily. Thus once a white dwarf cools below the Debye temperature, its luminosity plummets faster than the Mestel law alone would predict. In this Debye cooling phase, a white dwarf at a given luminosity is probably a bit younger than the Mestel law would suggest. Figure 23, shows a sketch of the white dwarf cooling curve. The solid line represents the standard Mestel cooling (the equation above). The dashed line at high luminosities (and young ages) represents the effects of neutrino cooling as making the curve more shallow. At the low luminosity end, a bump (dotted line) shows the effects of crystallization at the appropriate luminosity. The lowest luminosity portion shows the effects of Debye cooling.

6.3 Realistic Cooling Calculations

In constructing realistic cooling curves, modelers of white dwarf stars include all of the known physics within the model. The physical properties enter the cooling curve as the constitutive relations that provide the coefficients of the equa-

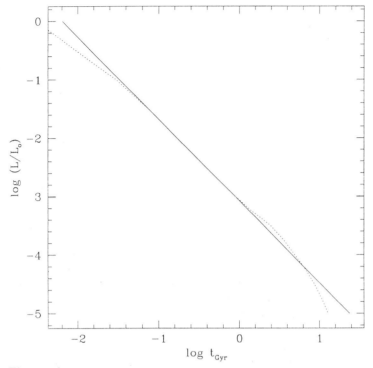

Fig. 23. A representative cooling curve for an average $0.60M_\odot$ C/O white dwarf based on the Mestel law. The dashed lines show the modifications to the Mestel law resulting from inclusion of neutrinos and of crystallization effects.

tions of stellar structure and evolution. These models use realistic conductive and radiative opacities, sophisticated equations–of–state, including the effects of degeneracy, Coulomb interactions, and crystallization. Thus the simplified analysis by Mestel (1952) has in practice been superseded in recent years. While the agreement between the best current models and the analytic result from the Mestel theory is remarkable, the new models show trends that are seen in the white dwarf luminosity function that confirm that the physics contained in the models is indeed accurately modeled.

The most complete models of cooling white dwarfs, and the white dwarf luminosity function, is the work of Wood (1992), who computed many sequences of cooling white dwarf models with many different combinations of parameters. Since white dwarf stars at the cool end of the luminosity function have largely forgotten their initial conditions, Wood used a homologous set of starting models. Convergence of the evolutionary tracks was essentially complete by the time the models reached a luminosity of roughly $0.1L_\odot$.

Three representative sequences from the calculations of Wood (as quoted by Winget et al. 1987) are shown in Figure 24. This figure shows that several of the basic scalings of the Mestel law hold for these complete models. First note that

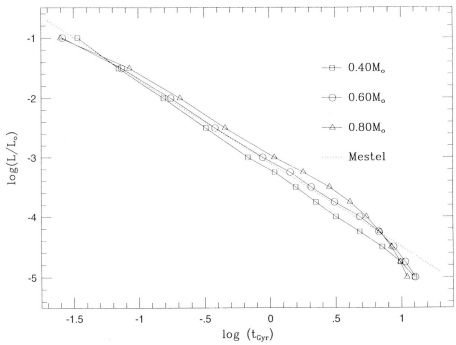

Fig. 24. Cooling curves for representative white dwarf evolutionary sequences. Data from Winget et al. (1987); figure from Hansen & Kawaler (1994)

at a given luminosity the more massive models are older, as expected from the above equation. All of the tracks parallel the Mestel law; that is, they show a general power–law slope of -1.4. Note that the luminosity of the $0.80 M_\odot$ model shown begins to drop quickly at a luminosity below $10^{-4} L_\odot$. This is the effect of crystallization; these models have already largely crystallized, and they are showing the effects of Debye cooling discussed in a previous section. The $0.60 M_\odot$ model begins Debye cooling at a lower luminosity and later time; this implies that more massive models cool more slowly, but that they crystallize earlier.

Before the downturn, the cooling curve flattens compared to the Mestel law. This is consistent with the expectations, described above, that as crystallization begins, the specific heat rises with the release of latent heat of crystallization. This "extra" luminosity source slows the evolution. Once largely crystallized, the models enter the Debye cooling stage resulting in the acceleration of the luminosity drop of the white dwarf models. Because of this, even though they cool more slowly initially, we expect that the faintest white dwarfs in the galaxy have, on average, higher mass than slightly more luminous (and therefore younger still) white dwarfs.

6.4 Construction of Theoretical Luminosity Functions

With theoretical cooling curves, we need a way to see if the theoretical inputs to the models are realistic. The principal observational contact of the theory of white dwarf cooling is the observed *luminosity function* of white dwarf stars: the number of white dwarfs per cubic parsec per unit luminosity (or bolometric magnitude), usually denoted as Φ. The cooling rate of white dwarfs is one input into the construction of a theoretical luminosity function. As described by Wood (1992), a theoretical determination of the luminosity function requires evaluation of the expression

$$\Phi(L) = \int_{M_{\text{low}}}^{M_{\text{hi}}} \int_{L_{\text{low}}}^{L_{\text{hi}}} \psi(t)\, \phi(M)\, \frac{dt_{\text{cool}}}{d\log(L/L_\odot)}\, \frac{dm}{dM}\, dL\, dM \tag{79}$$

at a given value for the galactic age. In this expression, $\psi(t)$ is the star formation rate at the time of the birth of the white dwarf progenitor, $\phi(M)$ is the initial mass function (usually assumed to be the Salpeter IMF) for stars with initial mass M, and dm/dM is the initial–final mass relation described in the first section. The mass limits of the integration cover the mass range of main sequence stars that produce white dwarfs, with the lower mass being the main–sequence turn–off age for the population age and the upper mass limit being about $8M_\odot$. The luminosity interval of integration corresponds to the faintest white dwarfs at the given age and some upper value (such as $10L_\odot$). The remaining number is the derivative of the cooling age, which is taken at the chosen value of L from the white dwarf cooling models.

The observed luminosity function, in principle, provides information that constrains the star formation history of the galaxy (in the age, $\phi(M)$, and $\psi(t)$), the $M_i - M_f$ relation (in dm/dM), and the white dwarf cooling history. The observed white dwarf luminosity function of a population can constrain one or all of these important parameters. In the case of a globular cluster, measurement of the luminosity function allows, in principle, precise calibration of the white dwarf cooling rates and the initial–final mass relations, since the age is (presumably) known. In the case of the solar neighborhood and, by extension, the galactic disk, reliable white dwarf cooling models can constrain the age of the galactic disk.

Iben & Laughlin (1989) show how the luminosity function appears for several simplified relations for the star formation rate, etc.; this paper is an excellent starting point for those who wish to work with the white dwarf luminosity function for various purposes. Figure 25, shown in lecture, shows an example of a simplified luminosity function and the effect of the population age on it. Quite simply, the older the population the lower the luminosity of the cutoff. The stars populating the lowest luminosity bin are the oldest white dwarfs in the sample; with increasing age, they reach lower luminosity and pile up because of the lengthening cooling time scale.

Recall the effects of neutrino emission on the cooling curve; at high luminosities, neutrinos accelerate cooling. Thus at the high luminosity side of the luminosity function, Φ should fall below the Mestel curve. When crystallization

White Dwarf Stars

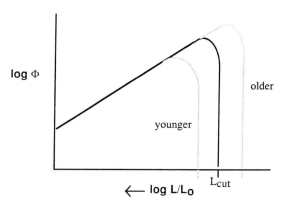

Fig. 25. Simplified luminosity functions for populations of three different ages. In this case, we assume a constant star formation rate with time, a Mestel cooling law ($t_{\rm cool} \propto L^{-5/7}$), a single–mass population, and three disk ages. Note that the slope of the luminosity function matches the cooling function when $\log \Phi$ is plotted versus $\log L$. This schematic plot follows this convention, and the additional convention of decreasing luminosity to the right.

begins, the cooling rate temporarily slows, and the luminosity function will rise above the Mestel curve. Debye cooling will cause a drop in Φ with the associated accelerated cooling. Of course, realistic calculations of the luminosity function are needed for detailed comparison with the observed luminosity function.

6.5 The Age of the Galactic Disk

Figure 26 shows a sample of realistic luminosity functions from Wood (1992) that employ a distribution of white dwarf masses (through the dm/dM, $\psi(t)$, and $\phi(M)$ terms) and theoretical cooling curves. Note that the general slope of this cooling curve is nearly $-5/7$, but with a depression at high luminosities and a bump near the final cutoff. Also shown in Figure 26 is the luminosity function of Liebert et al. (1988) as adapted by Wood.

Agreement between the theoretical and observed luminosity functions above the drop–off is very good; those portions that disagree may result from changes in the star formation rate and/or the IMF over time. Such possibilities provide an intriguing use of the white dwarf luminosity function to explore the history of star formation in our galaxy (or, in the near future, in other stellar populations such as globular clusters).

Figure 25 clearly shows the effect of increasing age on the white dwarf luminosity function; older populations have fainter white dwarf stars. The low–luminosity cutoff decreases in luminosity with increasing population age; the figure shows luminosity functions for ages from 6 Gyr to 13 Gyr. Inspection of this figure reveals that the observed luminosity function is consistent with an age for the disk of the Milky Way of between 8 and 11 Gyr, with a best value of about 9 Gyr.

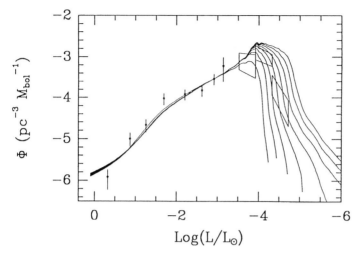

Fig. 26. Theoretical white dwarf luminosity functions from Wood (1992, Figure 26) for several input disk ages. (Note that the ordinate is actually $\log \Phi$. The drop–off in luminosity corresponds to disk ages from 6 Gyr (at the highest luminosity) to 13 Gyr. Also shown is the observed white dwarf luminosity function, with the lowest luminosity points represented by error boxes.

The derived age of the galactic disk depends on the input physics used in the computation of the white dwarf cooling rates. Given an observed, or otherwise fixed, luminosity function for comparison, the uncertainties in the derived ages follow from uncertainties in various properties of the white dwarf models. These properties, the sensitivity of the age to them, and the real range of possible ages, are summarized in Table 2. This table uses information from Winget & Van Horn (1987) and Wood (1992). It lists how the time for a white dwarf with a mass of $0.60 M_\odot$ to drop to a luminosity of $10^{-4.4} L_\odot$ changes with various changes in the input physics.

For example, if the mass of the surface helium layer is increased by one decade, the cooling time for a $0.60 M_\odot$ white dwarf will decrease by 0.7 Gyr. In this tabulation, the values of m_{helium}, A_{core}, and Z_{env} are uncertain because of unknowns in the prior evolution of the white dwarf stars — they depend on how the star became a white dwarf star, on the $^{12}\mathrm{C}(\alpha,\gamma)^{16}\mathrm{O}$ nuclear reaction cross section, and on the trace metal content in white dwarf envelopes. The "real range" represents the current uncertainty in the cooling time given current uncertainties in the listed parameters. The dominant uncertainties arise from the thickness of the surface helium layer and on the core composition. The combination of these uncertainties yields a current best–estimate of the age of the galaxy of about 9.3 ± 1.5 Gyr.

Fortunately, through observations of the pulsating white dwarf stars, there is a way to measure these quantities that is independent of approaches that use the observed luminosity function. These observed pulsations allow us to measure the depth of subsurface transition zones and, therefore, the thickness of

Table 2. Dependence on $t_{\rm disk}$ on input physics

Input physics ("X")	$\frac{dt_{\rm disk}}{d\log X}$	real range
$m_{\rm helium}$	-0.7 Gyr	$\leq 1.4^*$ Gyr
$A_{\rm core}$ (C? O?)	-16 Gyr	$\approx 2^*$ Gyr
conductive opacity	$+7.6$ Gyr	≤ 0.3 Gyr
radiative opacity	$+1.4$ Gyr	≤ 0.06 Gyr
$Z_{\rm env}$	$+1.4$ Gyr	≈ 0.2 Gyr
other stuff	?	<1 Gyr ?

* measurable with pulsation observations

the surface helium layer in the pulsating DO and DB white dwarfs. As for the core composition, measurements of the rate of change of the pulsation period in DA (ZZ Ceti) white dwarf star will give us a precise measurement of the cooling time for those stars (as described below). Thus we will soon be able to determine the core composition. As will be seen in the last chapter, the current upper limits on the rate of change of the pulsation period in one star already rule out a core of pure oxygen, and suggest a mixed C/O core. In addition to improving the input physics into the white dwarf luminosity function, these observations allow us to empirically calibrate the luminosity function and rendering the deduced age of the galactic disk an empirical, rather than theoretical, determination.

7 Nonradial Oscillations of White Dwarfs: Theory

In the mid–1960s, Arlo Landolt was working with a new photoelectric photometer when he observed the white dwarf star HL Tau 76. In the testing process, he observed the brightness of this potential standard star as a function of time, and noticed a peculiar variation in the count rate with a period of about 720 seconds. After further checking, Landolt realized that this particular variation was caused by the star itself (Landolt 1968). The problem was that he was observing a white dwarf star, and (at the time) everyone knew that white dwarfs did not pulsate. What Landolt had discovered was the first of the ZZ Ceti stars, and the first of the pulsating white dwarf stars.

The period of 720 seconds was curious for another reason. In the mid–1960s, stars known to be pulsating variables were all radial pulsators, with pulsation periods that are comparable to the free–fall time. Being such compact stars, with a high mean density, white dwarfs have free–fall (or dynamical) times of order 10 seconds. The observed period was almost two orders of magnitude longer; how could a star pulsate with a period so much longer than its free–fall time? It was soon realized, by Chanmugam (1972) and Warner & Robinson (1972) that the pulsation periods were consistent with a new class of pulsation modes known as nonradial g–modes. Calculations by Osaki & Hansen (1973) using reasonable white dwarf models put this conclusion on a firm and quantitative basis. In this section we review the theory of nonradial pulsations in white dwarf stars, as background for understanding how these pulsations allow us to probe the interiors of white dwarf stars through stellar seismology.

White dwarf pulsation has been reviewed extensively in recent literature: see the reviews of white dwarf pulsations by Kawaler and Hansen (1989), Winget (1988), and Kawaler (1990a, 1995). For theoretical background on the equations of stellar pulsations, and the rich variety of their solutions, two texts are available: Cox (1980) covers both radial and nonradial pulsation, while Unno et al. (1989) covers nonradial pulsations in great detail.

7.1 Review of Observations

Nonradial g-mode pulsations are observed in white dwarf stars during three phases of their evolution spanning the time from their birth to middle age. During the hot phases of white dwarf evolution ($T_{\text{eff}} > 80,000\,\text{K}$), some DO white dwarfs that show apparent carbon and oxygen enrichment are observed to pulsate with periods of 300–1000 seconds. Along with these "DOV" stars, several planetary nebula nuclei with spectra similar to the DOV stars show variations with period of 1000–2500 seconds (Bond et al. 1993). At cooler effective temperatures near about 27,000 K lie the pulsating DB stars. Finally, in a narrow strip near $T_e \approx 12,000\,\text{K}$ we find the DAV (or ZZ Ceti) stars, which are pulsating white dwarfs with hydrogen dominated spectra. Most, if not all white dwarfs in the DAV instability strip are variables (Winget 1988); hence it is probable that the DAV stars are the most numerous type of pulsating star in the Galaxy.

Figure 27 shows a sample light curve for a pulsating white dwarf. This figure, from Winget et al. (1985), shows the light curve of the prototype of the pulsating DO white dwarf stars, PG 1159–035 (hereafter simply PG1159). It shows the characteristic period of about 500 seconds. At the middle of the figure, the pulsation amplitude temporarily drops to zero. This is the result of beating between two simultaneous pulsation modes of slightly different periods. In fact, PG 1159 pulsates in hundreds of independent modes (Winget et al. 1991) which require extensive observations to resolve. On a given night, however, it typically shows the effects of 8 temporal frequency bands; Figure 28 shows a representative power spectrum for PG 1159 as observed for about 5 hours.

It is the multiperiodicity of the pulsating white dwarf stars that provides the key to studying their interiors through stellar seismology. Each independent period gives a new piece of information about the star; since these are global oscillation modes, this information concerns the interior of the star. The pulsation data, in concert with simple adiabatic pulsation theory, provide accurate determinations of the mass, rotation rate, cooling/contraction rate, compositional stratification, and magnetic field strength. More chancy nonadiabatic pulsation theory provides a window on the timing of nuclear shell shutdown, limits on the convective efficiency, and tests of numerical computations of stellar opacities.

7.2 Hydrodynamic Equations

A complete description of the nonradial pulsations of a star requires solution of the full set of equations of 3-D hydrodynamics, coupled to equations that describe

White Dwarf Stars

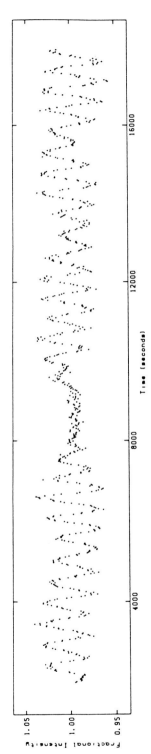

Fig. 27. Light curve for a 4.4 hour observation of the pulsating DO white dwarf PG 1159. The vertical scale gives the fractional amplitude of the variation in the brightness of the star. From Winget et al. (1985).

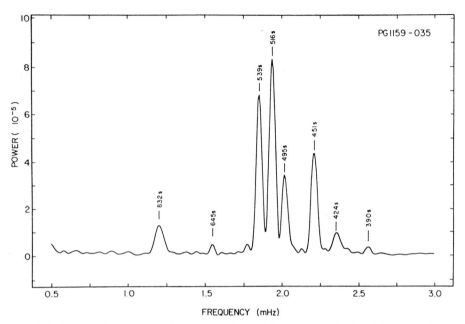

Fig. 28. Averaged power spectrum of PG 1159 for several single–night runs. From Winget et al. (1985)

the flow of radiation. Fortunately, the problem can be reduced through a linear perturbation analysis to a tractable level without losing too much accuracy. Here, I briefly outline the reduction of the problem to a few simple relations between the equilibrium structure of a star and the oscillation frequencies of nonradial modes. The equations derived here provide important keys to understanding the observed pulsation properties of white dwarfs. Much more thorough discussions of this material can be found in the excellent texts of Unno et al. (1989) and Cox (1980), as well as in Smeyers and Tassoul (1987), Tassoul (1980), and Hansen & Kawaler (1994).

As a starting point, the equations to be solved include: the Poisson equation,

$$\nabla^2 \phi = 4\pi G \rho, \tag{80}$$

the continuity equation,

$$\frac{\partial \rho}{\partial t} + \nabla \cdot (\rho \mathbf{v}) = 0, \tag{81}$$

and the equation of motion

$$\rho \left(\frac{\partial}{\partial t} + \mathbf{v} \cdot \nabla \right) \mathbf{v} = -\nabla P - \rho \nabla \phi, \tag{82}$$

where ϕ is the local gravitational potential. We begin by assuming that the pulsations are linear fluctuations about a static equilibrium state (i.e. one in which the time derivatives of all quantities are zero). The perturbations themselves can then be separated into a position dependent amplitude multiplied by a time dependent factor, that is,

$$\delta x(\mathbf{r}, t) = \delta x(\mathbf{r}) e^{i\sigma t}, \tag{83}$$

where σ is the frequency of the variation and δx represents the Lagrangian perturbation of the quantity x. The full set of equations is thus reduced to a set of equations relating the perturbations, with the equilibrium quantities appearing as coefficients and including the frequency of the perturbation σ.

Some reasonable physical assumptions make the equations even simpler. For g-modes in white dwarfs, we can usually ignore perturbations to the gravitational potential. This approximation is called the Cowling approximation. Another major simplification results from assuming that the motions are adiabatic; that is, that no heat is gained or lost anywhere in the star as a result of the fluctuations about equilibrium. If Γ_1 is the adiabatic exponent relating pressure and density, then the adiabatic approximation is given by

$$\frac{\delta P}{P} = \Gamma_1 \frac{\delta \rho}{\rho}. \tag{84}$$

In applying this assumption we relinquish the opportunity to study the energetics of pulsations, because driving and damping of oscillatory motion results from exchange of energy through the pulsation cycle. However, the periods of nonradial modes are determined by the adiabatic (i.e. mechanical) properties of the star. Hence an adiabatic analysis allows us to explore the pulsations in great detail with relatively simple physics.

These assumptions leave only two unknowns at each point in the star: the displacement in the radial direction and the perturbation of the pressure. The only dependence of these perturbations on angular coordinates are produced by a horizontal gradient term in one of the two remaining first order equations. We can therefore separate the perturbations into radial and angular components. A natural way of performing this separation can be found in any introductory quantum mechanics or electrodynamics text: the term containing the horizontal gradient is proportional to the Laplacian. We therefore express the angular dependence of the perturbations in terms of the set of spherical harmonics $Y_\ell^m(\theta, \phi)$, giving the perturbed quantities the form

$$\delta x(\mathbf{r}, t) = \delta x(r) Y_\ell^m(\theta, \phi) e^{i\sigma t} \tag{85}$$

The equations of adiabatic nonradial pulsations are now

$$\frac{1}{r^2} \frac{\partial}{\partial r}(r^2 \xi_r) = \frac{1}{c^2}\left(-1 + \left[\frac{\ell(\ell+1)\Gamma_1 P}{r^2 \rho}\right]\frac{1}{\sigma^2}\right)\frac{P'}{\rho} - \frac{1}{\Gamma_1}\xi_r \frac{d\ln P}{dr} \tag{86}$$

and

$$\frac{1}{\rho}\frac{\partial P'}{\partial r} = \xi_r\left\{\sigma^2 + \left[\left(\frac{d\ln\rho}{dr} - \frac{1}{\Gamma_1}\frac{d\ln P}{dr}\right)g\right]\right\} - \frac{1}{\Gamma_1}\frac{P'}{P}g. \tag{87}$$

where ξ_r is the perturbation displacement in the radial direction, g is the acceleration of gravity, and P' is the Eulerian pressure perturbation [1]. The terms in square brackets have the dimensions of the square of a frequency. These are the two characteristic frequencies of stellar pulsation. The frequency in equation (86), usually denoted S_ℓ, is the Lamb or acoustic frequency. It represents the local frequency of oscillation when pressure is the restoring force, and is therefore proportional to the square of the sound speed c. The second frequency, from equation (87), denoted N, is the Brunt-Väisälä or buoyancy frequency. This is the frequency of oscillation of a fluid element when buoyancy is the restoring force. This frequency represents the frequency of oscillation about the equilibrium position in regions that are stable against convection. In convective regions, the Brunt-Väisälä frequency is imaginary.

7.3 Local Analysis and the Dispersion Relation

We have now simplified the equations beyond the point where they are easily solved by numerical integration. However, it is instructive to continue with a local analysis to obtain some insight into the character of various solutions. If we assume that the characteristic frequencies change slowly over small ranges in r, then the perturbations take the form

$$P'(r) = P'(r_o) \exp(ik_r t) \qquad (88)$$

where k_r is the *local* radial wave number. The equations then provide a dispersion relation for k_r in terms of σ and the characteristic frequencies:

$$k_r^2 = \frac{1}{\sigma^2 c^2}(\sigma^2 - N^2)(\sigma^2 - S_\ell^2) \qquad (89)$$

Examination of this relation reveals that a mode is oscillatory (i.e. k_r is real) when the oscillation frequency is much smaller than both S_ℓ and N and when it is much greater than both. The case of large σ represents p-modes, such as the 5-minute solar oscillations. The case where σ is small represents the g-modes; it is these that are the modes of interest for white dwarf stars. For a time–independent solution to the equations, solutions are required to satisfy the standing wave condition that integral of k_r over the propagating regions is equal to an integer multiple of π.

7.4 g-mode Period Spacings

The set of possible standing wave solutions to the pulsation equations are the perturbation eigenfunctions, with eigenvalues equal to the frequency of the mode. For the case of g-modes, the standing wave condition then implies that the periods of the g-modes can be expressed as

$$\Pi_{n,\ell} = \frac{\Pi_o}{\sqrt{\ell(\ell+1)}}(n+\epsilon), \qquad (90)$$

---

[1] The horizontal perturbation in position, ξ_h, is $\frac{P'}{\sigma^2 r \rho}$.

where
$$\Pi_o = (2\pi)^2 \left[\int \frac{N}{r} dr \right]^{-1} \tag{91}$$

and n is the number of nodes in the perturbation eigenfunctions in the radial direction, and ϵ is a small constant. Therefore g-modes with consecutive values of n are equally spaced in period for modes with the same value of ℓ. The period spacing of the modes is determined by Π_o and by ℓ, where Π_o is a function of the equilibrium structure of the star. Thus from a given stellar model, one can determine the approximate g-mode period spectrum by simply determining the value of Π_o.

Identification of uniform period spacings in pulsating stars places immediate and important constraints on the stellar properties. As shown by Kawaler (1987b, 1988), and Kawaler and Bradley (1994) the value of Π_o for hot white dwarfs depends primarily on the mass of the star. At luminosities above about $10L_\odot$, the pulsation periods for DOV models are largely determined in the degenerate core (Kawaler et al. 1985). Since the mechanical properties (and equation of state) of degenerate material depends only on the amount of overlaying mass and not on the thermal structure, the mass is the principal parameter in determining Π_o for DOV stars. For white dwarfs between $0.4 M_\odot$ and $1.0 M_\odot$ and $L \geq 10 L_\odot$,

$$\Pi_o = 15.5 \left(\frac{M}{M_\odot} \right)^{-1.3} \left(\frac{L}{100 L_\odot} \right)^{-0.035} \text{ seconds} \tag{92}$$

with very weak dependence on the thickness of the surface helium layer.

In cooler white dwarfs, where the nondegenerate surface layers determine the pulsation periods, Π_o does depend on T_{eff} (i.e. Kawaler 1987a, Tassoul et al. 1990); however, the instability strips within which the pulsating stars are found are very narrow in temperature, and thus the period spacing is more-or-less dependent again only on mass (Bradley & Winget 1994).

7.5 Mode Trapping

In the ideal case of a uniform–composition equilibrium model, g-modes should show period spacings that are exact from one mode to the next. As we saw in earlier sections, real white dwarfs are compositionally stratified. The steep composition gradients produce steep gradients in the density, which in turn lead to locally rapid changes in the Brunt–Väisälä frequency and the Lamb frequency. These "glitches" in the characteristic frequencies lead to a phenomenon called mode trapping. Modes that happen to have nodes within the composition transition zones can have the amplitude of their eigenfunctions greatly reduced in the interior. The eigenfunctions are, in effect, pinched off below the surface at the composition transition zones (Winget et al. 1981).

With small amplitudes in the interior, trapped modes require less energy to excite and maintain. The kinetic energy of a nonradial mode can be expressed as

$$E_K \propto \frac{\sigma^2}{2} \int_0^R \rho r^2 dr \left[\xi_r^2 + \ell(\ell+1)\xi_h^2 \right]. \tag{93}$$

Note that the integral above is weighted by the density. Hence small amplitude oscillations in the degenerate interior, where the density exceeds $10^6 {\rm g/cm}^3$, contribute significantly to the kinetic energy. Trapped modes, which have greatly reduced interior amplitudes, have smaller kinetic energies than untrapped normal modes. Since the growth rate for the amplitude of a mode is inversely proportional to E_K, trapped modes are the most easily excited in a mode spectrum. Hence the filtering property of mode trapping is really an *adiabatic* effect.

Winget et al. (1981) demonstrated that mode trapping provides a natural mode selection mechanism for the ZZ Ceti stars, and is responsible for the selection of a few modes from the full g-mode spectrum in those stars. More recently, Hansen (1987, private communication) and Brassard et al. (1992a) performed an analytic analysis of mode trapping in conditions relevant to white dwarfs. In this analytic work, modes trapped in the surface hydrogen layer follow the relation

$$\Pi_i^2 = 4\pi^2 \lambda_i^2 \left\{ \left[1 - \frac{r_H}{R}\right] \ell(\ell+1) \frac{GM}{R^3} \right\}^{-1}, \qquad (94)$$

(Hansen 1987) where i is an index corresponding to the number of nodes between the surface and the H/He discontinuity, which occurs at a radius r_H. The numbers λ_i are constants related to the zeros of Bessel functions.

Numerical work confirms the basic analysis by Hansen. Several groups Kawaler & Weiss (1990), Brassard et al. (1992a), and Kawaler & Bradley (1994) computed grids of white dwarf models to test the Hansen relation. They showed that the periods of trapped modes in "realistic" numerical models of white dwarfs obey a Hansen relationship, with values of the λ_i that differ little from model to model and from investigator to investigator. This semi-analytic treatment allows us to determine (to at least the order–of–magnitude) the thickness of the surface composition transition zones in white dwarfs if we can determine the value of ℓ and assure ourselves that the observed modes are indeed trapped modes. The periods of trapped modes of the same ℓ should show fixed ratios determined by the values of λ_i.

The period difference between trapped modes is inversely proportional to the thickness of the surface layer; thin layers show trapped modes that are much farther apart in period than the normal g-modes. In the case of thin layers, several untrapped modes separate the trapped modes. If a surface layer is sufficiently thick, then the period difference between trapped modes is comparable to the natural period spacing of g-modes and mode trapping is not at all effective as a mode selection mechanism.

If the only consequence of mode trapping was reduced kinetic energy and, therefore, a higher likelihood of excitation to observable amplitude, mode trapping would be of little but academic interest. Many other nonadiabatic and nonlinear effects play important roles in amplitude and visibility of oscillation modes. However, what makes mode trapping so useful is that it affects the observed periods of the modes; specifically, mode trapping affects the period spacings in models. Because of the altered shapes of the eigenfunctions for trapped modes, their periods depart significantly from the uniform spacing described in the previous subsection (Kawaler 1987a). In effect, the trapped modes are nudged away

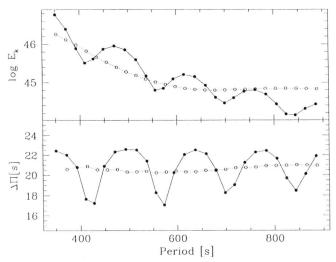

Fig. 29. Log of the pulsational kinetic energy, and period spacings, in two white dwarf models. The first model (open circles connected by dotted lines) has a homogeneous composition, while the second model has an outer layer of $10^{-4} M_\odot$ of helium. Note that the minima in kinetic energy correspond to minima in the consecutive-n period spacings.

from uniform spacing in the direction of shorter periods. Thus another signature of trapped modes is that they show smaller spacings from previous modes in a constant ℓ sequence of increasing n. Figure 29 shows the way in which mode trapping affects the kinetic energy and period spacings in some simple models of DOV stars.

The period spacing $\Delta\Pi$ as a function of period Π is a critical diagnostic of mode trapping. Minima in $\Delta\Pi$ usually correspond directly to trapped modes. Once trapped modes are identified, the Hansen relation, along with more detailed models, determines the position below the surface at which a mode sees a reflective boundary. For many stars that have been observed with the Whole Earth Telescope, trapped modes have been identified either from minima in the period spacing diagram or from comparison with the trapping period ratios. For more details about the phenomenon of mode trapping and applications to DA, DB, and DOV models see Kawaler & Weiss (1990), Kawaler (1990a), Bradley & Winget (1991), Brassard et al. (1992a, 1992b), Bradley et al. (1993), Kawaler & Bradley (1994), and Bradley & Winget (1994).

7.6 Rotational and Magnetic Splitting

The equations outlined above describe nonradial oscillations in nonrotating, axially symmetric stars. Nowhere in those equations is there any dependence on the azimuthal quantum number m. Hence the frequencies of modes with the same n and ℓ, but different values of m are identical in such stars. Any departures from spherical symmetry will break this m degeneracy, and lead to modes with different m having different frequencies. Rotation produces fine splitting of oscillation frequencies in a way analogous to Zeeman splitting of atomic spectral lines.

For slow ($\Omega \ll \sigma$) uniform rotation,

$$\sigma_{n\ell m} = \sigma_{n\ell} - m\Omega_{rot}(1 - C_{n\ell}) \qquad (95)$$

where $\sigma_{n\ell}$ is the frequency of a mode for the nonrotating case, $\sigma_{n\ell m}$ is the observed frequency, and $C_{n\ell}$ is a number that is a function of the perturbation eigenfunctions. For white dwarfs, where the amplitude of oscillatory motion in the horizontal direction is much larger than in the vertical,

$$C_{n\ell} \approx \frac{1}{\ell(\ell+1)} \qquad (96)$$

(Brickhill 1975). Therefore, slow uniform rotation causes splitting of pulsation frequencies into m components (assuming all modes with all available values of m are excited) that are equally spaced in frequency by $\Omega_{rot}(1 - C_{n\ell})$. Since for a given ℓ there are $2\ell + 1$ possible values of m, this means that, for example, $\ell = 1$ modes would appear as triplets in a power spectrum, $\ell = 2$ modes would appear as quintuplets, etc. While one expects equal splittings for all modes in a uniformly rotating star, it is probable that the rotation rate changes with depth within the star. In this case, modes which are concentrated near the surface (such as modes with large n) may show a different frequency splitting than modes in the same star that are concentrated further below. Thus measurement of rotational splitting over the entire observed frequency range can explore differential rotation with depth.

Another process that lifts the pure spherical symmetry of the problem is the effect of magnetic fields. If the magnetic pressure ($B^2/8\pi$) is comparable in magnitude to the amplitude of the pressure perturbation, then magnetic fields can influence the pulsation frequencies. The frequency changes depend on the field strength and on the field geometry, and will differ for different values of ℓ and m of the modes in question. One example of the influence of magnetic fields is that for force–free magnetic fields, as explored by Jones et al. (1989), in which the magnetic splitting is proportional to the absolute value of m. In this case, the splittings are not constant, and result in an offset of all m values to higher frequency. Such an effect has been observed in the pulsating DB white dwarf GD 358 (Winget et al. 1994), to be described in the next section.

7.7 The Seismological Toolbox

The table below summarizes the set of observed quantities that we use to explore beneath the surfaces of the pulsating white dwarf stars. This is the "seismological toolbox" for g-modes in white dwarf stars [2].

---

[2] The tools in the toolbox are similar to those used by solar seismologists, though they concern themselves with p-modes and the associated frequency spacings with changing n and ℓ.

Table 3. The white dwarf seismology toolbox.

measured quantity	symbol	information content
Period [s]	Π	class of modes (p, g, radial)
Period spacing [s]	$<\Delta\Pi>$, Π_o	stellar mass, luminosity
	$\Delta\Pi(\Pi)$	surface layer thickness
Frequency splitting [Hz]	$<\Delta\nu>$	rotation period, magnetic fields
	$\Delta\nu(\Pi)$	differential rotation, mag. fields
secular period change	$d\Pi/dt$	cooling/contraction rate

Note that all of the quantities above, and their interpretation, follow from simple linear adiabatic analysis; that is, they come from the mechanical structure of the star. They do not involve thermal effects (to first order), and so the simplicity of the modeling renders the conclusions particularly robust. This is the power of the methods of stellar seismology; you can learn a great deal about stars using simple pulsation theory, and even asymptotic approximations.

8 Pulsating White Dwarfs

The previous section describing the theory behind nonradial oscillations of stars reviews developments that were necessitated by the discovery of pulsating white dwarf stars. The study of these stars is really a story of close interplay between theory and observation. The scientific potential of the white dwarf pulsators were recognized early in the history of this discipline. Most notably, the pulsating white dwarfs are otherwise ordinary stars. Therefore, the pulsators are representative of all white dwarfs, and what we can learn about them can be applied to the entire class.

This section presents some selected highlights of what we have learned about the evolution of white dwarfs through study of the pulsators. It is not intended to be an exhaustive review; things are currently changing very quickly in this area and to try to make a comprehensive review is beyond the scope of this course. For a more complete sampling, I recommend obtaining the conference proceedings of the last two Whole Earth Telescope workshops (Meištas & Solheim 1993, 1995) and reading the many reports therein.

8.1 The Whole Earth Telescope

The significant observational challenges of white dwarf seismology involve technical challenges (such as very faint stars, low amplitude variations, etc.) as well as logistical challenges. The logistical difficulties arise because of the periods seen in these stars.

A large number of modes are available for a pulsating white dwarf; even if it chooses to pulsate in a relatively small fraction of those modes, very complex light curves can still result. If more than one mode is present in a star, then those modes will beat against one another. The light curve will show variations on time scales of the individual modes and on time scales corresponding to the frequency *differences* between the modes. The resulting light curves are extremely difficult, if not impossible, to fully decode from a single observing site. This is true of most nonradial pulsators: the large number of pulsation modes leads to beating effects with frequencies on the order of 1 cycle per day. Because a single observing site cannot observe its target when clouds, trees, or the Earth gets in the way, diurnal aliases necessarily appear in the Fourier power spectra of the data, and can hide true features of the star's power spectrum.

To solve this problem requires observations that are free from the 1 cycle per day aliases, and the only way to do that is to observe them 24 hours per day. This necessitates coordinated observations at observatories at widely distributed longitudes. Such observations are obtained for pulsating white dwarfs using an extremely remarkable instrument, the Whole Earth Telescope ("WET"). As described in detail by Nather et al. (1990), the WET is a global network of telescopes designed for highly coordinated interactive photometric observations of variable stars. The power of observations obtained in this way is illustrated in Figure 30, from Nather et al. (1990); in this case 9 observatories participated. The bottom–left panel shows the power spectrum of a singly–periodic sinusoid sampled at the same times as observations were obtained at a single site. The pattern of aliases is clearly visible. Also shown is the same sinusoid sampled at the same times as all participating sites. Because of the nearly complete coverage, the aliasing pattern disappears almost entirely. The WET is an extremely powerful instrument; one that has obtained data of extremely high quality on several pulsating white dwarfs. WET data allow us to use the tools of stellar seismology at nearly their full potential.

8.2 PG 1159 Stars and Pulsating PNNs

The PG 1159 stars are hot and luminous pre–white dwarfs. The designation corresponds to the spectral class characterized by CIV and OVI spectral lines with emission cores. PG 1159-035 was discovered to be a variable star by McGraw et al. (1979) and is the prototype of the DOV (or GW Vir) class of pulsating white dwarfs. McGraw et al. (1979) immediately recognized the astrophysical importance of these objects: they are rapidly evolving from the planetary nebula phase to the white dwarf cooling track. Thus they are among the hottest, and most rapidly evolving, stars in the Galaxy. NLTE model atmosphere analysis by Werner et al. (1991) shows that its effective temperature is close to 140,000 K; the effective temperatures of the PG 1159 spectral class ranges from 75,000 K to 160,000 K.

Central stars of some planetary nebulae show spectra very similar to the "naked" PG 1159 stars; some of these too are pulsating stars, but with periods that are significantly longer than the pulsation periods seen in the naked PG 1159 stars. The first of this class was the central star of the planetary nebula K1-16, discovered to be a variable by Grauer & Bond (1984). All of the other known pulsating central stars (except for RX J2117, see below) have been discovered and studied by Howard Bond and his collaborators. They have obtained globally

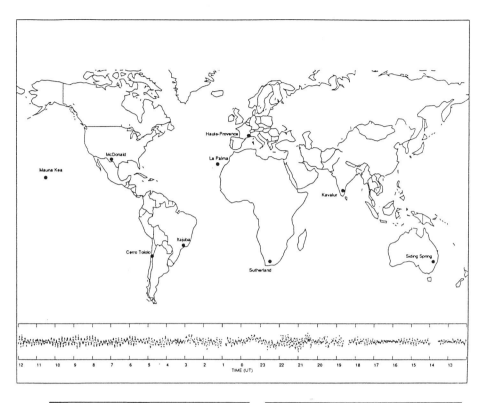
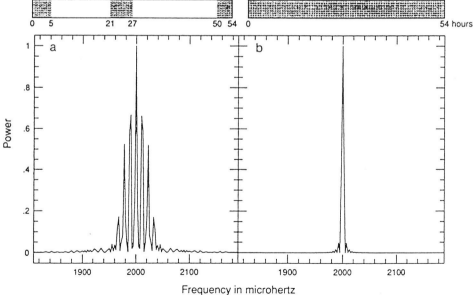

Fig. 30. (Top) Distribution of observatories participating in a Whole Earth Telescope observing campaign. (Bottom) Power spectra of a single sinusoid, sampled at the same time as observations at a single observatory, and over the entire network. Note 1 cycle/day alias pattern in the single–site data and the very clean signal in the multi–site data. Adapted from Nather et al. (1990).

coordinated CCD fast photometry of the central star of the planetary nebula NGC 1501 and Sand 3 (see Bond et al. 1993, 1995).

Figure 31 shows the power spectra of several hot white dwarfs in order of decreasing dominant period. Note the striking similarities (and differences) between these objects in terms of dominant period and distribution of peak amplitudes. It is left to the reader to derive conclusions based on the temperatures of these objects. The values of T_{eff}, in units of 10^3 K, are 96 ± 3 for NGC 1501 (Shaw & Kaler 1985), 130 for Sand 3 (Koesterke & Hamann 1995), 150 ± 15 for RX J2117 (Motch et al. 1993), 140 ± 10 for PG 1159, 100 ± 10 for PG 1707 (Werner et al. 1991), and 84 ± 2 for PG 2131 (Kawaler et al. 1995).

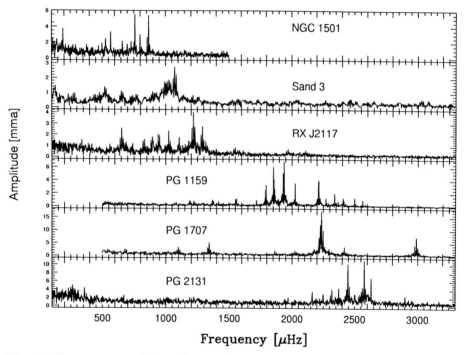

Fig. 31. Power spectra of hot pulsating degenerate stars. The data for PG 1159 are from Winget et al. (1991) and for PG 2131 from Kawaler et al. (1995). The other data are from works in preparation: data for NGC 1501 and Sand 3 were obtained by Bond et al. (1995); data for RX J2117 and PG 1707 data are from unpublished WET data.

PG 1159–035 Through temporal spectroscopy with the WET, the prototype of the pulsating hot white dwarfs, PG 1159, has yielded a wealth of information about what one of these stars look like on the inside. The reports of these observations and their analysis (Winget et al. 1991, Kawaler & Bradley 1994) show what such stars can teach us. PG 1159 is wonderful in that it shows all that we expect to see (and more) in these stars. An unbroken sequence of dozens of periodicities match very nicely with the expectations from stellar evolution and pulsation theory.

The noise level in the region of greatest power is less than 7×10^{-4} relative amplitude. This sensitivity allows identification of much lower amplitude peaks in the power spectrum that had been possible with single–site data; Winget et al. (1991) completely resolve all of the peaks (or, more precisely, bands of power) present in earlier data into groups of individual peaks. Some of these groups are triplets of individual peaks separated equally in frequency. Other groups show 4 or 5 component peaks, also equally spaced in frequency. The frequency spacing between components in the triplets is 4.2 μHz, while the spacing in the quintuplets is 7.0 μHz. The expected ratio of frequency splitting for $\ell = 1$ modes to $\ell = 2$ modes is 0.60 (see equations [95] and [96]), which is precisely what is observed! Hence PG 1159 pulsates in $\ell = 1$ and $\ell = 2$ modes; its rotation period is 1.38 days.

The data show a complete string of 20 $\ell = 1$ modes between 430 and 840 seconds. The mean period spacing for the $\ell = 1$ modes is 21.5 seconds, and 12.5 seconds for the $\ell = 2$ modes. The observed ratio of the period spacings is 1.72; the theoretical value is 1.73, from equation (91). Again there is excellent agreement between theory and observation. From these period spacings equation (92) gives the mass of PG 1159 as $0.590 \pm 0.01 M_\odot$.

Deviations from uniform spacing of the $m = 0$ periods are clearly apparent, and strongly suggestive of mode trapping by a subsurface composition transition. Kawaler and Bradley (1994) use the period spacing diagram to determine the thickness of the surface helium layer in PG 1159 as $3 \times 10^{-3} M_*$. The model that best fits the mean period spacing and the observed periods had a luminosity of $220 L_\odot$, an effective temperature of 136,000 K, $\log g = 7.4$, and a surface helium abundance of 30%. Figure 32 shows the final fit by Kawaler & Bradley (1994) between the period spacings in the model and in PG 1159. From traditional spectroscopy, Werner et al. (1991) find an effective temperature of 140,000 K and $\log g = 7.0 \pm 0.5$, in excellent agreement with the results from temporal spectroscopy.

PG 2131+066 Another PG 1159 star examined with WET was PG 2131+066 (hereafter PG 2131). PG 2131 was little–observed (photometrically) following its discovery as a variable star by Bond et al. (1984). Data obtained at McDonald Observatory in 1984 indicated at least 6 periodicities, but the limitations of single–site observing rendered positive period determinations impossible. The Kiel group has analyzed the spectrum of PG 2131, and report an effective temperature of about $80,000 \pm 10,000$ K, and $\log g = 7.5 \pm 0.5$ (Dreizler et al. 1995): they indicate that PG 2131 might be more helium–rich than PG 1159.

PG 2131 was the target of a WET run in September 1992. The star is significantly fainter than PG 1159; therefore the data obtained have a lower S/N level. Despite this, the coverage of PG 2131 was sufficient for Kawaler et al. (1995) to decompose the observed pulsation spectrum of PG 2131. PG 2131 pul-

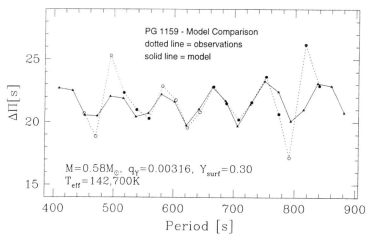

Fig. 32. Comparison between period spacings in PG 1159 and in the best model from Kawaler & Bradley (1994).

sates in $\ell = 1$ modes with several consecutive overtones, and a period spacing of 21.7 seconds. The power spectrum shows these modes are (roughly) equally spaced triplets, giving a rotation period of 5.07 hours. PG 2131 has a mass of $0.61 \pm 0.02 M_\odot$, and there is a hint of a discontinuity in the composition near $6 \times 10^{-3} M_*$ below the surface. The best-fit $0.61 M_\odot$ models have $T_{\text{eff}} = 84,000 \pm 2,000$ K and $L = 10 \pm 4 L_\odot$.

PG 2131 shows some similarities to PG 1159, but also some important differences. The largest amplitude periods are at higher frequency than PG 1159. There are fewer modes present, however, and there is no evidence for $\ell = 2$ modes. The amplitude of the pulsations is slightly larger in PG 2131 than in PG 1159. The differences in the pulsation behavior are probably correlated to the difference in temperature between PG 1159 and PG 2131.

Other PG 1159 Stars Additional WET campaigns of other hot white dwarfs are currently being analyzed. PG 1707 is another naked PG 1159 star for which WET data exist and are being analyzed, though that star had a gap in coverage (corresponding to central Asia) and therefore has a significant alias problem. WET data have also been obtained for RX J2117+34, the pulsating central star of an enormous old planetary nebula. This star may prove to be extremely important, as it shows periods that are shorter than any other central star, but longer than for the naked PG 1159 stars (Vauclair et al. 1993). It may provide a direct link between the planetary nebula nuclei and hot DO white dwarfs (Appleton, Kawaler, and Eitter 1993).

8.3 GD 358: A Pulsating DB White Dwarf

The prototype of the pulsating DB white dwarfs is GD 358, which was the first member of the class to be discovered (Winget et al. 1982b) following the theoretical prediction by Winget et al. (1982a) that DB white dwarfs should show analogous nonradial pulsations to the ZZ Ceti stars. GD 358 has an effective temperature of about 25,000 K (Thejl, Vennes, & Shipman 1991). It has an extremely complex light curve that made it an obvious target for study with the WET.

Winget et al. (1994) report on WET observations of GD 358, which have an extremely high duty cycle and clean power spectrum. They found over 180 separate pulsation frequencies in this star, including many triplets. The triplet spacing is measurably non–uniform, with the triplets at long period being systematically wider than at short period. Winget et al. (1994) interpret this as evidence of differential rotation, with the outer layers of GD 358 rotating faster than interior layers. The rotation period near the surface is 0.88 days, and 1.6 days at the deeper levels. In addition, Winget et al. (1994) find a systematic shift of the $m = 0$ peak to lower frequencies, which is the signature of magnetic fields (Jones et al. 1989). Measurement of this shift allows Winget et al. (1994) to determine the magnetic field strength of 1300 ± 300 G. Other peaks in the power spectrum lie precisely at the sums and/or differences of the dominant "normal" peaks, demonstrating that important nonlinear effects are operating in this star.

Model analysis of GD 358 by Bradley & Winget (1994) successfully reproduces in detail the pulsation frequencies. They determine a mass of GD 358 of $0.61 \pm 0.03 M_\odot$, $T_{\text{eff}} = 24 \pm 1 \times 10^3$ K, and a luminosity of $0.050 \pm 0.012 L_\odot$. The resulting distance to GD 358 is 42 ± 3 pc, meaning that the pulsations of GD 358 (with a dose of theory) are a distance indicator. In principle, the same distance determination technique can be applied to the PG 1159 stars, but in those hotter stars the bolometric correction introduces a much larger uncertainty.

Perhaps the most striking result of the modeling by Bradley & Winget (1994) is that the observed periods are best matched with a surface layer of pure helium of approximately $1.2 \times 10^{-6} M_\odot$. This is three order of magnitude smaller than PG 1159. This difference in helium layer mass suggests that establishing an evolutionary connection between the PG 1159 stars and DB white dwarfs requires finely–tuned mass loss beyond the PG 1159 stage. Alternatively, these results challenge the notion of there being a direct evolutionary relationship between these objects.

Exploration of this dilemma is another example of how temporal spectroscopy of white dwarf stars drives our growing understanding of their origins and chemical evolution. Dehner (1995) and Dehner & Kawaler (1995) show that the two classes of stars *do* appear to have a direct evolutionary connection. They computed the evolution of an initial model based on the PG 1159 pulsational data, including time–dependent diffusion. As expected, diffusion causes a separation of the elements; a thickening surface layer of nearly pure helium overlays a deepening transition zone where the composition changes to the surface composition of the original model. In the temperature range inhabited by GD 358 and the pulsating DB white dwarfs, this pure helium surface layer is $\sim 10^{-5.5} M_*$ thick; see Figure 21. The pulsation periods of the resulting model show a good fit to the WET observations. The model is very similar to the model used by Bradley & Winget (1994) to match the pulsation observations of GD 358. These results demonstrate the plausibility of a direct evolutionary path from PG 1159 stars to the much cooler DB white dwarfs.

8.4 The ZZ Ceti Stars

Though they are historically the first of the pulsating white dwarfs to be discovered, and they are the most numerous of the broad class of pulsating white dwarfs (and of pulsating stars in general), this review does not attempt to cover the subject of the ZZ Ceti stars in any detail. These stars are, however, an important test bed for several areas of astrophysics, and will be discussed in the final section in the context of measurement of the rate of evolution of white dwarfs, and of convective efficiency. The reader is referred to several reviews of the pulsating white dwarfs; perhaps the most complete is the Ph.D. dissertation by J. C. Clemens (Clemens 1993a), but see the more readily available review Clemens (1993b). Another a useful review is still that of Winget (1988).

9 Astrophysical Applications of White Dwarfs

In this final section, I'd like to describe some of the areas of astrophysics where white dwarfs are useful probes. I'll begin with a description of how the pulsating white dwarfs have been used to demonstrate that stars do indeed evolve with time, and that the rate of evolution of white dwarfs provides a direct test of some of the fundamental inputs into stellar evolution theory, and a test of some basic physics. The next section will collect some of the results of analysis of the white dwarf luminosity function. White dwarfs can be used to study star formation in globular clusters, and planetary nebulae (that contain newly formed white dwarfs) can be used as a powerful and accurate extragalactic distance indicator. Cool pulsating white dwarfs are teaching us, in some detail, about how convection works.

9.1 Stellar Evolution as a Spectator Sport

As discussed in the previous sections, pulsations probe the interiors of white dwarfs; the pulsation periods are determined by the global properties of the stars. As stars evolve and change (by cooling in the case of white dwarf stars) their global properties, including the periods of the nonradial g−modes, change. The period changes that reflect evolutionary changes occur on a time scale that is comparable to the cooling time scale. That is, the e−folding time for the pulsation periods is comparable to the e−folding time for the luminosity or effective temperature. For example, if a star cools at a rate that reduces its luminosity by a factor of two in 10^7 years, then the pulsation period will change significantly in 10^7 years. Assuming that the pulsation itself remains stable (that is, transients and other effects do not change the pulsation period or amplitude over time scales of months or years), it is possible to measure the pulsation period to a precision of 1 part in 10^7 in one year of time. Therefore, in principle, a star evolving as fast as a hot white dwarf can show measurable period changes on "human" time scales.

While measuring the period with great accuracy is a direct way of measuring secular period changes, there is a much better, and more stable, method. This is one familiar to pulsar specialists. What is measured is not the period per se, but the phase of the pulsation... that is, the time of the first maximum after a given time. Assuming that the period is strictly constant, the time of maximum can be predicted into the near (or distant) future. Observation at those later dates will show maxima that may or may not arrive at the predicted times. Call the observed time of maximum O and the predicted time C, we can write an equation for the difference, $(O - C)$, in terms of the assumed pulsation period and its time derivatives as follows:

$$(O - C) \approx \frac{1}{\Pi} \frac{d\Pi}{dt} (t - t_o)^2 + ... \tag{97}$$

The above equation, valid when the period at time t_o is known precisely, shows that the values of $(O - C)$ change with time squared. Thus if a secular period change is present within the star, it will result in a parabolic $(O - C)$ diagram. Samples of this effect are sketched in Figure 33.

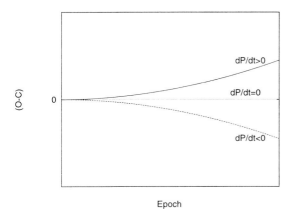

Fig. 33. Schematic $(O - C)$ diagram. If there is a secular period increase, then maxima in the observed light curve will arrive later and later, with the delay proportional to the square of the elapsed time, resulting in the parabolic curve upwards.

How does the value of the period change depend on the changes in the structure of the star? For white dwarfs pulsating in nonradial g-modes, this question has been addressed by Winget et al. (1983), and followed up by Kawaler et al. (1985). They showed that the rate of period change can be represented by

$$\frac{1}{\Pi} \frac{d\Pi}{dt} = -\frac{1}{T} \frac{dT}{dt} + \frac{1}{R} \frac{dR}{dt} \tag{98}$$

where the time derivatives on the right–hand side are for the temperature and radius in the region of the star that is important for period formation. Kawaler et al. (1985) show that for the hot PG 1159 stars, the period is determined by the degenerate interior, while in cooler white dwarfs, the outer layers are most important in determining the periods. Therefore period changes measure changes in the core in hot PG 1159 stars, and in the envelope in the DB and DA white dwarfs.

PG 1159 Stars For PG 1159 stars, Kawaler et al. (1985) show that cooling is more important than contraction as an evolutionary process; these stars are already largely degenerate and have nearly reached the constant–radius white dwarf cooling track. Thus the expected period changes can be estimated as the e-folding time for the surface temperature. This gives an estimated rate of period *increase* of

$$\frac{1}{\Pi}\frac{d\Pi}{dt} \approx -\frac{1}{T}\frac{dT}{dt} \approx \frac{1}{10^6 \text{yr}} \approx +10^{-14}\text{s}^{-1}. \tag{99}$$

This is an exciting possibility; most white dwarfs in the PG 1159 region are cooling primarily because of neutrino emission. Therefore, the rate of period change is a direct measurement of the plasmon (and/or bremstrahlung) neutrino emission rates. These rates have never been measured experimentally.

With a period derivative at this level, the change in the time of maximum in an $(O - C)$ diagram after one year will be 10 seconds, and after two years, 40 seconds. The timing accuracy for the dominant modes of pulsation in PG 1159 stars is much better than this for a typical observing run, and so the secular period increase in these stars is, in principle, easy to measure. This fact was noted by McGraw et al. (1979) in the discovery paper of PG 1159. The measurement was finally made by Winget et al. (1985), and confirmed to high precision by Winget et al. (1991). Paradoxically, the measured value of $\dot{\Pi}/\Pi$ was *negative*: $-2 \times 10^{-14}\text{s}^{-1}$ for the 516 second mode! This suggests that contraction was playing a larger role than anticipated in the evolution of the pulsation modes in PG 1159. At first, this large rate of decrease suggested that the neutrino rates used in the evolutionary calculations were incorrect, or that an additional agent was responsible for accelerated contraction (such as axions).

Evolutionary models computed by Kawaler & Bradley (1994) explain why the period of PG1159 is decreasing with time: the 516 s mode is a trapped mode. They found that, in PG 1159 models, trapped modes do show a negative $\dot{\Pi}/\Pi$. Recall that the period of a given mode will change in response to changes occurring in the region of the model that are sampled by that mode. Since each mode samples a slightly different part of the model, the value of $\dot{\Pi}/\Pi$ can vary from one mode to the next. Untrapped modes sample most of the star in these models, so that they all have roughly the same value of $\dot{\Pi}/\Pi$. However, modes which are trapped in the outer layers sample regions where contraction plays a large role. Thus trapped modes occasionally show period decreases with time while most modes have increasing periods.

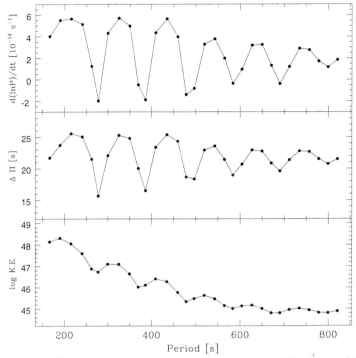

Fig. 34. The various effects of mode trapping in a PG 1159 model. From Kawaler & Bradley (1994).

Figure 34, adapted from Kawaler & Bradley (1994) illustrates this effect. The top panel shows $\dot{\Pi}/\Pi$ as a function of period for $\ell = 1$ modes in a $0.59 M_\odot$ PG 1159 model. The middle panel shows the period spacing as a function of period for the same model, and the bottom panel shows the log of the pulsational kinetic energy. Trapped modes are identified using the bottom two panels as those modes near a local minimum in the energy plot, which correspond to modes at minima in the period spacing diagram. Clearly, the top panel shows that trapped modes show minimum rates of period change. In fact, in this model trapped modes experience decreasing periods with time. This is because they sample only the outermost layers, where gravitational contraction occurs on a significantly shorter time scale than in the rest of the star, and on a time scale shorter than the local cooling time scale. It is clear that mode trapping provides the answer to why $\dot{\Pi}/\Pi$ is negative for the observed mode in PG 1159.

ZZ Ceti Stars While $\dot{\Pi}/\Pi$ can be large, and therefore observable, in PG 1159 stars, it is much smaller in ZZ Ceti stars. These stars are much cooler, and therefore cool and evolve more slowly than their hot cousins. However, two factors combine to make ZZ Ceti stars another prime class of stars for measuring rates of period change. First, there are several ZZ Ceti stars with very short periods (of order 100 to 200 seconds), and therefore they show more cycles on a given night, allowing precision timing of times–of–maxima and amplifying the effects of $\dot{\Pi}/\Pi$ on the $(O-C)$ diagram. The second reason is that the available time base on some of the ZZ Ceti stars is extremely long, approaching 20 years; recall that the precision of the measurement of $\dot{\Pi}/\Pi$ improves as time–squared goes by.

For ZZ Ceti stars,

$$\frac{1}{\Pi}\frac{d\Pi}{dt} \approx -\frac{1}{L}\frac{dL}{dt} \approx \frac{1}{7 \times 10^8 \text{yr}} \approx +10^{-17}\text{s}^{-1}. \quad (100)$$

which gives an advance in the measured time of maximum of about 1 second in 10 years, or 4 seconds in 20 years. The principal star used in the attempt to measure $\dot{\Pi}/\Pi$ is G117–B15A, with the analysis led by S. O. Kepler and his colleagues. A recent determination, using Whole Earth Telescope data and all available earlier data, results in a measurement of $\dot{\Pi}/\Pi$ of $(+1.5\pm1.3)\times10^{-17}\text{s}^{-1}$ (Kepler 1993).

This is, effectively, only an upper limit on the magnitude of $\dot{\Pi}/\Pi$ in this star, but it is still of significant interest. Recall the Mestel cooling curve for white dwarfs, and the dependence of the cooling rate on the atomic mass of the core and envelope. The Mestel law can be manipulated to give the time scale for luminosity decrease in terms of the core composition for the mass of choice $(0.60 M_\odot)$, or the luminosity of the ZZ Ceti stars $(10^{-3} L_\odot)$, to give:

$$\frac{1}{\Pi}\frac{d\Pi}{dt} \approx 1.4 \times 10^{-17} \frac{A}{12} \quad (101)$$

Thus the observed upper limit to the rate of period change in G 117–B15A already rules out the possibility that the star has a core heavier than oxygen, and in fact appears to require that the core is mostly carbon. This is a remarkable result, as it confirms some of the basic assumptions made in modeling white dwarfs: namely, that the core is carbon and oxygen!

9.2 The White Dwarf Luminosity Function and Our Galaxy

Much has already been discussed about the white dwarf luminosity function. Matt Wood's results (1992) indicate that, using the best available white dwarf models and observed luminosity functions, the age of the galactic disk in the vicinity of the Sun is 9.3±1.5 Gyr. The uncertainty in this number will be reduced in the near future with further observation and analysis of the pulsating white dwarfs. Pulsations will reduce the uncertain value of the thickness of the surface helium layer, and as the core composition becomes more tightly constrained by measurements of $\dot{\Pi}/\Pi$. While the uncertainties will become smaller, the value of the age itself probably will not change significantly. From the shape of the luminosity function, it is apparent that the star formation rate has been roughly constant over the history of the disk. Had there been bursts of star formation at some times, these bursts would have produced bumps along the observed luminosity function that are not seen in the data (Wood 1992, Iben & Laughlan 1989).

This age for the age of the disk of the galaxy is consistent with the current measurements of the age of the Universe based measurements of the Hubble constant. Recent measurements hover around $H_o \approx 77$ km/s/Mpc. Assuming $\Lambda = 0$ and $\Omega = 1$, this corresponds to a Hubble age of about 8 to 9 Gyr, which is entirely consistent with the age of the disk as determined with the white dwarf luminosity function so all is well. However, there are things called globular clusters which currently appear to have an age of about 14 to 16 Gyr. This is consistent with the white dwarf age if the globular clusters formed first in the formation of the galaxy, and the disk formed at a later time. This model of galaxy formation is a leading contender ..., but only if the globular cluster age can be reconciled with the cosmological age (Chaboyer 1995).

9.3 White Dwarfs and Cluster Ages

A related use of white dwarfs is in studying stellar populations in clusters. The low–luminosity cutoff of nearby white dwarfs at an M_V of about $+16$ yields an age for the disk of the Galaxy. If we look at a young cluster that formed, say, a 1 Gyr ago, then the oldest white dwarfs in that cluster should be at most about 1 Gyr old (minus the pre–white dwarf life time of the first stars to produce white dwarfs, or about 20 million years), and will be at $M_V \approx 13$. Clusters at a distance of about 630 pc should therefore not have white dwarfs fainter than about $V = 22$ if they are less than 10^9 years old. Of course, identification of such faint white dwarfs poses a challenge; demonstration that the sample is complete and that there is indeed a cutoff is even more difficult. Still, much can be learned from ground based observations of faint white dwarfs in clusters. Chuck Claver (Claver 1995) and others are in the process of surveying nearby clusters to very faint magnitudes to try and identify white dwarfs in clusters.

The determination of a white dwarf cooling age in a cluster would provide an age estimate that is nearly independent of other techniques for estimating cluster ages, such as main–sequence turnoff ages or dynamical ages. Alternatively, given the age of the cluster from other methods, the white dwarf sequence can be used to provide an accurate distance for the cluster.

Clearly, the best instrument for probing the luminosity function of white dwarfs in clusters is the Hubble Space Telescope. Indeed, HST has been used in a number of studies of clusters in efforts to study their white dwarf population (see, for example, Paresce et al. 1995 on 47 Tuc, von Hippel et al. 1995, and Richer et al. 1995 on M4). The very recent results of Richer et al. (1995) are truly remarkable; they report detection of a very distinct white dwarf sequence in the nearby globular cluster M4. Figure 35 shows their color–magnitude diagram for M4, along with the white dwarf sequence. The sequence is very well defined, and stretches over 4 magnitudes, and is consistent with a very small mass dispersion. They do not indicate the low luminosity cutoff in this cluster – it lies below the limiting magnitude for their observations. However, they

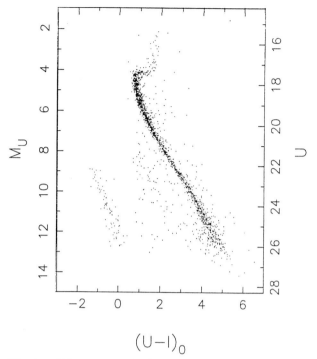

Fig. 35. Color magnitude diagram for the globular cluster M4, obtained by Richer et al. (1995) with the Hubble Space Telescope

convincingly demonstrate that the observed white dwarf luminosity function is consistent with the theoretical determination by Wood (1992).

9.4 The Planetary Nebula Luminosity Function and Galaxy Distances

As mentioned in previous sections, the white dwarf luminosity function has proven to be useful in many ways beyond the study of white dwarf structure and evolution. It is a probe of the star formation history of the Galaxy, of the age of the galactic disk, the ages of globular and open clusters, etc. Another luminosity function involving white dwarfs is the luminosity function of planetary nebulae (or "PNLF"). The PNLF reflects the rate of evolution of the central star, coupled to the rate of evolution of the surrounding nebulae. With some basic simplifying assumptions about radiation interaction with the nebula, and the dynamics of the expanding nebular shell, the PNLF can give useful information about the rate at which stars cross the H–R diagram as they evolve from the AGB to the white dwarf cooling track.

Unfortunately, while in principle the PNLF is a useful tool, its implementation is hampered by severe observational selection effects when using planetary nebulae in our galaxy. The chance of detecting planetary nebulae are affected not only by the vagaries of nebular formation and expansion, but also by the surrounding interstellar dust (frequently contributed by the AGB star which produced the nebula), dust in the local environment, etc. Perhaps the most debilitating factor, though, is the lack of reliable distances to the nebulae. Measuring the distance to planetary nebula is a thorny problem indeed; the standard distance measurement techniques frequently quote errors of 50% or more.

One solution to the distance (and completeness) problem is to consider a population of planetary nebula at a fixed distance, such as the planetary nebulae associated with the galactic bulge, or the Magellanic Clouds. This removes the distance uncertainty, as distances to these locations are available by other means. For these reasons, there are concentrated efforts now underway to study the central stars of planetary nebula in near the galactic center and in the LMC and SMC using the Hubble Space Telescope and large ground-based instruments. In addition, huge numbers of planetary nebula can be detected in other galaxies (out to the Virgo Cluster) by narrow-band observations at 5007Å, corresponding to the brightest emission line seen in planetary nebulae (an [OIII] line). As described by Jacoby (1989), subtracting images in [OIII] and the surrounding continuum makes planetary nebulae identification in distant galaxies relatively easy.

One of the remarkable results of studies of the PNLF in external galaxies is that the shape of the PNLF is relatively constant from one galaxy to the next. Furthermore, the PNLF drops rapidly at a fixed absolute V magnitude (actually, a V magnitude equivalent for the 5007Å [OIII] line). Jacoby (1989) details the construction of theoretical luminosity functions, and the determination of the PNLF from observations of several galaxies. The rapid drop-off is the result of the sharply peaked mass distribution of planetary nebula central stars, coupled with the shape of their evolutionary track and the steep dependence of the rate of fading on the mass. Figure 36 shows two plots from Jacoby (1989) that illustrate this remarkable property This figure shows the luminosity function for several galaxies, along with theoretical PNLF calculations using models in the literature. The data points are indicated in the figure legend; the observed PNLFs have been shifted horizontally by amounts corresponding to their observed distance moduli; they have been shifted vertically to meet at 1 magnitude below the brightest PN to correct for different numbers of detected planetaries in each galaxy.

The galaxies M31 and M81 are similar spiral galaxies, while the points labeled "Leo" are for three galaxies in the Leo group (two ellipticals and one S0). These PNLFs, obtained for galaxies of different morphological types and therefore representing different stellar populations, have an extremely close overlap. Another remarkable observation is that they all show the same cutoff at the bright end of $V = -4.48 \pm 0.04$. Jacoby (1989) points out that measurement of the *bright* end of the PNLF is sufficient to measure the apparent magnitude of the cutoff; assuming that the absolute magnitude of this cutoff is the same in all galaxies therefore allows an independent determination of the distance. This technique has been refined and exploited in determination of accurate ($\pm 10\%$) distances to a galaxies out to the Virgo cluster; the results of this research, and comparison of several different distance indicators is summarized in the excellent review of cosmological distance determination by Jacoby et al. (1992).

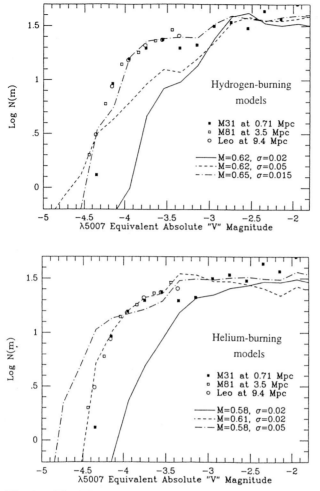

Fig. 36. The PNLF for external galaxies, from Jacoby (1989). In both panels, data points represent the PNLF for the indicated galaxy, shifted horizontally by an amount corresponding to the distance indicated in the legend. Lines represent theoretical models of the PNLF with mean masses, and mass dispersions, as indicated. The top panel uses theoretical models that are powered by hydrogen shell burning; the models used in the bottom panel are powered by the helium–burning shell.

A further use of the extragalactic PNLF is to work the problem in reverse: how does the mean *observed* PNLF compare to the theoretical PNLF using various flavors of pre–white dwarf models? This question has been addressed by Jacoby (1989) and, in further detail, by Dopita et al. (1992). The basic conclusion follows from the Figure 36; the top panel shows the observed extragalactic PNLF compared to theoretical ones computed using central star models powered by hydrogen burning, while the lower panel shows the same data compared to

White Dwarf Stars

helium–burning models. In the theoretical models, the free parameters of the synthetic population are the mean mass and the spread in that mean mass; both parameters can be adjusted to give improved fits with the data. The best–fit parameter pair for the hydrogen–burning models is $(<M>,\sigma) = (0.65, 0.015)$, while for helium–burning models it is $(0.60, 0.020)$. Changing the mean mass by a small amount degrades the fit significantly, as does a small change in σ. Based on measurements of white dwarf masses and the width of the white dwarf (and planetary central star) mass distributions, it appears that the extragalactic PNLF favors those PN central star models that burn helium in their outer layers, rather than the "standard model" which requires hydrogen–burning central stars.

While the distance determination technique is an empirical determination based on observed calibrations, accurate stellar models can be used to provide an absolute calibration. Dopita et al (1992) have attempted this calibration; they conclude that progress in an absolute calibration requires a better understanding of the physics of PN central star evolution. By assuming the veracity of the empirical calibration, the physics of white dwarf formation can indeed be constrained by galaxies at 10 Mpc!

9.5 Driving and Damping of Pulsations and Convective Efficiency in White Dwarfs Ceti Stars

Returning again to the pulsating white dwarfs, the positions of the unstable stars in the H–R diagram are determined by the details of the operation of the driving mechanism. In at least the pulsating DB and DA white dwarfs, the driving mechanism is related to the surface composition. When the material near the surface becomes partially ionized, then the so-called κ and γ mechanisms can operate to make the star vibrationally unstable. These mechanisms are the same as those that operate in the more classical variable stars. To illustrate this effect, consider the linear perturbation of the radiative flux in a way analogous to the way we considered perturbations to the other physical quantities when discussing the pulsating stars. The perturbation of L_r is

$$\frac{\delta L_r}{L_r} = 4\frac{\delta r}{r} - \frac{\delta \kappa}{\kappa} - 4\frac{\delta T}{T} + \left(\frac{dT}{dr}\right)^{-1}\frac{d}{dr}\left(\frac{\delta T}{T}\right). \tag{102}$$

Writing the opacity in the form of a Kramers opacity $\kappa = \kappa_o \rho^n T^{-s}$ and employing a thermodynamic identity or two, this can be rewritten to give the radiative luminosity perturbation in terms of the temperature perturbation alone:

$$\frac{\delta L_r}{L_r} = \frac{\delta T}{T}\left\{s + 4 - \frac{\frac{4}{3} + n}{\Gamma_3 - 1}\right\}. \tag{103}$$

Recalling that $\Gamma_3 - 1$ is 2/3 for a normal perfect gas, and that under normal circumstances in an ionized stellar plasma $s = 3.5$ and $n = 1$, we see that the term in braces on the right hand side is normally of order $+4$. That is, an increase in temperature results in a corresponding increase in the radiative flux.

If a portion of the star is compressed and heated, then this excess heat is removed more quickly through radiation, and therefore the energy of compression is not available to augment the pulsations. However, if the term in braces becomes negative, then the increase in temperature resulting from compression causes a *reduction* in the radiative flux, and the energy is effectively bottled up; it is then available to assist the dynamical expansion that follows compression, and thus the pulsation is given an additional kick which can lead to increasing amplitude. Under what physical conditions can this driving occur?

If the material is undergoing ionization, then the opacity of the material can be quite large. In this case, an increase in the temperature does not necessarily result in a large decrease in the opacity, and the value of s can become quite small or even become negative. Similarly, Γ_3 drops in an ionization zone. Together, these two effects can cause the term in braces in equation (103) to drop below zero, and the region therefore contributes to pulsational instability. However, the question of pulsational stability is not a local question; pulsation is a global phenomenon. Therefore, one needs to add up all of the local contributions to stability and/or instability; if the star as a whole is to be vibrationally unstable, the narrow driving regions must overcome damping by the rest of the star.

Computation of vibrational stability of realistic white dwarf models is extremely difficult, and plagued by numerical uncertainties of opacity tables and the like. Fortunately, the condition of vibrational stability can be understood relatively simply and quantitatively. Simply put, if the region where driving occurs is too deep within the star, then the driving will be insufficiently strong to overcome damping; that is, the time scale for heat flow from the driving zone is too long compared with the time scale over which driving is occurring. Similarly, if the driving zone is too close to the surface, then the modulation of the heat flux over the pulsation cycle is too slow compared to the response time of the overlying layers, and the star adjusts instantaneously. Therefore, a condition for driving to overcome damping is that the pulsation period (the time scale over which the heat flux is being perturbed) must be comparable to the thermal response time of the surrounding material. Put another way,

$$\Pi \approx \tau_{\text{th}} \approx \frac{<c_v T> \Delta m}{L} \tag{104}$$

where c_v is the specific heat, Δm is the mass above the driving zone, and L is the luminosity of the star. When the driving zone has a thermal time scale comparable to the pulsation period, then instability may occur.

For the $\kappa - \gamma$ effect, the position at which driving occurs can be identified with the base of the surface convection zone; the large radiative opacities that generate convective instabilities are intimately related to partial ionization of the dominant chemical species. Thus, we can track the thermal time scale of the driving region by examining the thermal time scale at the base of the surface partial ionization zone. Figure 37 shows the position (in mass fraction) of the base of the surface convection zone in DA white dwarf models that employ different models for the convective flux, all variants of the standard mixing length theory. Also shown is the thermal time scale at the base of the surface convection

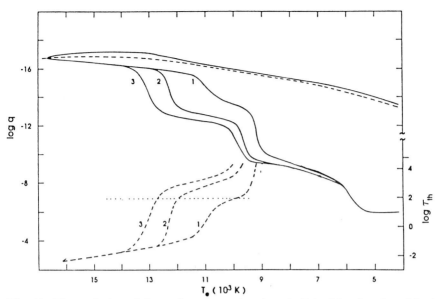

Fig. 37. The evolution of the surface convection zone in DA white dwarf models, from Tassoul et al. (1990). The position of the boundaries of the convective zone, in terms of fractional mass below the surface, is indicated by solid lines for several different formulation of the mixing length theory ranging from very efficient convection (ML3) to standard MLT. Dashed lines show the thermal time scale at the base of the surface convection zone.

zone. Notice that the convection zone deepens rapidly with decreasing effective temperature in all cases, but it deepens soonest for more efficient convection. When the thermal time scale increases to about 100 seconds, the models become unstable to g−mode pulsations; these first unstable models represent the blue edge of the ZZ Ceti instability strip.

Agreement of full–scale computations of driving and damping of white dwarf pulsations and the simple scaling described above is very close. In 1982, this theory was demonstrated to be a predictive theory. Donald Winget had investigated the driving mechanism of the ZZ Ceti stars, and described it in detail in his Ph.D. dissertation done at the University of Rochester (Winget 1981). His analysis demonstrated the accuracy of the above relation; he then reasoned that DB white dwarfs should also be unstable to the same sort of pulsations known to occur in the DA white dwarfs, but at higher effective temperatures. The scaling law above allowed him to make a prediction of the effective temperatures at which DB white dwarfs should pulsate (Winget 1981, Winget et al. 1982a). He then undertook a search for such pulsating DB white dwarfs at McDonald Observatory. The discovery of the pulsating DB white dwarf GD358 by Winget et al. (1982b) was the first instance of a variable star class to be predicted theoretically before the discovery of the first member.

Note also that accurate determination of the blue edge of the ZZ Ceti instability strip can give an excellent measurement of the efficiency of convection in white dwarf stars. The most recent determination of T_{eff} of the blue edge of the ZZ Ceti instability strip can be found in Bergeron et al. (1995); they find that optical spectrophotometry alone gives a blue edge for the ZZ Ceti strip at $T_{\text{eff}} = 13,650$, which is in rough agreement with earlier studies. From inspection of Figure 37, it is clear that this hot blue edge requires that convection in DA white dwarfs is much more efficient than standard mixing length theory allows. The ML3 formulation (see Tassoul et al. 1990 for details about ML2 and ML3) gives a blue edge at this temperature. However, Bergeron et al. (1995) also employ additional data from the ultraviolet in a more accurate spectroscopic determination of the effective temperatures of the ZZ Ceti stars. Atmospheric models that are consistent with the UV and optical spectra require less efficient convection, and give a value for the blue edge of $T_{\text{eff}} = 12,460$. This blue edge is also consistent with our simple model for driving of the pulsations; Figure 37 shows that this temperature for the blue edge corresponds to the ML2 convective efficiency. Still, the ZZ Ceti stars show us that convection in white dwarfs is more efficient that standard mixing length theory.

Convection in white dwarf stars has been the subject of a remarkable series of numerical hydrodynamic simulations, reported by Ludwig et al. (1994). These 3-D simulations include realistic atmospheric boundary conditions, and display all of the features of turbulent convective flow seen in models of the convective zones of main sequence stars. However, the action occurs on a much faster time scale, and over a greatly reduced spatial scale, as a consequence of the high gravity of these stars. Ludwig et al. (1994) have also produced time–averaged and angular–averaged thermal profiles for comparison with atmospheric and interior models. As described by Bergeron, these numerical simulations produce thermal profiles that are somewhat consistent with ML2 models with a reduced mixing length of 0.6; however, the thermal profiles of the hydrodynamic models cannot be reproduced in detail by any form of simple mixing length theory. Gautschy (1995) explores the consequences of the thermal profile on the vibrational instability of ZZ Ceti models; he uses the structural properties of the Ludwig et al. (1994) models in concert with white dwarf interior models. Gautschy (1995) is unable to produce vibrational instability in models that lie within the ZZ Ceti instability strip that have this realistic structure. This problem poses an interesting dilemma for the study of white dwarf pulsations and convection

9.6 Final Thoughts

These lectures have described various topics in the structure and evolution of white dwarf stars. The field of white dwarfs is a big field; there are a lot of people working on these little stars. I hope that it is clear that there is far more to the study of white dwarfs than I have been able to cover in these lectures; indeed, there are many things that have been left out. The uses are growing with time; I leave it to the reader to imagine ways to apply white dwarfs to new areas.

References

Aannestad, P. and Sion, E. (1985): *A.J.* **90**, 1832.
Adams, W.S. (1914): *P.A.S.P.* **26**, 198
Adams, W.S. (1915): *P.A.S.P.* **27**, 236
Angel, J.R.P., Borra, E.F., & Landstreet, J.D. (1981): *Ap.J.Supp.* **45**, 457
Appleton, P.N., Kawaler, S.D., & Eitter, J.J. (1993): *A.J.* **106**, 1973
Barstow, M.A., Fleming, T.A., Diamond, CJ., Finley, D.S., Sansom, A.E., Rosen, S.R., Koester, D., Marsh, M.C., Holberg, J.B., & Kidder, K. (1993): *M.N.R.A.S.* **264**, 16
Beaudet, G., Petrosian, V., & Salpeter, E.E. (1967): *Ap.J.* **150**, 979
Bergeron, P., Saffer, R., & Liebert, J. (1992): *Ap.J.* **394** 228.
Bergeron, P., Wesemael, F., Lamontagne, R., Fontaine, G., Saffer, R., & Allard, N. (1995): *Ap.J.* **449**, 258
Bessel, F.W. (1844): *M.N.R.A.S.* **6**, 136
Blöcker, T. (1993): *Acta Astr.* **43**, 305
Bond, H.E., Grauer, A.D., Green, R.F., & Liebert, J. (1984): *Ap.J.* **279**, 751
Bond, H.E., Ciardullo, R., & Kawaler, S.D. (1993): *Acta Astr.* **43** 425
Bond, H.E. et al. (1995): *Ap.J.* (in preparation)
Bowen, G. & Willson, L.A. (1991): *Ap.J.Lett.* **375**, L53
Bradley, P.A., & Winget, D.E. (1991): *Ap.J.Supp.* **75**, 463
Bradley, P.A. & Winget, D.E. (1994): *Ap.J.* **430**, 850.
Bradley, P.A., Winget, D.E., & Wood, M.A. (1993): *Ap.J.* **406**, 661
Brassard, P., Fontaine, G., Wesemael, F. & Hansen, C.J. (1992a): *Ap.J.Supp.* **80**, 369
Brassard, P., Fontaine, G., Wesemael, F. & Tassoul, M. (1992b): *Ap.J.Supp.* **81**, 747
Brickhill, A.J. (1975): *M.N.R.A.S.* **170**, 404
Chaboyer, B. (1995): *Ap.J.Lett.* 444, L9
Chandrasekhar, S. (1939): *An Introduction to the Study of Stellar Structure*, (Chicago: University of Chicago Press).
Chanmugam, G. (1972): *Nat. Phys. Sci.* **236**, 83
Chapman, J. (1995): in *I.A.U. Colloquium #155: Astrophysical Applications of Stellar Pulsation*, ed. R. Stobie and P. Whitelock, (Provo, PASP Conference Series), (in press)
Chayer, P., Fontaine, G., & Wesemael, F. (1995a): *Ap.J.Supp.* **98**, 180
Chayer, P., Vennes, S., Pradhan, A., Thejl, P., Beauchamp, A., Fontaine, G., & Wesemael, F. (1995b): *Ap.J.* (in press)
Ciardullo, R., Jacoby, G., & Harris, W. (1991): *Ap.J.* **383**, 487
Claver, C. (1995): Ph. D. Dissertation, University of Texas at Austin
Clemens, J.C. (1993a): Ph. D. Dissertation, University of Texas at Austin
Clemens, J.C. (1993b): *Baltic Astron.* **2**, 407
Cox, J.P. (1968): *Principles of Stellar Structure*, (New York, Gordon and Breach)
Cox, J.P. (1980): *Theory of Stellar Pulsation*, (Princeton, Princeton U. Press)
D'Antona, F. (1990): *Ann.Rev.A.Ap.* **28**, 139
Dehner, B.T. (1995), Ph.D. Dissertation, Iowa State University
Dehner, B.T. & Kawaler, S.D. (1995): *Ap.J.Lett.* **445**, L141
Demers, S., Kibblewhite, E., Irwin, M., Nithakorn, D.S., Beland, S., Fontaine, G., & Wesemael, F. (1986): *A.J.* **92**, 878
Dopita, M., Jacoby, G., Vassiliadis, E. (1992): *Ap.J.* **389**, 27
Dorman, B., Rood, R., & O'Connell, R. (1993): *Ap.J.* **419**, 596

Dreizler, S. et al. (1995): in *9th European Workshop on White Dwarf Stars*, ed. D. Koester & K. Werner (Dordrecht: Kluwer), (in press)

Eggleton, P., Faulkner, J., & Flannery, B. (1973): *Astron. Astrophys.* **23**, 325

Fleming, T.A., Liebert, J. & Green, R.F. (1986): *Ap.J.* **308**, 176

Fontaine, G., & Wesemael, F. (1987): in *IAU Colloq. 95, The Second Conference on Faint Blue Stars*, ed. A.G.D. Philip, D.S. Hayes, & J. Liebert, (Schenectady: Davis), 319

Gautschy, A. (1995): *Astron. Astrophys.* (in press)

Giclas, H.L., Burnham, R.,Jr., & Thomas, N. G. (1971): *Lowell Proper Motion Survey, The G-Numbered Stars*

Giclas, H.L., Burnham, R., Jr., & Thomas, N. G. (1978): *Lowell Obs.Bull.* No.163

Grauer, A.D. & Bond, H.E. (1984): *Ap.J.* **277**, 211

Green, R.F. (1980): *Ap.J.* **238**, 685

Green, R., Schmidt, M., & Liebert, J. (1986): *Ap.J.Supp.* **61**, 305

Greenstein, J.L. (1976): *Ap.J.* **227**, 224

Greenstein, J.L. (1984): *Ap.J.* **276**, 602

Greenstein, J.L. (1986): *Ap.J.* **304**, 334

Greenstein, J. & Liebert, J.L. (1990): *Ap.J.* **360**, 662

Hamada, T. & Salpeter, E. (1961): *Ap.J.* **134**, 683

Hansen, C.J. & Kawaler, S.D. (1994): *Stellar Interiors: Physical Principles, Structure, and Evolution*, (New York, Springer–Verlag)

Herschel, F.W. (1782): *Phil.Trans.R.Soc.Lon.* **72**, 112

Herschel, F.W. (1785): *Phil.Trans.R.Soc.Lon.* **75**, 40

Hertzsprung, E. (1915): *Ap.J.* **42**, 115

Hubbard, W. & Lampe, M. (1969): *Ap.J.Supp.* **18**, 297

Iben, I. Jr. (1984): *Ap.J.* 333.

Iben, I. Jr. (1991): *Ap.J.Supp.* **76**, 55

Iben, I. Jr. & Tutukov, A. (1984): *Ap.J.* **282**, 615

Iben, I. Jr. & MacDonald, J. (1985): *Ap.J.* **296**, 540

Iben, I. Jr. & Laughlin, G. (1989): *Ap.J.* **341**, 312

Itoh, N., Mitake, S., Iyetomi, H., & Ichimaru, S. (1983): *Ap.J.* **273**, 774

Itoh, N., Kahyama, Y., Matsumoto, N., & Seki, M. (1984): *Ap.J.* **285**, 758

Jacoby, G. (1989): *Ap.J.* **339**, 39

Jacoby, G. et al. (1992): *P.A.S.P.* **104**, 599

Jones, P.W., Pesnell, W.D., Hansen, C.J., & Kawaler, S.D. (1989): *Ap.J.* **336**, 403

Kawaler, S.D. (1986): Ph.D. Dissertation, University of Texas at Austin

Kawaler, S.D. (1987a): in *Stellar Pulsation*, ed. A.N. Cox, W.M. Sparks, & S.G. Starrfield, (Berlin: Springer), 367

Kawaler, S.D. (1987b): in *IAU Colloq. 95, The Second Conference on Faint Blue Stars*, ed. A.G.D. Philip, D.S. Hayes, & J. Liebert (Schenectady: Davis), 297

Kawaler, S.D. (1988): in *IAU Symposium 123, Advances in Helio- and Asteroseismology*, ed. J. Christiansen-Dalsgaard & S. Frandsen, (Dordrecht: Reidel), 329

Kawaler, S.D. (1990a): in *Confrontation Between Stellar Pulsation and Evolution*, ed. C. Cacciari & G. Clementini, (Provo: Astr. Soc. of the Pacific), 494

Kawaler, S.D. (1990b): *B.A.A.S.* **21**, 1077

Kawaler, S.D. (1995): in *I.A.U. Colloquium #155: Astrophysical Applications of Stellar Pulsation*, ed. R. Stobie and P. Whitelock, (Provo, PASP Conference Series), (in press)

Kawaler, S.D. & Bradley, P.A. (1994): *Ap.J.* **427**, 415

Kawaler, S.D., Hansen, C.J., & Winget, D.E. (1985): *Ap.J.* **295**, 547

Kawaler, S.D. & Hansen, C.J. 1989, in *IAU Colloquium #114: White Dwarfs*, ed. G. Wegner (Berlin: Springer-Verlag), 97

Kawaler, S.D., & Weiss, P. (1990): in *Proc. of the Oji International Seminar, Progress of Seismology of the Sun and Stars*, ed. Osaki, Y., & Shibahashi, H., (Berlin: Springer), 431

Kawaler, S.D. et al. (1995): *Ap.J.* **450**, 350

Kepler, S.O. (1993): *Baltic Astron.* **2**, 444

Kilkenny, D., O'Donoghue, D., & Stobie, R. S. (1991): *M.N.R.A.S.* **248**, 664

Kippenhahn, R. & Weigert, A. (1990): *Stellar Structure and Evolution*, (Berlin, Springer-Verlag)

Koesterke, L. & Hamann, E. (1995): in *9th European Workshop on White Dwarf Stars*, ed. D. Koester & K. Werner (Dordrecht: Kluwer), (in press)

Landau, L. & Lifschitz, E. (1958): *Statistical Physics*, (London, Pergamon Press)

Landolt, A. (1968): *Ap.J.* **153**, 151

Ledoux, P. J., and Sauvenier-Goffin, E. (1950): *Ap.J.* **111**, 611

Liebert, J. (1980): *Ann.Rev.A.Ap.* **18**, 363

Liebert, J. (1986): in *Hydrogen Deficient Stars and Related Objects*, ed. K. Hunger et al. (Dordrecht: Reidel) 367

Liebert, J., Dahn, C., Monet, D. (1988): *Ap.J.* **332**, 891.

Ludwig, H.-G., Jordan, S., & Steffen, M. (1994): *Astron. Astrophys.* **284**, 105

Luyten, W.J. (1969): *Proper Motion Survey with the Forty-eight Inch Schmidt Telescope. XVIII. Binaries with White Dwarf Components*, (Minneapolis: University of Minnesota)

Luyten, W.J. (1979): *NLTT Catalogue*, (Minneapolis: University of Minnesota)

MacDonald, J. (1992): *Ap.J.* **394**, 619

McCook, K. & Sion, E.M. (1987): *Ap.J.Supp.* **65**,

McGraw, J.T., Starrfield, S.G., Liebert, J., & Green, R.F. (1979): in *IAU Colloq. 53, White Dwarfs and Variable Degenerate Stars*, ed. H.M. Van Horn & V. Wiedemann, (Rochester: Univ. of Rochester), 377

Meištas, E. & Solheim, J.-E., eds. (1993): *The Second WET Workshop Proceedings*, *Baltic Astron.* **2**, 357

Meištas, E. & Solheim, J.-E., eds. (1995): *The Third WET Workshop Proceedings*, *Baltic Astron.* (in press)

Mestel, L. (1952): *M.N.R.A.S.* **112**, 583

Motch, C., Werner, K., & Pakull, M.W. (1993): *Astron. Astrophys.* **268**, 561

Nather, R.E., Winget, D.E., Clemens, J.C., Hansen, C.J., & Hine, B.P. (1990): *Ap.J.* **361**, 309

Osaki, Y. & Hansen, C.J. (1973): *Ap.J.* **185**, 277

Oswalt, T. & Smith, J.A. (1995): in *Proceedings of the 9th European Workshop on White Dwarfs*, ed. D. Koester & K. Werner (Heidelberg: Springer-Verlag), (in press)

Paczynski, B. (1970): *Acta Astron.* **6**, 426

Paresce, F., De Marchi, G. & Romaniello, M. (1995): *Ap.J.* **440**, 216

Pelltier, C., Fontaine, G., Wesemael, F., Michaud, G., & Wegner, G., (1986): *Ap.J.* **307**, 242.

Pippard, A.B. (1957): *The Elements of Classical Thermodynamics*, (Cambridge, Cambridge University Press)

Renzini, A. (1981): in *Physical Processes in Red Giants*, ed. I. Iben, Jr. & A. Renzini, (Dordrecht, Reidel), 431

Richer, H. et al. (1995): *Ap.J.Lett.* **451**, L17

Russell, H.N. (1944): *A.J.* **51**, 13

Salpeter, E.E. (1955): *Ap.J.* **121**, 161
Schatzman, E. (1950): *Publ. Obs. Copenhagen,* #149.
Schatzman, E. (1958): *White Dwarfs,* (Amsterdam, N. Holland)
Schmidt, G. (1987): in *IAU Colloq. 95, The Second Conference on Faint Blue Stars,* ed. A.G.D. Philip, D.S. Hayes, & J. Liebert, (Schenectady: Davis), 377
Schmidt, G. & Smith, P. (1994): *Ap.J.Lett.* **423**, L63
Schmidt, G. & Smith, P. (1995): *Ap.J.* (in press)
Schönberner, D. (1983): *Ap.J.* **272**, 708.
Shapiro, S. & Teukolsky, S. (1983): *Black Holes, White Dwarfs, and Neutron Stars: The Physics of Compact Objects,* (New York: Wiley)
Shaw, R. & Kaler, J. (1985): *Ap.J.* **295**, 537
Sion, E.M. (1986): *P.A.S.P.* **98**, 821.
Sion, E. & Liebert, J. (1977): *Ap.J.* **213**, 468
Sion, E.M., Greenstein, J., Landstreet, J., Liebert, J., Shipman, H., & Wegner, G. (1983): *Ap.J.* **269**, 253
Smeyers, P. & Tassoul, M. (1987): *Ap.J.Supp.* **65**, 429
Stobie, R.S., Morgan, D.H., Bhatia, R.K., Kilkenny, D., & O'Donoghue, D. (1987): in *IAU Colloq. 95, The Second Conference on Faint Blue Stars,* ed. A.G.D. Philip, D.S. Hayes, & J. Liebert (Schenectady: Davis), 493
Tassoul, M., (1980): *Ap.J.Supp.* **43**, 469
Tassoul, M., Fontaine, G., & Winget, D.E. (1990): *Ap.J.Supp.* **72**, 335
Thejl, P., Vennes, S., & Shipman, H.L. (1991): *Ap.J.* **370**, 355
Truran, J.T., & Livio, M. (1986): *Ap.J.* **308**, 721
Unno, W., Osaki, Y., Ando, A., Saio, H., & Shibahashi, H. (1989): *Nonradial Oscillations of Stars (2nd Edition),* (Tokyo, Univ. of Tokyo Press).
Uus, U. (1970): *Nauch. Inform. Akad. Nauk. USSR,* **17**, 30
Van Horn, H.M. (1971): in *White Dwarfs,* ed. W. Luyten (Dordrecht, Reidel), 97
van Maanen, A. (1913): *P.A.S.P.* **29**, 258
Vassiliadis, E. & Wood, P.R. (1993): *Ap.J.* **413**, 641
Vauclair, G., Belmonte, J.A., Pfeiffer, B., Chevreton, M., Dolez, N., & Motch, C. (1993): *Astron. Astrophys.* **267**, L35
von Hippel, T., Gilmore, G., & Jones, D. (1995): *Astron. Astrophys.* (in press)
Warner, B. & Robinson, E.L. (1972): *Nat. Phys. Sci.,* **234**, 2
Weidemann, V. (1990): *Ann.Rev.A.Ap.* **28**, 103
Werner, K. (1993): in *White Dwarfs: Advances in Theory and Observation,* ed. M.A. Barstow, (Dordrecht: Kluwer), 67
Werner, K., Heber, U., & Hunger K. (1991): *Astron. Astrophys.* **244**, 437
Wesemael, F., Greenstein, J., Liebert, J., Lamontagne, R., Fontaine, G., Bergeron, P., & Glaspey, J. (1993): *P.A.S.P.* **105**, 761 (W93)
Winget, D.E. (1981): Ph.D. dissertation, University of Rochester
Winget, D.E. (1988): in *IAU Symposium 123, Advances in Helio- and Asteroseismology,* ed. J. Christensen-Dalsgaard & S. Frandsen, (Dordrecht: Reidel), 305
Winget, D.E., Van Horn, H.M., & Hansen, C.J. (1981): *Ap.J.Lett.* **245**, L33
Winget, D.E., Van Horn, H.M., Tassoul, M., Hansen, C.J., Fontaine, G., & Carroll, B. (1982a): *Ap.J.Lett.* **252**, L65
Winget, D.E., Hansen, C.J., and Van Horn, H.M. (1983): *Nature,* **303**, 781
Winget, D.E., Hansen, C.J., Liebert, J., Van Horn, H.M., Fontaine, G., Nather, R.E., Kepler, S.O., & Lamb, D.Q. (1987): *Ap.J.Lett.* **315**, L77

Winget, D.E. & Van Horn, H.M. (1987): in *IAU Colloq. 95, The Second Conference on Faint Blue Stars*, ed. A.G.D. Philip, D.S. Hayes, & J. Liebert, (Schenectady: Davis), 363

Winget, D.E., Kepler, S.O., Robinson, E.L., Nather, R.E., & O'Donoghue, D. (1985): *Ap.J.* **292**, 606

Winget, D.E., Robinson, E.L., Nather, R.E., & Fontaine, G. (1982b): *Ap.J.Lett.* **262**, L11

Winget, D.E., et al. (1991): *Ap.J.* **378**, 326

Winget, D.E., et al. (1994): *Ap.J.* **430**, 839

Wood, M. (1992): *Ap.J.* **386**, 539

Wood, P.R. & Faulkner, D.J. (1986): *Ap.J.* **307**, 659

Neutron Stars

G. Srinivasan

Raman Research Institute, Bangalore 560 080, India

There is a very rich literature on the subject of neutron stars. They were predicted as soon as the neutron itself was discovered, and many of their properties and manifestations elucidated long before their discovery. In the ensuing chapters we have tried to strike a balance between the physics of neutron stars themselves on the one hand, and their origin and evolution on the other. We have deliberately chosen not to be exhaustive in citing original references since this very often affects the flow of arguments in a pedagogical article. We have, however, attempted to cite as many text books and review articles as possible.

After a historical introduction to the subject, which we feel is very revealing, we turn to a discussion of the nature of pulsars in Chapter 2. Since the overwhelming majority of neutron stars that have been detected are radio pulsars it is appropriate to begin with a discussion of them. Chapter 3 is devoted to a discussion of the physics of neutron stars. This is followed by a discussion of the question concerning the progenitors of neutron stars (Chapter 4). In Chapter 5 we take up the important topic of pulsars in binary systems: their origin and evolution. Millisecond pulsars are the subject of Chapter 6. One of the intriguing questions concerning neutron stars is the evolution of their magnetic fields. Some recent ideas concerning this question are discussed in Chapter 7. The phenomenon of glitches in neutron stars is discussed in Chapter 8. The last Chapter reviews some recent and exciting ideas about plate tectonics in neutron stars.

1 A Historical Introduction

The main objective of this introductory chapter is to point out not only the historical beginnings of the subject but also to remind ourselves that many concepts and ideas that we now consider as well established were anticipated on very general grounds nearly sixty years ago.

1.1 What Are the Stars?

Every now and then there occurs a great revolution in science. By scientific revolution we mean the following: Very often there are questions that arise in science which appear meaningless or even frivolous within the premise of science, and suddenly they acquire a meaning. This constitutes a revolution. One of the great revolutions in science occurred in the middle of last century. To the positivist philosophers who influenced European thinking so much in the 18th and 19th century *it was in the nature of things that we shall never know what the stars are.* And yet, with Fraunhofer's discovery of the absorption lines in the spectrum of the Sun, and their explanation by Kirchoff, a great revolution had occurred. It suddenly became clear that at least the outer layers of the Sun and the stars were gaseous. Soon there were numerous attempts to understand stars as gaseous masses.

Lane was one of the first persons to argue that stars were gaseous masses that were stable because the gravitational pressure was balanced by the pressure of a *perfect gas* viz., Boyle's law. This picture was amplified by Kelvin and Helmholtz. What Lane, Kelvin and Helmholtz argued was that the stability of the Sun must be understood in terms of the equation of hydrostatic equilibrium .

$$\frac{dP(r)}{dr} = -\frac{GM(r)\rho(r)}{r^2}$$

where the pressure is to be calculated from the ideal gas equation of state viz.,

$$P_{gas} = \frac{\rho k T}{\mu m_p}$$

where k is Boltzmann's constant, μ the mean molecular weight, and m_p the mass of the proton.

Regarding the source of the heat, Kelvin and Helmholtz argued that it was due to gravitational contraction. However, the serious difficulty with this picture was quickly appreciated. From Virial theorem one could estimate the kinetic energy of the plasma and it was clear that at the present luminosity the Sun will radiate it away in a mere 10^7 years. But various biological and geological evidences pointed to the fact that the Sun must have been radiating roughly at its present luminosity for billions of years. Therefore there must be a continuing source of energy in the Sun. Eddington was the first to suggest (around 1920) that the source of energy may be subatomic or nuclear energy, as we call it today. When he encountered skepticism to this remarkably prescient idea he is known to have remarked "what is possible in the Cavendish Laboratory cannot be too complicated for a star", and "how else can we produce the observed helium in the Universe?".

In the early 1920s Eddington made a major modification to the 19th century principles of stellar structure. He invoked the idea of *radiative equilibrium* (a concept introduced earlier by Karl Schwarzschild), and argued that one must take into account not only the gas pressure but also the pressure of the outward flowing radiation. Thus the total pressure that balances gravity is the sum of gas pressure and radiation pressure. It is worthwhile reminding ourselves of the basic equations of stellar structure for stars in radiative equilibrium.

$$\frac{d}{dr}\left[\frac{\rho k T}{\mu m_p} + \frac{1}{3}aT^4\right] = -\frac{GM(r)\rho}{r^2}$$

$$\frac{dP_{rad}(r)}{dr} = -\left(\frac{L(r)}{4\pi r^2 c}\right) \cdot \frac{1}{l}$$
$$\frac{dL(r)}{dr} = 4\pi r^2 \epsilon \rho$$

where l is the mean free path of the photons, L the luminosity, and ϵ the energy generated per gram of material per unit time.

This theory of the stars due to Eddington was enormously successful (Eddington 1926).

1.2 Why Are the Stars as They Are?

Following Eddington we may ask 'why are the stars as they are?'. To explain the meaning of this cryptic question we may ask in parallel the following question viz., 'why are the atoms as they are?', by which we mean: (i) why do the atoms have the sizes they have? (ii) why atoms occur only in a finite range of atomic masses and charges? We may answer these questions by saying the following: Quantum theory is at the base of our understanding of atoms. And this theory naturally provides us with a length scale with which to measure atoms and so on. In a similar fashion we wish to now ask 'why are the stars as they are?' viz., (1) why do the luminosities of stars depend only on their masses and not on their radii? (2) why do stars occur only in an incredibly narrow range of masses? (3) why is there a limit to the luminosities of stars?

Eddington's theory was able to answer all these questions in a convincing manner. Perhaps the most intriguing of all is the second question which concerns the range of masses of stars. Eddington addressed this question and answered it in his famous parable of a physicist on a cloud-bound planet (Eddington 1926). "The outward flowing radiation may be compared to a wind blowing through the star and helping to distend it against gravity. The formulae to be developed later enable us to calculate what proportion of the weight of the material is borne by this wind, the remainder being supported by the gas pressure. To a first approximation the proportion is the same at all parts of the star. It does not depend on the density nor on the opacity of the star. It depends only on the mass and molecular weight. Moreover, the physical constants employed in the calculation have all been measured in the Laboratory, and no astronomical data are required. We can imagine a physicist on a cloud-bound planet who has never heard tell of the star calculating the ratio of radiation pressure to gas pressure for a series of globes of gas of various sizes, starting, say, with a globe of mass 10 g, then 100 g, 1000 g and so on, so that his nth globe contains 10^n g. Table I shows the more interesting part of his results."

"The rest of the Table would consist mainly of long strings of 9's and 0's. Just for the particular range of mass about the 33rd to 35th globes the Table becomes interesting, and then lapses back into 9's and 0's again. Regarded as a tussle between matter and aether (gas pressure and radiation pressure) the contest is overwhelmingly one-sided except between numbers 33–35, where we may expect something to happen."

Table I

No. of globes	Radiation pressure	Gas pressure
..	.....	.....
..	.....	.....
..	.....	.....
32	0.0016	0.9984
33	0.106	0.894
34	0.570	0.430
35	0.850	0.150
36	0.951	0.049
37	0.984	0.016
38	0.9951	0.0049
39	0.9984	0.0016
..	.....	.....
..	.....	.....
..	.....	.....

"What 'happens' is the stars."

"We draw aside the veil of cloud beneath which our physicist has been working and let him look up at the sky. There he will find a thousand million globes of gas nearly all of mass between his 33rd and 35th globes — that is to say, between 1/2 and 50 times the Sun's mass. The lightest known star is about 3.10^{32} g, and the heaviest about 2.10^{35} g. The majority are between 10^{33} and 10^{34} g, where the serious challenge of radiation pressure to compete with gas pressure is beginning."

But why is this tussle important for 'stars to happen'? Eddington did not elaborate. A deeper understanding of why the masses of stars lie in such a narrow range, although their luminosities vary by million or more, was provided by Chandrasekhar (Chandrasekhar 1937).

Let us introduce the parameter β which is the ratio of the gas pressure to the total pressure

$$P_{tot} = p_{gas} + p_{rad}$$
$$= \frac{1}{\beta} p_{gas} = \frac{1}{1-\beta} p_{rad}$$

One can now write the total pressure as follows:

$$P_{tot} = \left[\left(\frac{k}{\mu m_p} \right)^4 \cdot \frac{3}{a} \cdot \frac{1-\beta}{\beta^4} \right]^{1/3} \rho^{4/3}$$

Chandrasekhar argued that for a star to be mechanically stable the following inequality has to be satisfied.

$$\frac{1}{2}G\left(\frac{4\pi}{3}\right)^{1/3} M^{2/3} \bar{\rho}^{4/3} \leq P_c \leq \frac{1}{2}G\left(\frac{4\pi}{3}\right)^{1/3} M^{2/3} \rho_c^{4/3}$$

Here, $\bar{\rho}$ is the mean density and ρ_c the central density. P_c is the sum of gas pressure plus radiation pressure at the centre. The right hand side of the inequality gives the necessary condition for the stability of a star, i.e.

$$\mu^2 M \left(\frac{\beta_c^4}{1-\beta_c}\right)^{1/2} \geq 0.19 \left[\left(\frac{hc}{G}\right)^{3/2} \cdot \frac{1}{m_p^2}\right]$$

The combination of fundamental constants on the right hand side has the dimension of MASS. Its value is $\approx 29.2 M_\odot$. We thus see that in a theory of stars where gravity is balanced by the sum of gas pressure and radiation pressure, the natural *scale of mass* is of stellar magnitude. This is why the stars are as they are! Or, as Chandrasekhar put it:

> "We conclude that to the extent the above theory is at the base of the equilibrium of actual stars, to that extent the above combination of natural constants, providing a mass of proper magnitude for the measurement of stellar masses, is at the base of a physical theory of stellar structure."

1.3 White Dwarfs

Despite its sensational success the standard model soon received a severe jolt. The difficulty arose with the discovery of a new class of stars called the White Dwarfs. These were unusual for the following reason: although their masses were comparable to the Sun they were no bigger than the Earth, implying a mean density $\sim 10^6$ gm cm^{-3}. Eddington was the first to realize that these highly condensed stars posed a serious problem. As he put it,

> "I do not see how a star which has once got into this highly compressed condition is ever going to get out of it ... It would seem that the star will be in awkward predicament when its supply of subatomic energy fails ... a star will need energy to cool."

This paradox was better stated by R.H. Fowler in 1926:

> "The stellar material will have radiated so much that it has less energy than the same matter in normal atoms expanded at the absolute zero temperature. If part of it were removed from the star and the pressure taken off, what could it do?"

What Fowler meant was the following. The electrostatic energy E_{es} per unit volume of completely ionized atoms is given by

$$E_{es} = 1.32 \times 10^{11} Z^2 \rho^{4/3}$$

The kinetic energy E_{kin} of the particles in this volume under the assumption that the gas was a *perfect gas*, is given by

$$E_{kin} = \frac{3}{2}\frac{\rho kT}{\mu m_p} = \frac{1.24 \times 10^8}{\mu}\rho T$$

Now, if one removed the pressure the matter can expand and resume the state of ordinary atoms only if

$$E_{kin} > E_{es}$$

of equivalently, only if

$$\rho < \left(10^{-3}T/\mu Z^2\right)^3$$

It turns out that for densities that obtain in white dwarfs, $\sim 10^6 \text{gm cm}^{-3}$, this condition is violated. This is what Eddington meant when he said "a star will need energy to cool": it can only get out of the compressed state by expanding, but it doesn't have enough energy to expand! So it is doomed!

This major problem was solved in 1926 by R.H.Fowler who argued that at such high densities Boyle's law will break down and that one should calculate the pressure due to the electrons according to the newly discovered statistics of Fermi and Dirac (Fowler 1926). It is worth mentioning that this was the first application of the new quantum statistics which had been published just three months earlier!

Let us elaborate on Fowler's idea: In a white dwarf the pressure that balances gravity is the quantum mechanical or *degeneracy pressure of the electrons*, and not the classical pressure given by Boyle's law. The degeneracy pressure is determined by the number density of electrons and to a good approximation *independent of temperature*. This immediately resolved Eddington's paradox. One can pack matter into such high densities only at the expense of giving a tremendous kinetic energy to the electrons. The moment one removes the external pressure, the matter will have no difficulty in expanding to the state of normal atoms. In Fowler's words:

> "The black dwarf material is best likened to a single gigantic molecule in its lowest quantum state. On the Fermi-Dirac statistics, its high density can be achieved in one and only way, in virtue of a correspondingly great energy content. But this energy can no more be expended in radiation than the energy of a normal atom or molecule. The only difference between black dwarf material and a normal molecule is that the molecule can exist in a free state while the black dwarf matter can only so exist under very high external pressure."

In the summer of 1930 Chandrasekhar worked out the complete theory of white dwarfs based on the seminal idea of R.H.Fowler (Chandrasekhar 1931a). Let us have a brief look at the salient steps in the theory.

The quantum mechanical pressure of a high density electron gas is given by the simple expression

$$P \propto n^{5/3} = K_1 \, \rho^{5/3}$$

Neutron Stars

The mass-radius relation for white dwarfs can now be worked out from the equations of hydrostatic equilibrium. Since the pressure is independent of temperature, the problem of mechanical stability separates out from the thermodynamics of the star. From

$$\frac{dP}{dr} = -\frac{GM(r)\rho}{r^2}$$

or

$$\frac{P}{R} \propto \frac{GM^2}{R^5}$$

which will suffice for dimensional arguments, one immediately gets

$$R \propto \frac{1}{M^{1/3}}$$

This is a remarkable result. A more massive white dwarf will have a smaller radius, contrary to normal matter! It follows from the above relation that the mean-density

$$\bar{\rho} \propto M^2$$

This theory predicted that a stable configuration of nonzero radius is possible for all masses. The significance of this may be seen as follows. A star is stable because it generates energy by fusion. When it runs out of fuel it will contract and contract till the degeneracy pressure of the electrons arrests the gravitational collapse. It will still shine, though dimly, because of some fossil heat. The heat itself plays no role in supporting it. Eventually it will radiate away all the heat and the star will die peacefully as a black dwarf. *All* stars will find their peace eventually this way. Or so it seemed for a while.

In July of 1930 Chandrasekhar set off on a long voyage to Cambridge to work with R.H.Fowler. During this voyage he had a second look at his theory and this was to change our picture of stellar evolution radically. The fallacy in the theory just outlined can be traced to a very simple fact. We saw that the mean density of a white dwarf increases as the square of the mass. We also saw that according to quantum statistics the maximum momentum of the electrons increases as $\rho^{1/3}$. So, at sufficiently high density, or in a sufficiently massive white dwarf, the velocity of the electrons will approach the velocity of light and effects of special relativity, or the variation of mass with velocity, will become important. This has an important consequence for the pressure of the gas. It comes about for the following reason. For small velocities

$$E = p^2/2m$$

whereas in special relativity

$$E^2 = p^2c^2 + m^2c^4$$

In the extreme relativistic case $E \sim pc$. This simple modification alters the expression for the pressure as follows

$$P_{rel} = K_2 \rho^{4/3}$$

instead of $\rho^{5/3}$. One can now recalculate the mass-radius relation using the equation of hydrostatic equilibrium:

$$\frac{dP}{dr} = -\frac{GM\rho}{r^2}$$

With the modified equation of state, $P \propto \rho^{4/3}$, we get

$$\frac{M^{4/3}}{R^5} \propto \frac{GM^2}{R^5}$$

Since the radius cancels out, this relations can be satisfied only for a *unique mass*! An exact calculation gives for this unique mass

$$M = 0.197 \left[\left(\frac{hc}{G}\right)^3 \frac{1}{m_p^2}\right] \frac{1}{\mu_e^2} = 1.4 M_\odot$$

where μ_e is the mean molecular weight of the electrons.

Again we see the same combination of fundamental constants, but this time this is an exact result. A completely relativistically degenerate star can have this and only this mass. This is one of the most beautiful results in physics. Chandrasekhar interpreted this mass as the *limiting mass* for white dwarfs; they cannot have a mass greater than this (Chandrasekhar 1931b).

One now had a theory of white dwarfs in the nonrelativistic approximation and in the extreme relativistic case. The next step was to work out an exact theory without making these assumptions. This Chandrasekhar proceeded to do. The mass-radius relation he derived is shown in Fig. 1. For white dwarfs of small mass the previous theory predicted the same radius as the exact theory. But for larger masses the exact theory predicted a smaller radius and as M approached the limiting mass the radius went to zero! This limiting mass has come to be known as the *Chandrasekhar limit*.

1.4 Can All Stars Find Peace?

Let us take stock of what we have discussed so far. If a star has a mass less than $1.4 M_\odot$, it will eventually find peace as a black dwarf. But what is the ultimate fate of gaseous stars more massive than M_{Ch}? The first question to be resolved was the condition under which a star, initially gaseous, can develop a degenerate core. It might appear that we have already answered this question, but we have not. In a gaseous star there is radiation pressure in addition to gas pressure. Whether or not degeneracy can set in in a gaseous star will be determined by the amount of radiation pressure, as was shown by Chandrasekhar in 1932. Since it is such an elegant and powerful argument it is worth going through it (Chandrasekhar 1932). The gas cannot be regarded as degenerate if for the given density and temperature the pressure calculated according to Boyle's law

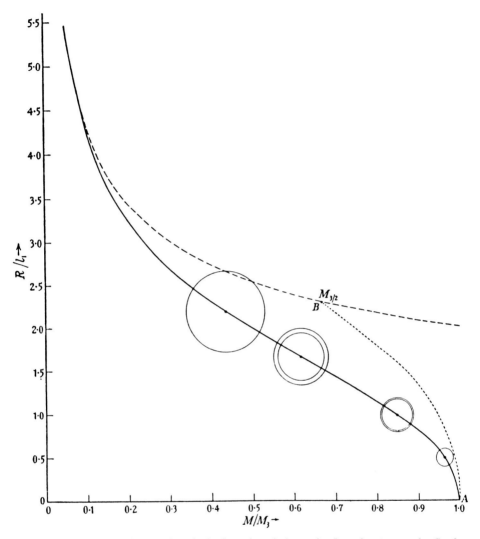

Fig. 1. Mass-radius relation for ideal white dwarfs from *An Introduction to the Study of Stellar Structure* by Chandrasekhar (1939a). The solid-line represents the exact relation for the completely degenerate configurations.

exceeds the pressure calculated using the degeneracy formula. The expression for the classical pressure of the electron gas can be cast in the following form:

$$p_{e,\,class} = \frac{\rho kT}{\mu_e m_p} = \left[\left(\frac{k}{\mu_e m_p}\right)^4 \cdot \frac{3}{a} \cdot \frac{1-\beta}{\beta}\right]^{1/3} \rho^{4/3}$$

where $(1 - \beta)$ is the fraction of the total pressure contributed by radiation. The condition that degeneracy cannot set in, viz.,

$$p_{e,\, class} > p_{e\, deg}$$

or

$$\left[\left(\frac{k}{\mu_e m_p}\right)^4 \cdot \frac{3}{a} \cdot \frac{1-\beta}{\beta}\right]^{1/3} \rho^{4/3} > K_2 \rho^{4/3}$$

reduces to the inequality

$$(1 - \beta) > 0.092$$

Thus if the radiation pressure is more than 9.2% of the total pressure degeneracy cannot set in. Now, in the standard model of Eddington radiation pressure becomes more and more important as we go to more massive stars. Using this model Chandrasekhar derived the critical mass of the star in which radiation pressure is precisely 9.2% of the total pressure:

$$M_{crit} \sim \frac{6.6}{\mu^2} M_\odot$$

The evolutionary significance of this critical mass is the following. Stars more massive than this cannot, during the course of their evolution, develop degeneracy in their interiors; and accordingly an eventual white dwarf stage is impossible without a substantial ejection of the mass. The above argument is so powerful, and convincing, that about 60 years ago Chandrasehar made the following confident assertions:

> "For all stars of mass greater than M_{crit} the perfect gas equation of state does not break down however high the density may become, and the matter does not become degenerate. An appeal to Fermi-Dirac statistics to avoid the central singularity cannot be made." (1932)

> "The life-history of a star of small mass must be essentially different from the life-history of a star of large mass. For a star of small mass the natural white dwarf stage is an initial step towards complete extinction. A star of large mass ($> M_{crit}$) cannot pass into a white dwarf stage, and one is left speculating on other possibilities." (1934)

Now a few remarks about a sad chapter in modern astrophysics – the controversy between Eddington and Chandrasekhar. Eddington did not like the limiting mass; perhaps because he, more than anyone else, saw the full implications of it (!). As he put it (Eddington 1935):

> "Chandrasekhar, using the relativistic formula which has been accepted for the last five years, shows that a star of mass greater than a certain limit M_{crit} remains a perfect gas and can never cool down. The star has to go on radiating and radiating, and contracting and contracting until, I suppose, it gets down to a few km radius, when gravity becomes strong enough to hold in the radiation, and the star can at last find peace."

> "Various accidents may intervene to save the star, but I want more protection than that. I think there should be a law of Nature to prevent the star from behaving in this absurd way!"

To put it in modern terminology, Eddington feared the star will become a BLACK HOLE. Surprisingly he did regard this as a fatal difficulty and this made him reject the Chandrasekhar formula for the pressure of a relativistic fermi gas. But it is worth appreciating the following point: Eddington was not disturbed by a star becoming a black hole for, after all, as he put it that way "the star can at last find peace". What must have shocked him was the inevitable singularity that must result. Most people are still unhappy with that. In the late sixties some exact theorems were proved by Penrose and Hawking which established the inevitability of a singularity once a black hole is formed. About this exact result of Penrose and Hawking, Chandrasekhar has remarked that for the first time in the history of physics the acceptance or rejection of a conclusion reached entirely within the premise of Science, is based on purely philosophical grounds. Penrose himself has said that he is agnostic about the singularity, though he himself proved it. Eddington, too, did not like the inevitable singularity (that Chandrasekhar's theory implied) on philosophical grounds.

To summarize, stars with mass less than $1.4 M_\odot$ will die peacefully as white dwarfs. Stars more massive than M_{crit} cannot develop degeneracy, however much they contract. These stars will collapse indefinitely and in the process become black holes, as Eddington had feared.

1.5 Neutron Stars and Supernovae

What about stars of intermediate mass? In a paper published in 1939 Chandrasekhar conjectured that these stars *will* develop degenerate cores, and that the cores will be relativistically degenerate. Sooner or later the mass of these cores will grow to $1.4 M_\odot$ at which point they will collapse to become *neutron stars*.

> "If the degenerate cores attain sufficiently high densities (as is possible in these stars) the protons and electrons will combine to form neutrons. This would cause a sudden diminution of pressure resulting in the collapse of the star to a neutron core giving rise to an enormous liberation of gravitational energy. This may be the origin of the supernova phenomenon."

Chandrasekhar (1939b)

An implication of this prediction is that the masses of neutron stars should be close to $1.4 M_\odot$ – the maximum mass for white dwarfs.

This brings us to the developments till 1939 in one quarter. But there were some developments in parallel. For example, in 1934 Baade and Zwicky wrote one of the most prophetic papers in the entire astronomical literature and they ended it as follows:

"With all reserve we advance the view that supernovae represent the transitions from ordinary stars into neutron stars which in their final stages consist of extremely closely packed neutrons."

In this paper in a single stroke Baade and Zwicky not only invented neutron stars but they gave a theory for supernova explosions, as well as the origin of cosmic rays. By 1939 Oppenheimer and Volkoff had constructed models of neutron stars and also estimated the maximum possible mass for neutron stars in analogy with the maximum mass of white dwarfs. These exotic objects were not taken too seriously by the astronomers for many decades. However, some physicists, particularly Landau's school in Moscow, pursued the understanding of the interiors of such dense stars and predicted many remarkable things such as superfluidity of their interiors.

In parallel to these theoretical developments there was a remarkable drama unfolding concerning the nature of the Crab nebula. [We refer to the book entitled *Supernovae* by Murdin and Murdin (1985) for an account of the unveiling of the mystery of the Crab Nebula.] It is worth recalling a few historical facts. By 1854 it was established by Lassell that there was genuine diffuse emission from the Crab nebula and that it was not just unresolved stars. By 1916 Sliphar had concluded from his spectroscopic observations that the Crab nebula was expanding at $\sim 1000 \text{ km s}^{-1}$. By 1920 strong arguments were put forward that the Crab must in some way be related to the supernova of 1054 AD. Lundmark had noted that the position of the Chinese guest star of 1054 was very close to the Crab nebula. In 1928 Hubble took the remarkable step of arguing that the expansion age of the Crab nebula strongly supported the idea that it must be the remnant of the supernova of 1054 AD. This remarkable conjecture by Hubble was published in the Popular Essays and promptly went unnoticed by professional astronomers. It was left to Jan Oort to make a formal identification of the Crab nebula as the remnant of the historic supernova by a scholarly analysis of the Chinese historical records. This was in 1942. Around the same time Walter Baade and Minkowski came to the conclusion that the "southwest star" in the nebula must be the stellar remnant of the supernova explosion, and that the star had a large proper motion. In 1942 Baade also concluded that the nebula was not only expanding but it was *accelerated* after the supernova explosion. This meant that there must have been a post-supernova injection of energy into the nebula. Baade conjectured that the southwest star which he and Minkowski had earlier identified as the stellar remnant of the supernova must in some ways be responsible for its post-supernova acceleration. In 1949 John Bolton used the cliff-top interferometer in Australia to conclude that there was radio emission from the Crab. Apart from the Sun, this was the first identification of radio emission with an optical object. In a characteristically brilliant prediction Shklovski argued in

1952 that the optical as well as radio continuum radiation from the Crab nebula must be synchrotron radiation and that it must be polarized. Two years later the polarized nature of the diffuse optical emission was confirmed by the historical observations of Dombrowsky. By 1957 the polarization of the radio emission was also established. The outstanding theoretical question at that time was "what is the source of relativistic particles and the magnetic field in which they radiate?" In 1957, Piddington in Australia wrote a short and remarkable paper in which he argued that the magnetic field of the Crab nebula could neither be the interstellar field nor a magnetic field produced in a supernova explosion. According to him the magnetic field had the appearance of a "wound up" field like the windings of the spring of a watch! He went on to conjecture that such a field could, in principle, have been produced by a central star and went on to deduce the number of times the star must have turned around during the age of the nebula to produce the observed field!! Considering how scanty the radio maps of the Crab nebula were in 1957 this was an extraordinary conclusion. In the meantime Shklovskii argued that the cosmic rays produced in the supernova explosion cannot account for the present day optical radiation, but that there must be a continuous injection of the relativistic particles into the nebula. The argument was simple: the radiative lifetime of high energy electrons are very short compared to the age of the nebula. So in the beginning of 1960s one was left with the question of understanding the nature of the central engine in the Crab nebula for the supply of relativistic particles and magnetic field. By 1964 Woltjer and Ginzburg had independently conjectured that neutron stars must have magnetic field $\sim 10^{12}$ G. It was the privilege of Franco Pacini to argue in 1967 that a rapidly spinning, strongly magnetized neutron star could be the central engine in the Crab nebula. A few months later Jocelyn Bell made history by discovering pulsars. Soon a pulsar in the Crab nebula was found and its energy loss rate confirmed that it was, indeed, the powerhouse energising the Crab nebula. This is one of the most remarkable stories of extraordinary observations and bold theoretical predictions going hand in hand and culminating in a truely great discovery.

1.6 Black Holes

Let us now go back to the 1930s and to the paper by Oppenheimer and Volkoff (1939). Having concluded that neutron cores cannot have masses greater than $\sim 0.7 M_\odot$, they rejected Landau's idea that static neutron cores buried in the cores of stars can be the source of stellar energy. And what about the fate of very massive stars? Like Chandrasekhar a few years earlier they, too, were left speculating: *"Either the Fermi equation of state must fail at very high densities, or that the star will continue to contract indefinitely never reaching equilibrium"*. In 1939 Oppenheimer and Snyder chose between these alternatives. They concluded that *"when all thermonuclear sources of energy are exhausted a sufficiently heavy star will collapse. This contraction will continue indefinitely till the radius of the star approaches asymptotically its gravitational radius. Light from the surface of the star will be progressively reddened and can escape over a progressively nar-*

rower range of angles till eventually the star tends to close itself off from any communicaton with a distant observer. Only its gravitational field persists."

This historical introduction should give a flavour of the truly pioneering efforts made nearly sixty years ago which gave us our first understanding of how stars ultimately find peace.

2 The Nature of Pulsars

2.1 Phenomenology

We now turn to a discussion of Pulsars. For a detailed exposition we refer to the excellent texts by Manchester and Taylor (1977) and Lyne and Graham-Smith (1990), as well as the review articles by Taylor and Stinebring (1986) and Srinivasan (1989).

The first observations of a pulsar was through its radio emission at a wavelength of a few metres (Hewish et al. 1968). Extraordinarily, the radiation arrived in bursts of pulses with great regularity. Figure 2 shows the pulses from pulsar PSR 1919 discovered by Jocelyn Bell. The first thing one notices is that the amplitude varies greatly from pulse to pulse. When observed with high time resolution the individual pulses show subpulse structure. The figure also shows that the emission from pulsars is fairly wide band over the radio spectrum. The actual arrival time of any given pulse is different at different frequencies with the lower frequency pulses arriving later (Fig. 3). This delay was noticed in the very first observation itself and was correctly accounted for as due to the dispersion of the signal in the interstellar plasma. The dispersion measure which is the difference in arrival times at adjacent frequencies is directly proportional to the column density of electrons along the line of sight to the pulsar. Thus, given a model for the distribution of free electrons in the interstellar medium this dispersion can be used to estimate the distances to pulsars.

The duty cycle of pulsars or the fractional time for which the radiation is received is typically $\sim 3\%$, and is independent of the pulsar period. Individual pulses show a great variety of intensity variations. Some pulsars show "microstructure" within a pulse. Others have subpulses which "drift" with remarkable regularity within the average pulse profile (Fig. 4). The time averaged or integrated pulse profiles show incredible variety (Fig. 5), but remarkably, each is a signature of a particular pulsar, and this average profile sometimes remains stable for more than a decade! Another remarkable property exhibited by many pulsars is "nulling" which represents a series of missing pulses. The duration of nulling can be as long as eight hours as in the case of PSR 0826-B4. Till recently it was thought that nulling may represent the "last gasps" of dying pulsars. But it now appears that pulsars do not null by simple virtue of their age. The subpulses show a fascinating "memory" during nulling. Although the train of pulses is interrupted during a null, after the pulsar resumes functioning the subpulse reappears at a phase as though no nulling had occurred and continues the drift within the pulse window. The pulse radiation is highly polarized. The position

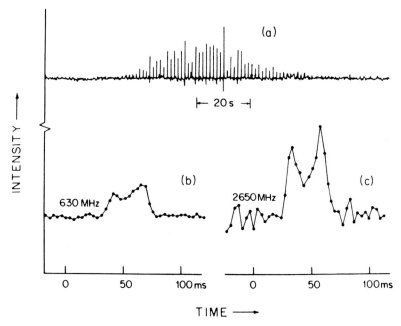

Fig. 2. (a) A train of pulses from PSR 1919, the first pulsar to be found. The period is about 1.3 sec. and the amplitude is seen to vary with time. (b) and (c) show the structure within the pulse as observed at two widely separated frequencies.

angle of linear polarization varies smoothly within a pulse window, and the instantaneous direction of polarization is independent of frequency after allowance is made for Faraday rotation in the interstellar medium (Fig. 6). Many pulsars also show weak circular polarization.

2.2 The Pulsar Population

Nearly 600 pulsars have been detected to date in the Galaxy. In addition, two have been detected in the large Magellanic cloud and one in the SMC. Distribution of pulsars in our Galaxy is shown in the Fig. 7. Most of these are within a couple of kiloparsec of the Sun, and their average distance from the galactic plane is ~ 400 parsec. The distribution of the observed period and period derivatives is shown in Fig. 8. The overwhelming majority of pulsars have periods ~ 1 sec and period derivatives $\sim 10^{-15}$ s s^{-1}. We shall return to a detailed discussion of this population.

2.3 The Emission Mechanism

The first direct observational evidence which suggested that pulsars must be rotating magnetized neutron stars came from the smooth variation of the direction

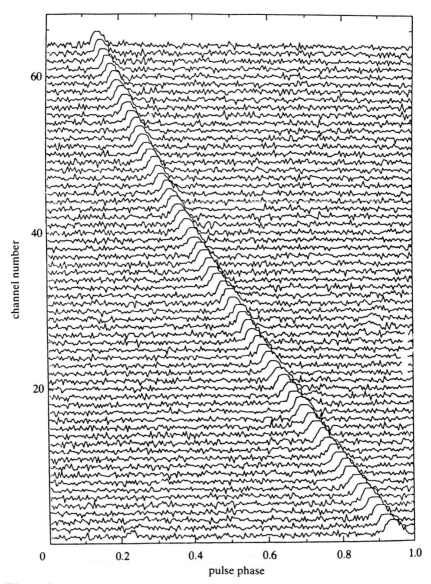

Fig. 3. Dispersive delay of the pulsed signal from PSR 1641-45, which has a dispersion measure of 475 cm^{-3} pc. The vertical axis represents the frequency: channel 1 corresponds to 1400 MHz and channel 64 to 1720 MHz (from Manchester 1993).

of linear polarization within the pulse. This was first observed in the Vela pulsar by Radhakrishnan and Cooke (1969). This led them to advance the now widely accepted "polar cap model" or the "lighthouse model" of pulsars (Fig. 9). According to this model if particles are accelerated near the polar cap then because of the strong magnetic field they will be constrained to move along the open field

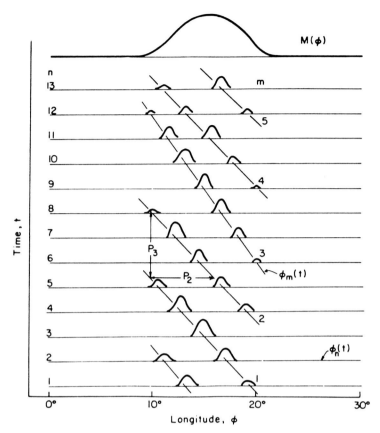

Fig. 4. Description of drifting subpulse phenomenon. The observer's path is $\phi_n(t)$ which crosses 360° of longitude in one pulse period. Time is then measured in pulse number n. The pulsed emission is found in a narrow region of longitude with an average level given by $M(\phi)$, the pulse profile. Narrow subpulses are found to drift along paths $\phi_m(t)$, with m indexing successive paths. These are separated, nominally, P_2 in longitude and P_3 in time (from Backer 1973).

lines like "beads on a string". Since the field lines diverge from the polar cap these particles will emit *curvature radiation*. If the particle is non-relativistic then this radiation will not be beamed, and one would expect to see elliptically polarized radiation in general. But observations clearly show that the radiation is very highly linearly polarized. The only way to get purely linear polarization is if the particles that are sliding along the curved field lines were highly relativistic. In this case the radiation will be beamed and linearly polarized in a plane containing the particular field line under consideration. Seen in projection the radiation will be linearly polarized parallel to the magnetic field projected on to the plane of the sky. In this picture the sweep of the plane of polarization within the pulse merely reflects the geometry of the projected field lines near the magnetic pole. The degree of the sweep of the plane of polarization within the

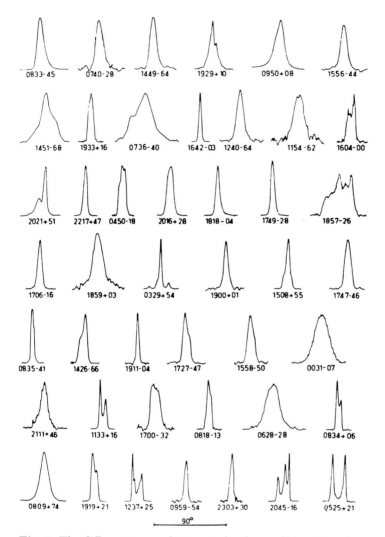

Fig. 5. The different types of integrated pulse profiles of 45 pulsars arranged in order of increasing period. All profiles are plotted on the same longitude scale where 360° longitude equals the pulsar period. A bar indicating 90° of longitude is shown at the bottom of the figure (from Manchester and Taylor 1977).

pulse will depend upon how the locus of the line of sight intersects the diverging bundle of field lines from the magnetic pole, or the angular separation of the observer's line of sight from the magnetic axis. This simple model was able to elucidate several things in one stroke:

1. One is dealing with a rotating magnetized neutron star.
2. That the radio radiation emanates very close to the polar cap.

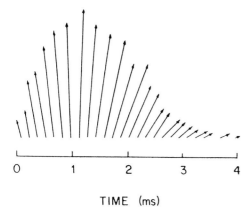

Fig. 6. The sweep of the plane of linear polarization within the pulse window of the Vela pulsar, PSR 0833-45 (from Radhakrishnan and Cooke 1969).

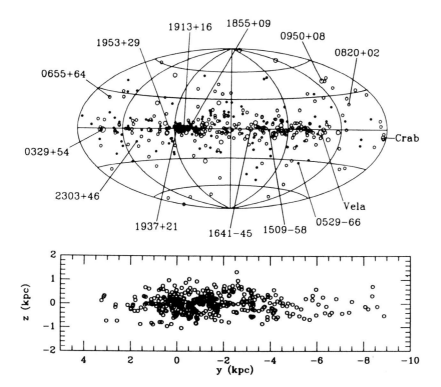

Fig. 7. The upper figure shows the distribution of 398 pulsars in galactic coordinates. Four different sizes of circles correspond to 400 MHz average flux densities <10, 10 to 100, 100 to 1000, and >1000 mJy; the lower figure shows the approximate location of 391 pulsars, projected on to $y-z$ plane of the Galaxy. The Sun lies at $(0,0)$, and the galactic centre is assumed to be at $(-10,0)$ (from Taylor and Stinebring 1986).

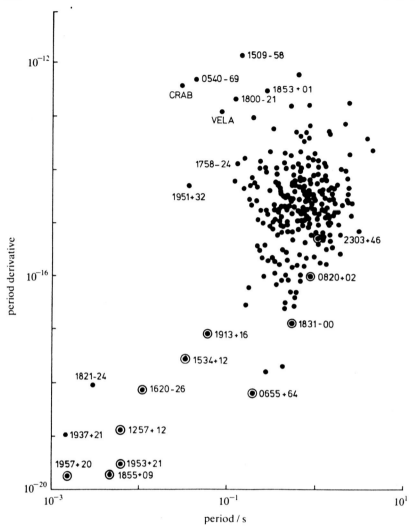

Fig. 8. The distribution of periods and period derivatives of approximately 370 pulsars.

3. The radiation that we observe must be curvature radiation by relativistic particles.

The richness of this polar cap model was soon demonstrated in a classic paper by Komesaroff (1970) who showed that if the radio emission is due to curvature radiation then the pulsar beam must be a *hollow cone*. Since the field lines very close to the magnetic axis have very little curvature there will be no radiation from particles accelerated along these field lines. The double pulse observed in some pulsars, the pulse width, polarization, as well as the spectrum of pulses found natural explanation in this model.

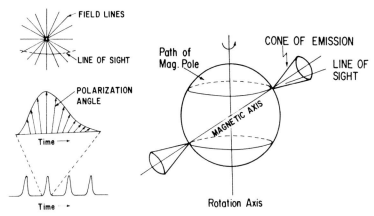

Fig. 9. The polar cap model of Radhakrishnan and Cooke (1969). The observational inspiration for this model was the sweep of the position angle of linear polarization within the pulse window (schematically shown by the figure on the left). The length and orientation of the *arrows* represent the polarized flux and position angle. This led to the 'light house model' for pulsars shown on the right; the observed radiation is believed to originate very close to the magnetic poles, and beamed into a narrow cone. The figure at the top left shows the geometry of the projected field lines and the locus of line of sight.

2.4 Pulsars as a Dynamo

The Radhakrishnan-Cooke model did not attempt to provide an explanation for how relativistic particles are accelerated along the field lines. But this is precisely what had been postulated by Goldreich and Julian in a classic paper published around the same time. Goldreich and Julian (1969) considered an axisymmetric case for simplicity viz., the magnetic axis parallel to the rotation axis, and demonstrated that such a neutron star will function as a *unipolar inductor* or a *homopolar generator*. Consider a uniformly magnetized, conducting sphere. Let us spin it about an axis which coincides with the magnetization axis. Because the sphere is a conductor and its material is moving in a magnetic field, a magnetic force $\frac{e}{c}(\mathbf{\Omega} \times \mathbf{r}) \times \mathbf{B}$ is exerted on each free charge. The charges therefore redistribute themselves within the sphere and on its surface in such a way as to set up an electric field inside the sphere given by

$$\mathbf{E} = -\frac{1}{c}(\mathbf{\Omega} \times \mathbf{r}) \times \mathbf{B}, \quad r < R$$

so that the total Lorentz force on each charge vanishes.

Let us assume that there is a vacuum outside the star. One can now solve Laplace's equation to derive the electric field outside the sphere by matching potentials etc. If the magnetic field is a pure dipole, the external potential is quadrupolar

$$\Phi = -\left(\frac{B\Omega R^5}{3cr^3}\right) P_2(\cos\theta), \quad r > R$$

The surface charge density would be

$$\sigma = -\frac{B\Omega R}{4\pi c}\cos^2\theta$$

Inside the star $\mathbf{E}\cdot\mathbf{B} = 0$, whereas outside the sphere

$$\mathbf{E}\cdot\mathbf{B} = -\left(\frac{\Omega R}{c}\right)\left(\frac{R}{r}\right)^7 B^2 \cos^3\theta, \quad r > R$$

In other words, outside the sphere there is a component of the electric field parallel to the magnetic field. To get a feeling for numbers, if $R = 10$ cm, $\Omega = 10^3$ s^{-1} and $B = 10^4$ G one would generate ~ 5 volts between the pole and the equator (Fig. 10). If one examines the current path through the sphere one will discover that there is a $(\mathbf{j}\times\mathbf{B})$ force, which will brake the rotation. According to Goldreich and Julian this is roughly how pulsars work. If we now make the above estimate for a pulsar one will get

$$\Delta\Phi = 3\times 10^{16}\left(\frac{B_{12}}{P}\right) \text{ Volts}$$

where B_{12} is the magnetic field is units of 10^{12} G and P is its period in seconds. A truly astronomical voltage! Our discussion so far has been based on a vacuum exterior. Since $\mathbf{E}\cdot\mathbf{B} = 0$ in the interior and nonzero outside, it must change continuously through the charge layer. If this layer is to be stable then the electric force along the magnetic field must be balanced by the gravitational force. Unfortunately, this is not possible. The component of $\mathbf{E}$ along $\mathbf{B}$ is 10^{10} V cm^{-1} and thus the electrostatic force is $\sim 10^{10}$ times stronger than gravity. Therefore the charges will shoot out along B and the exterior will no longer be a vacuum.

2.5 The Near and Wind Zone

The charges drawn off the surface will distribute themselves so as to make the field lines equipotentials. Then $\mathbf{E}\cdot\mathbf{B} = 0$ both inside and outside. The moral to be drawn is that a rotating magnetized neutron star cannot be surrounded by a vacuum.

Near the pulsar the magnetic-energy density greatly exceeds the particle energy density so that the charged particles will be constrained to slide along the field lines which rigidly rotate with the star. Any transverse motion is very quickly damped out because of radiation reaction. Before discussing the motion of these charges, let us estimate the plasma density. Since the field lines are equipotentials we can solve for the potential everywhere outside. From this one can calculate E and hence the charge density. Goldreich and Julian gave the following estimate for the number density.

$$n = 7\times 10^{-2}\left(\frac{B_z}{P}\right) \text{ cm}^{-3}$$

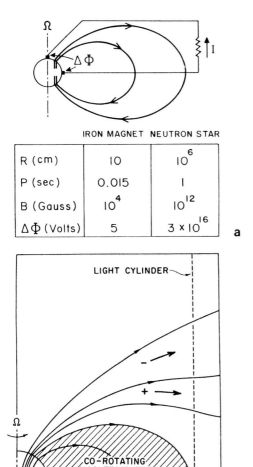

Fig. 10. a A rotating, conducting magnetized sphere as a homopolar generator. The typical voltages that will be generated by an iron magnet and a neutron star are given in the table; **b** The Goldreich-Julian model predicts that positive and negative charges will be pulled out of the surface and accelerated along the open field lines (from Goldreich and Julian 1969).

where B_z is in Gauss. One's first reaction would be to say that these charges will corotate as they are trapped in the field lines. This is not possible for all the charges. Here one must introduce an important length in the problem viz., the light cylinder radius, $r \sin \theta = c/\Omega$. This is the distance at which a charge corotating with the star will be travelling with the velocity of light. Therefore, charges attached to those field lines which close within the light cylinder will corotate. This defines the pulsar magnetosphere. The corotation probably has no observable consequences for the pulsar. The toroidal current due to the rotation of the space charge is of the order $\left(\frac{\Omega r}{c}\right)^2 \frac{B}{r}$ and becomes an important source for the magnetic field only near the light cylinder. Thus the poloidal magnetic field is an approximate dipole close to the star and becomes quite distorted near the light cylinder.

The picture is very different along the field lines that cross the light cylinder, the so called open field lines. Clearly corotation is not possible for particles trapped on these lines. If you admit this then one will have a discontinuity in the space charge at the light cylinder, and these charges cannot be in dynamic equilibrium for the same reason we had for the surface charges.

Thus particles must stream out along these open field lines. If Ω and $\mathbf{B}$ are parallel, the electric potential on the surface is highest at the equator and decreases towards the poles. There will be a critical field line at the base of which the potential will be the same as that in the interstellar medium. Thus electrons will stream out along the higher latitude lines and protons will escape along the lower latitude open field lines. The location of the critical field line which separates the charges will be determined by the condition that the star loses no net charge. If Ω and $\mathbf{B}$ are antiparallel then the location of the electron and proton streams will be interchanged; the protons will now flow along the higher latitude lines.

What happens to these particles? How do they get accelerated? This is a very difficult problem and no satisfactory solution is available. The following simple-minded approach is what has been possible. One takes the attitude that these open field lines are not really equipotentials. The star does not manage to achieve this and the particles are blasted out by the electric field parallel to B. Let us look at the region of the polar cap from which the open field lines emanate. The last open field line makes an angle of θ_0 with the rotation axis given by

$$\sin\theta_0 \approx \left(\frac{\Omega R}{c}\right)^{1/2} = \left(\frac{R}{R_{LC}}\right)^{1/2}$$

Using the general expression given earlier one can evaluate the potential difference between this outermost field line and the pole:

$$\Delta\Phi = \frac{1}{2}\left(\frac{\Omega R}{c}\right)^2 BR$$

One can now solve for the electric field in a region bounded by the polar cap, the rotation axis and the conducting magnetosphere boundary. The most energetic particles escaping should have energies of about $\Delta\Phi$ or,

$$E_{max} = 3 \times 10^{12}\, \frac{Z R_6^3\, B_{12}}{P^2}\ \text{eV}$$

Here Z is the atomic charge. So if we consider a pulsar with $P = 1s$, $B = 10^{12}$ G and $R = 10^6$ cm, an electron will get accelerated to an energy of 3×10^{12} eV or a Lorentz factor $\gamma = 6 \times 10^6$ ($E = \gamma mc^2$). For fast pulsars like the Crab one can get a γ of about 10^9 or $E \sim 10^{15}$ eV. Thus one expects pulsars to be an important source for cosmic rays. Whereas one can think of many other acclerators for electrons (such as supernova remnants), pulsars can also accelerate protons and heavy nuclei.

Admittedly the above discussion was rather simplified – although deceptively simple it is an extremely difficult problem – but we now have a picture of why a rotating magnetized neutron star will emit relativistic particles. To recall, the polar cap model which set out to explain the sweep of the linear polarization within the pulse required a copious supply of relativistic particles which will emit curvature radiation. The aligned rotator model of Goldreich and Julian provides the required theoretical basis to the lighthouse model of pulsars.

2.6 The Dipole Radiator

Before proceeding further in trying to understand the radiation from pulsars let us discuss an alternative model viz., the magnetic dipole radiation model. It may be recaled that inclined magnetic rotator will emit magnetic dipole radiation whose luminosity will be given by

$$L = \frac{2}{3c^3} m_\perp^2 \Omega^4 = \frac{1}{6c^3} B^2 R^6 \Omega^4$$

Since the radiated energy can only come at the expense of the stored rotational energy

$$\frac{d}{dt}\left(\frac{1}{2} I \Omega^2\right) = -L$$

On simplifying this equation one finds that the period derivative $\dot{P}$ is proportional to B^2/P. This simple model tells us that an inclined radiator must slow down according to a very precise relation. Very soon after the discovery of pulsars it was established that their periods in fact are lengthening, and this gave added support to the idea that pulsars must be rotating magnetized stars. The measured period and period derivative enables one to estimate the magnetic field of the neutron star. Figure 11 shows the distribution of magnetic fields of the observed population of pulsars now numbering more than 500. It is seen that the majority of pulsars have magnetic fields in the range $10^{12.5} - 10^{13}$ G, as conjectured even prior to their discovery.

2.7 Gaps, Sparks and Pairs

The basic difficulty in understanding the pulsed radiation was in trying to understand the origin of a copious supply of relativistic particles. Although Goldreich and Julian provided the basic framework, much of the acceleration in this model occurs well outside the light cylinder. But as we saw, the sweep of the linear polarization interpreted as the geometry of the projected field lines near the polar cap requires a supply of relativistic particles fairly close to the stellar surface.

The next major step in this direction was taken by Sturrock (1971) who pointed out that the relativistic wind from the pulsar will consist mostly of electron-positron pairs – (the particles pulled out from the surface will be a

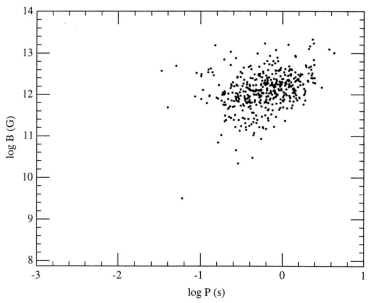

Fig. 11. The distribution of periods and the derived magnetic fields of pulsars. The majority of pulsars have periods between 10^{12} and 10^{13} Gauss.

minority). These primary particles accelerated along the open field lines will produce very high energy gamma-rays. These gamma-rays are unstable against the production of electron-positron pairs because of the strong magnetic field near the pulsar. Each of these particles in turn will produce more gamma-rays which produce more electron-positron pairs etc. Thus a cascade develops resulting in an *electron-positron wind* from the neutron star. This seminal idea was the basis for the "polar cap **gap model**" due to Ruderman and Sutherland (1975). To date, this model represents the most successful attempt in trying to understand the phenomenology of radio emission from pulsars.

In the Goldreich-Julian model, both electrons and positive ions are pulled out of the surface of the neutron star. Ruderman and Sutherland pointed out that whereas electrons can readily be pulled out of the surface, ions are unlikely to be pulled out despite the large electric field generated near the surface. This has to do with the large binding energy of the ions in the crust. Ruderman and his collaborators argued that this is a natural consequence of the very strong magnetic field near the surface. When $B \sim 10^{12}$ G one is in a regime where the Larmor radius is smaller than the Bohr radius. Therefore an atom near the surface will not be spherical but instead needle-like along a direction perpendicular to the surface. These needle-like atoms form long molecular chains consisting of a one-dimensional lattice of ions with an outer sheath of electrons. These long molecular chains form a highly anistropic, very dense, and strongly bound crustal matter. As we shall see, one expects matter in the outer layers of the crust to be mainly Fe^{56}. Ruderman and Sutherland argued that the binding energy per ion in such a lattice will be ~ 14 keV. They pointed out that the electric field required to pull these ions from such a lattice is very much greater than the maximum electric field available for most pulsars. Of course, in a very young and hot neutron star the ions can be emitted by "thermionic emission". Such a situation may obtain only in the case of a young Crab pulsar.

If one assumes that only electrons can flow out of the surface of the neutron star, but not the positive ions, then it has a profound implication for the structure of the magnetosphere. To be specific, Ruderman and Sutherland considered the model in which rotation axis and the magnetic axis were *antiparallel* rather than parallel. In this case the magnetosphere just above the polar caps has a net positive charge. If the spin and the magnetic momentum are parallel the magnetosphere above the polar cap is negative. Let us pursue the case where the base of the magnetosphere over the polar cap is positively charged. As the positively charged particles flow out along the open field lines and through the light cylinder the base of the magnetosphere will pull away from the surface producing a *vacuum gap* (Fig. 12). In this region $\mathbf{E}\cdot\mathbf{B} \neq 0$ i.e., there is component of the electric field parallel to the magnetic field in this gap, and at the pole the electric field will be normal to the surface given by

$$E_{gap} = 2\frac{\Omega B}{c}h$$

and the potential drop across this gap will be given by

$$\Delta\phi = \frac{\Omega B}{c}h^2$$

where h is the height of the gap. As the height of the gap begins to increase (essentially with the velocity of light) the voltage across the gap will grow very rapidly till at some stage the vacuum in the gap becomes unstable against electron-positron production.

Breakdown of the Gap

Imagine a stray gamma-ray of energy $> 2mc^2$ entering this vacuum gap and producing an electron-positron pair. These particles will be accelerated by the electric field parallel to the magnetic field. In the antiparallel configuration that we are considering, the electrons will move towards the positively charged stellar surface, and the positrons will move outwards. We saw earlier that the electric field is so strong that within a height which is of the order of the width of the polar cap these charges will be accelerated to energies ~ 10^{12} eV. These particles sliding along curved field lines will themselves radiate. The characteristic frequency of this radiation will be

$$\omega \sim \frac{3}{2}\gamma^3\left(\frac{c}{\rho}\right)$$

where ρ is the radius of curvature of the orbit. These photons will be beamed in the forward direction within an angle γ^{-1}. If the γ of the electron and postiron

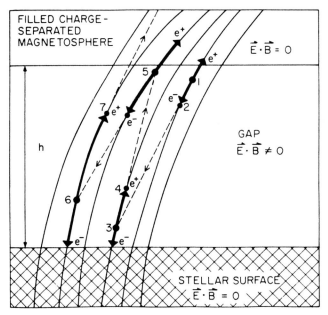

Fig. 12. Electron-positron cascade triggered by gamma rays in the polar cap GAP (from Ruderman and Sutherland 1975). Imagine an electron-positron pair being created in the GAP. The particles will be quickly accelerated to ultrarelativistic energies by the strong electric field (parallel to the magnetic field) in the gap. Because of the strong magnetic field the charges will be constrained to move along the curved field lines, and consequently emit curvature radiation. These photons, in turn, will create more pairs and thus a cascade will develop. In the Ruderman-Sutherland model, the gap accelerated electrons will flow towards the stellar surface and the positrons will flow out along the open field lines.

are sufficiently high, the frequency of the curvature photons will be large enough to produce $e^+ - e^-$ pairs, and so the process will cascade exponentially in time. The gap in the magnetosphere will now be shorted out by this avalanche process. At what stage will this breakdown occur? Roughly speaking, the gap height should be comparable to the mean free path of the photon in the magnetic field. The mean free path for this process is given by the well known expression

$$l = \frac{4.4}{(e^2/\hbar c)} \cdot \frac{\hbar}{mc} \cdot \frac{B_q}{B_\perp} \cdot \exp \frac{4}{3\chi},$$

$$\chi \equiv \frac{\hbar \omega}{2mc^2} \cdot \frac{B_\perp}{B_q}$$

$$B_q = m^2 c^3 / e\hbar = 4.4 \times 10^{13} \text{ G}$$

ω is the frequency of the photon, and $B_\perp = B \sin\theta$ where θ is the angle between the direction of the photon and the magnetic field. (The mean free path is infinite for photons with $B_\perp = 0$.) Once the gap is shorted out by the cascade the story repeats again, viz., as the positive charges flow out a vacuum gap will once again develop, discharge again and so on. So there is really no steady state process possible. If you like, a pulsar is a continuously sparking dynamo with a loose contact at the pole!

Pulsar Radiation

Let us call the electron-positron pairs produced in the gap as 'primaries'. As already remarked, of these two only the positrons will leave the gap; the electrons will flow towards the stellar surface. These positrons continue to emit curvature radiation even outside the gap. These photons can in turn produce further pairs which we shall call the 'secondaries'. In fact, conditions are more favourable here to produce pairs since the field lines are flaring out, and therefore a photon will see a larger $B_\perp$ and consequently have a much smaller mean free path. The copious production of secondaries will ensure that the field lines become equipotential, and therefore there will be no component of the electric field parallel to the magnetic field. Because of this, unlike in the gap both the electrons and positrons will stream outwards along gently curving field lines. The pulsed radiation that we see is the curvature radiation produced by these secondary particles. We remarked earlier that in the Goldreich-Julian model the acceleration region was beyond the light cylinder. Now we have acceleration inside the light cylinder. Thus phase stability and the polarization characteristics can all be accounted for.

This model due to Ruderman and Sutherland was also able to explain at a qualitative level several other things. For example, when a discharge begins at some point on the polar cap the electric field that has built up in the gap rapidly decreases at that location. This would be expected to inhibit the formation of another discharge within a distance of the order of the height of the gap. Thus the discharge of the gap may in fact be through a series of localised "sparks". This temporal behaviour may account for the microstructures observed in pulses. The particular location of the spark discharges within the polar cap could easily account for the subpulses within the pulse. The drifting of the subpulses within the average pulse was ascribed by Ruderman et al. to the movements of sparks around the polar cap due to $\mathbf{E} \times \mathbf{B}$ drift.

2.8 Coherence of the Radio Radiation

Soon after pulsars were discovered it became clear that the radio radiation must be coherent. This conclusion followed from the observed very high brightness temperature $10^{25} - 10^{30} K$. Komesaroff (1970), Sturrock (1971), Pacini and Rees (1970) among others were quick to point out that the observed coherence may be due to "bunching" of particles as they slide along curved field lines. It is not very difficult to estimate the number of particles in a bunch that must radiate in phase to explain the observed brightness temperature. Let us recall that if ρ is the radius of the curvature of the field line the characteristic frequency emitted in curvature radiation is given by

$$\nu_c = \left(\frac{c}{\rho}\right)\gamma^3$$

where γ is the Lorentz factor of the radiating particles. For typical radio frequencies particles must have a $\gamma \sim 10^2$. Therefore a brightness temperature $\sim 10^{25} K$ would require approximately 10^{14} particles radiating in phase. How these bunches are formed or even whether such bunches can be formed is still very unclear. Since we have just discussed the gap model due to Ruderman and Sutherland, we might mention the suggestion they made. According to them, bunching is a consequence of a 2-stream instability resulting from the streaming of the highly relativistic gap accelerated positrons through the much slower pair plasma produced above the gap. We shall not dwell any further on this topic because not much progress has been made during the last decade and this continues to be an outstanding challenge (for a modern assessment of this intriguing problem we refer to Melrose 1993).

2.9 Period Evolution

Death Line

Let us return to the distribution of the periods and the derived magnetic fields of pulsars. Figure 13 is in many ways the equivalent of the H-R diagram in the theory of stellar evolution. The magnetic field of pulsars plays essentially the same role as the mass of a star. While discussing the dipole radiator model of pulsars we showed that the period derivative of a pulsar is proportional to the square of the magnetic field and inversely proportional to the period. Therefore, as long as the magnetic fields remain constant during the evolution the trajectories of pulsars would be horizontal in this diagram. But it should be remembered that the "speed" with which a given pulsar transits in this diagram depends on its period; as the pulsar ages its spindown rate also decreases. Therefore, the paucity of very rapidly spinning pulsars, as well as the piling up of pulsars towards longer periods, is easily understood. From the observed period and period derivative one can define the characteristic age of the pulsar as $\tau = P/2\dot{P}$. The characteristic age defines an upper limit to the true age of the pulsar; if pulsars are born spinning extremely rapidly then it would in fact represent the true age provided the magnetic field remained constant. Lines of constant characteristic age are also shown in this figure. From this one readily sees that the majority of pulsars are typically a few million years old. If the magnetic field decays then the trajectories of pulsars will no longer be horizontal. They will begin to drop and eventually become vertical when the magnetic field decay dominates their evolution. The physical meaning of this is, of course, clear. Since the period derivative is proportional to the square of the magnetic field, as the field decays the period of pulsar does not lengthen significantly and the trajectory therefore becomes vertical. A remarkable property of this observed distribution is that there are no pulsars located to the right of a line marked as "death line". There

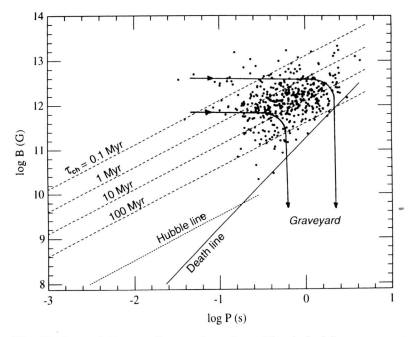

Fig. 13. An evolutionary diagram for pulsars. The dashed lines represent lines of constant characteristic age $\left(\tau = P/2\dot{P}\right)$. It will be seen that most pulsars are typically a few million years old. There are no pulsars to the right of the *Death line*; this line corresponds to the critical voltage the neutron star has to generate for the *polar cap gap* to break down due to electron-positron avalanche. The evolutionary track of a typical pulsar is also shown. Their trajectory will be horizontal as long as the magnetic field remains constant. As the field begins to decay the trajectory will droop, and eventually become vertical (a field decay timescale of 10^8 yr has been assumed here).

is a simple and elegant explanation for this. We saw that a pulsar is essentially a sparking dynamo. Crucial to pulsar activity is the development of large voltages in the polar cap region. As the period of pulsar lengthens the voltage generated by the dynamo begins to decrease since it is proportional to B/P^2. When this voltage drops below a certain critical value the breakdown of the gap due to electron-positron production no longer occurs, and pulsar activity turns off. The death line plotted in Fig. 13 is not the envelope of the observed population, but in fact the theoretically predicted death line.

For how long do neutron stars function as pulsars? The lifetime of a pulsar is obviously the time taken to cross the death line. This, of course, depends upon the strength of the magnetic field of individual pulsars. High field pulsars have shorter lifetime than low field pulsars. Again, the story can be more complicated if the magnetic fields of pulsars decay relatively fast. To leave you with a number, the present estimates of the average lifetime of pulsars is ~ 10 Myr.

Glitches and Timing Noise

In the discussions so far we have been concentrating on the secular evolution of the period. It is straightforward to understand why periods of pulsars lengthen. But superimposed on this well understood behaviour are two kind of deviations which are not yet fully understood. These are (i) fairly continuous but erratic variations in the period, also known as *timing noise*, and (ii) spectacular changes in the rotation period, known as *glitches*. Although we do not propose to discuss timing noise it is important to at least have a feeling for the kind of erratic variations that one is talking about. Figure 14 shows the timing residuals of a number of pulsars. Technically speaking, the residuals are the difference in the arrival time of pulses from the expected arrival time after allowing for known effects such as period lengthening etc. Such a timing noise is found predominantly in young pulsars which typically have large magnetic fields and relatively short periods and therefore large spindown rates. Occasionally pulsars suddenly spin up or glitch. The first such glitch was discovered in the Vela pulsar in 1969 very soon after its discovery (Radhakrishnan and Manchester 1969). And these glitches seem to recur (Fig. 15). For example, in the last twenty years about half of the 20 or so glitches observed so far were observed in the Crab and Vela pulsars. Thus, like timing noise it appears that glitches are also a phenomenon that are most common in young pulsars. In a glitch there is a sudden increase in the rotation rate followed by an exponential decay back to the pre-glitch rate. The glitches in the Vela pulsar tend to be much more spectacular than in the Crab. For example,

$$\Delta \Omega/\Omega \sim 2 \times 10^{-6}, \; \Delta \dot{\Omega}/\dot{\Omega} \sim 10^{-2}; \; \text{VELA}$$
$$\Delta \Omega/\Omega \sim 10^{-8}, \quad \Delta \dot{\Omega}/\dot{\Omega} \sim 10^{-3}; \; \text{CRAB}$$

The post-glitch recovery timescale also seems to vary all the way from a few weeks to a few years. There are also other intriguing things. For example, in some pulsars it never fully recovers from a glitch. In the case of the Crab pulsar there is an even more puzzling behaviour in that after each glitch the steady state value of the *spindown rate* seems to increase. We shall return later to this remarkable phenomenon of glitches.

2.10 High-Frequency Radiation

Optical and X-Ray Radiation

Our discussion so far has concentrated exclusively on radio emission from pulsars. Although only a very small fraction of the rotational energy lost by the pulsar can be accounted for by the radio radiation ($L_R/L \sim 10^{-4}$ in the case of the Crab pulsar) radio emission is a rather subtle and still ill-understood process. But it is relatively straightforward to understand at least the optical and X-ray emission from pulsars, although γ-ray emission is again somewhat complicated.

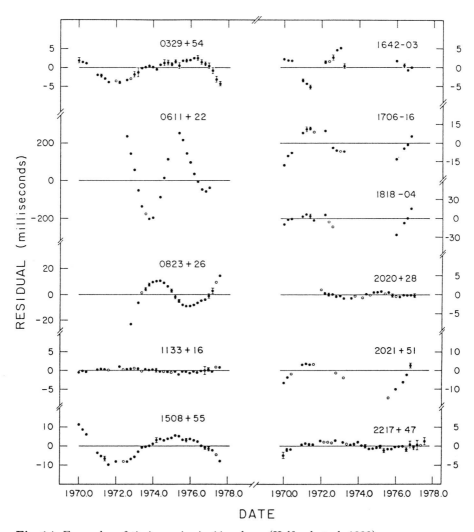

Fig. 14. Examples of timing noise in 11 pulsars (Helfand et al. 1980).

Optical emission has been detected to date from three pulsars, and X-rays from about half-a-dozen pulsars. Regarding γ-ray emission from pulsars the picture is just emerging thanks to observations from the GRO. The average pulse profiles in the optical, X-ray and γ-ray for the Crab pulsar is shown in Fig. 16. The most remarkable thing to be noticed is that in all cases there is a main pulse and a subpulse separated by 140 degrees, and these are precisely in phase. Having said this we should point out that the behaviour is very different in the case of the Vela pulsar! Quite remarkably, no pulsed X-ray emission has been detected so far although it is a strong γ-ray source. While the γ-ray spectrum is similar to that of the Crab pulsar with a 140 degree separation between the

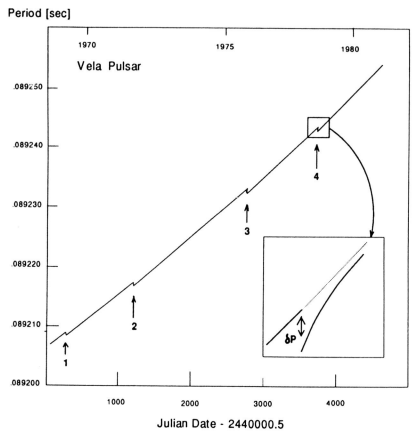

Fig. 15. The first four giant glitches of the Vela pulsar (from Downs 1981). The change in the period at each glitch is of the order $\Delta P/P \sim 10^{-6}$, and the timescale for the recovery after the glitch is of the order of a few months.

main pulse and the subpulse, the separation is only 90 degrees in the case of the optical emission. Returning to the Crab, about 10^{-3} of the total spindown energy loss rate is emitted as X-rays and γ-rays.

The first pulsar to be detected in the optical wavelength was, of course, the Crab pulsar. It turned out to be the very star – the "southwest star" – in the Crab nebula to which Walter Baade had pointed out way back in 1942 and conjectured that it must be the stellar remnant of the supernova of 1054 AD. Soon X-ray pulses were also detected from the Crab. Regarding the origin of the optical and X-ray emission, a consensus emerged straightaway that it must be incoherent synchrotron radiation arising due to the gyration of particles, perhaps near the light cylinder. Unlike in the radio one was not constrained by derived brightness temperatures to invoke any coherent mechanism. The observed intensities in X-ray and optical combined with models for incoherent synchrotron radiation straightaway implied that the number of radiating electrons and/or

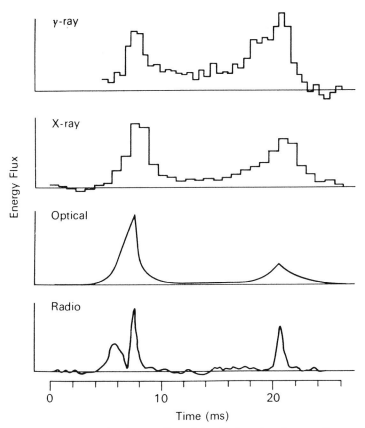

Fig. 16. Integrated pulse profiles of the Crab pulsar from radio to gamma-ray frequencies. The two widely separated components are seen to be precisely aligned over the entire range (from Manchester and Taylor 1977).

positrons exceeds $\sim 10^{37}$ s^{-1}. Pacini (1971) constructed a simple model to calculate the secular decrease of the optical and X-ray luminosity of pulsars, and showed that the bolometric synchrotron luminosity will be roughly proportional to P^{-10} where P is the period of the pulsar in question. The argument is rather simple and worth recalling. As is well known, the synchrotron luminosity is given by

$$L \sim \frac{e^2}{c} \left(\frac{eB_\perp}{mc}\right)^2 \gamma^2 n V$$

where $B_\perp$ is the perpendicular component of the magnetic field in the emitting region, γ the Lorentz factor of the particles, n their number density, and V the emitting volume. If one assumes that the volume of the emitting region is proportional to $R_{LC}^2 \Delta$ where R_{LC} is the radius of the light cylinder, and Δ is the measure of the thickness of the emitting region in the radial direction, then

$$L \propto \left(n R_{LC}^2 \gamma\right) \gamma \Delta B_\perp^2$$

The quantity in bracket in the above equation is related to the total energy flux across the light cylinder in the form of relativistic particles, and is therefore proportional to B_0^2/P^4 where B_0 is the surface magnetic field of the star and P the rotation period. Assuming a dipole law for the magnetic field, this immediately leads to the following law for the bolometric synchrotron luminosity of the pulsar

$$L \propto \frac{B_0^4}{P^{10}}(\gamma \Delta)$$

Pacini went on to argue that the product $(\gamma \Delta)$ cannot have a strong dependence on the period of the pulsar, and therefore the synchrotron luminosity should be proportional to P^{-10}. Such a law would imply that the optical luminosity of the Crab pulsar should decrease by about 1/2% every year. Since the observed optical luminosity of the Crab pulsar provides a normalization for the above equation, Pacini went on to predict that there might be a good chance to detect optical emission from the Vela pulsar and that its average magnitude might be ~ 25. Many years later, the Vela pulsar was indeed detected, and it is a 26th magnitude star. More recently, optical radiation has also been detected from the 50-millisecond pulsar 0540-69 in the Large Magellanic Cloud.

Although one readily concluded that the optical and X-ray emission must be incoherent synchrotron radiation, one had to simultaneously explain how a large flux of electrons and positrons are produced in order to account for the observed intensities. As we will see in our subsequent discussion of the Crab nebula, it is not enough to produce the required number of electrons *or* positrons but electron-positron *pairs* must be created. We shall defer the argument for this till a little later.

Pair-Production Mechanisms

The production of $e^+ - e^-$ pairs in the magnetosphere of young pulsars can be caused in a number of ways (Ruderman 1987). The most important among them are:

(i) The conversion of γ-rays into $e^+ - e^-$ pair in very strong magnetic fields. This mechanism which we have already discussed is expected to be important only within a few stellar radii of the surface.
(ii) The collision of a high energy γ-ray with another γ-ray or an X-ray can also lead to pair production.
(iii) The same process as above can occur when an ultra-energetic photon $\sim 10^{12} - 10^{13}$ eV collides with an optical or infrared photon.

Once one has created the $e^+ - e^-$ pairs there are a number of ways in which the γ-rays can be produced: (a) curvature radiation, (b) inverse Compton scattering of extremely relativistic $e^+ - e^-$ by soft photons, (c) synchrotron radiation due to gyration about the magnetic field.

The γ-rays needed to create an electron-positron wind, as well as those that are directly observed point to very efficient acceleration mechanisms. What these accelerators are and where they are located are still highly controversial. But from observations and theory the following kinds of constrains can be placed:

1. It is quite reasonable to suppose that the pair production mechanisms, as well as the acceleration mechanism, are not likely to be very effective at distances much greater than the light cylinder. This is because the mechanisms we have just enumerated for pair production require either large magnetic fields or large fluxes of infrared, optical or X-ray photons. These conditions are likely to be met only near a star.
2. The reported observations of 10^{12} eV γ-rays from the Crab and Vela implies that the emitting region must be located at $r > 1/3$ light cylinder radius. Otherwise such high energy γ-rays will be unstable against pair production in the strong magnetic fields, and can never escape. Therefore, it is reasonable to suppose that both the pair production, as well as the observed γ-ray emission must arise in the outer magnetosphere.
3. The radiation from optical frequencies to 10 GeV probably comes from the same region of the magnetosphere since no differences in the pulse arrival times have been detected.

The most comprehensive attempts to understand high frequency radiation from the Crab and Vela pulsars is due to Ruderman and his collaborators (Cheng, Ho and Ruderman 1986). Their arguments are far too complex to permit even a brief summary. We shall merely say that they give convincing arguments for the existence of 'gaps' such as those shown by the hatched regions in Fig. 17. We use the word 'gap' in the same sense that we used before, viz., that in these regions there is a component of the electric field parallel to the magnetic field. The details of how such a model is employed to produce the observed radiation are very elaborate and the themes are as complex as a Wagnerian opera! To put in a nutshell, an electron or a positron accelerated through the outer gap can radiate around 10^5 γ-rays during this passage through the gap. These γ-rays in turn create pairs from collisions with soft X-rays. When such pairs are created within the gap itself the electrons and positrons are separated, accelerated in opposite directions, and the process repeats. Beyond such an outer gap the pairs are unaccelerated and lose their energy essentially by synchrotron radiation. This is the origin of the observed optical to hard X-ray radiation. The γ-rays in the MeV to GeV region are thought to be produced by the inverse Compton scattering of the outward flowing electron and positron wind on some of the X-ray photons themselves. This model is able to explain at a qualitative level the observed spectrum of the Crab pulsar, and also able to explain the production of 10^{38} electron-positron pairs per second to account for the observed luminosity of the Crab nebula.

2.11 The Crab Nebula

As we remarked in the introductory chapter, the pursuit of an understanding of the Crab nebula eventually pointed inescapably to the presence of a central neutron star. In that context we were quite satisfied that the energy loss rate of

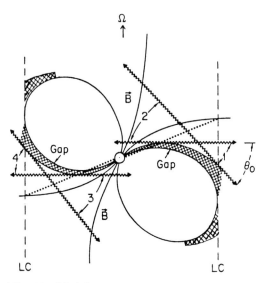

Fig. 17. Model outer magnetosphere with indicated radiation from four emission regions near the *outer gap* boundary (from Ruderman 1987).

the Crab pulsar could nicely account for the luminosity of the nebula. But many details need to be filled in. For a start, there are two things to be understood. First, 'how does the pulsar produce relativistic particles in the Crab nebula?'. Second, 'what is the origin of the wound up magnetic field for the Crab nebula?'.

The Goldreich-Julian model of the pulsar as a unipolar inductor gave us a very concrete picture of how relativistic particles are emitted by a pulsar. To recall briefly, a rapidly rotating strongly magnetized neutron star generates very large voltages, and as a consequence there is a non-vanishing component of electric field parallel to the magnetic field outside the star. Since the electrostatic force on a charged particle near the surface far exceeds all other forces, charges are drawn off the surface and accelerated to very high energies. We saw that in the case of the Crab pulsar electrons could be easily accelerated to a Lorentz factor $\sim 10^9$ or $E \sim 10^{15}$ eV. Thus, the relativistic wind from the pulsar is the source of the relativistic particles in the Crab nebula. But there remained a difficulty, viz., that the charge density of particles predicted by the Goldreich-Julian theory falls short of the number of particles required to explain the luminosity of the nebula by many orders of magnitude.

We already saw that even to explain the radio, optical and X-ray radiation one needs to supplement the primary particles pulled out from the surface with a large number of electron-positron pairs produced in the magnetosphere. The synchrotron, optical, and X-ray radiation from the Crab nebula puts a even stronger constraint on the injection rate of charged particles into the nebula. It turns out that approximately 10^{39} particles must be injected into the nebula per second with an average energy $\sim 10^{12}$ eV. But this current of particles cannot be exclusively electrons or positrons. The net electric current due to the flow of electrons must be almost exactly balanced in the nebula by one of opposite sign. Quite apart from this, if there was a net current as big as what we have just mentioned then it would generate a magnetic field exceeding that in the accelerating region itself. Therefore, if electrons are injected and accelerated into the nebula by the pulsar then an almost equal number of positively charged particles must flow with them. Since there is no mechanism of producing a flux of $\sim 10^{39}$ s^{-1} of positive ions from the stellar surface one must conclude that somewhere in the vicinity of the stellar surface there must be a copious production of $e^+ - e^-$ pairs. To summarize, a rapidly spinning strongly magnetized pulsar energising a synchrotron nebula such as the Crab nebula must emit a strong electron-positron wind.

The next thing that needs to be understood is the origin of the magnetic field in the Crab nebula. Let us recall the conjecture by Piddington nearly ten years before discovery of the pulsar that the magnetic field of the Crab nebula could neither be the interstellar field nor the magnetic field produced at the time of the explosion. He was led to the remarkable conclusion that the magnetic field of the nebula must be some sort of a "wound up" field (Piddington 1957). This finds a natural explanation in the Goldreich-Julian picture that we have been discussing (Fig. 18). At large distances from the pulsar the escaping charges provide a poloidal current source which gives rise to a toroidal magnetic field in the far zone. It turns out that it is possible to estimate the rigorous lower bound for this field in terms of the poloidal magnetic field

$$\frac{B_t}{B_p} \geq \frac{r}{R_{LC}}$$

Thus, the toroidal magnetic field produced by the relativistic wind dominates over the poloidal field beyond the light cylinder. This lower bound in turn sets a lower bound on the rate of escape of charges from the star.

At large distances the poloidal field becomes radial $B_p \propto \frac{1}{r^2}\Psi(\theta)$. Therefore the toroidal magnetic field drops off as $1/r$, a behaviour characteristic of *radiation fields*. This is the most remarkable feature of the Godreich-Julian wind viz. *although one is talking about a DC field produced by the wind, it has the most characteristic feature of a radiation field viz., it falls off as $1/r$*. The energy outflow from the pulsar can be calculated by integrating the radial component of the Poynting flux over a spherical cell. Similarly the angular momentum flux can be obtained by integrating the Maxwell stress tensor. When one does this one finds that the star loses energy at a rate

$$\frac{dW}{dt} \approx \frac{B^2 R^6 \Omega^4}{c^3}$$

This is a truly remarkable result. Apart from a numerical coefficient of order unity, this is the same formula we had for the energy loss rate when we used the description of dipole radiation. It may be recalled that the classical dipole radiation model is relevant only for an inclined radiator. Strictly speaking, an aligned radiator should not slow down because there is no radiation, but we now find that an aligned radiator will also slow down and at the same rate as one

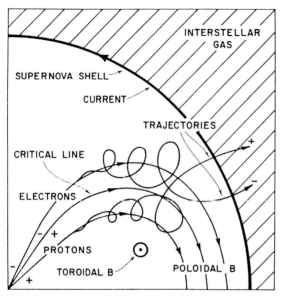

Fig. 18. Schematic diagram showing the supernova cavity, shell, and the ambient interstellar gas (from Goldreich and Julian 1969).

would have predicted in the dipole radiation model. But the energy and angular momentum are now carried away by the relativistic wind from the pulsar. Now that one understands why an aligned radiator slows down — i.e., via a relativistic wind — one can turn around and ask how in the dipole radiation model one can produce the relativistic wind which is crucial not only for pulsar activity but also to explain the activity of the Crab nebula. In an independent investigation Ostriker and Gunn (1969) came up with an ingenious way to achieve this. It is reasonable to assume that although Goldreich and Julian considered only the aligned rotator the results of the analysis are likely to be similar whether the dipole axis is aligned with the rotation axis or not, at least in the zone close to the surface. Therefore one can conclude with some confidence that in either picture there must be a plasma surrounding the pulsar. Now let us do the following thought experiment. Imagine that one is standing at the light cylinder and dropped an electron on the outgoing low frequency dipole wave. Because the frequency of the electromagnetic wave is very low i.e., it is the same as the frequency of rotation of the pulsar, the particle will find itself in a nearly static crossed electric and magnetic field. In a time short compared to the period of the wave the particle will be accelerated in this crossed electric and magnetic field to a relativistic speed in the direction of the wave vector. Thereafter the particle will essentially *surf ride* on the wave. Here one has a situation reminiscent of Cherenkov radiation or Landau damping viz., the velocity of the particle matches the phase velocity of the wave. Hence, in principle, resonant transfer of energy is possible between the particle and the wave. Ostriker and Gunn demonstrated that charged particles will be accelerated to ultrarelativistic energies by the low frequency radiation. Therefore, in either of the two descriptions there will be a strong relativistic wind emanating from the pulsar. One can think of this wind as being accelerated either by the large electric field generated near the surface or accelerated by the low frequency dipole radiation from the star.

It now remains to understand in some detail the observed properties of the Crab nebula. One now has a concrete picture of how the relativistic particles and the magnetic field in the nebula are generated. The description so far has been for a pulsar in vacuum. In this case the relativistic wind or the low frequency wave will reach arbitrarily large distances. The pulsar is in fact in a cavity swept up by the debris of the supernova explosion, and therefore the energy density in the form of relativistic particles and magnetic field has been accumulating in this cavity for nearly a thousand years. The question one has to now ask is 'how will this modify or simplify our description?'. On very general grounds one can expect that something drastic may happen at a characteristic radius R_s from the pulsar. This is the radius at which the ram pressure due to the energy flow from the pulsar $L/4\pi R_s^2 c$ balances the total magnetic plus particle pressure built up in the nebula. A rough estimate (Rees and Gunn 1974) suggests

$$\frac{R_s}{R_{neb}} \sim \left(\frac{\dot{R}_{neb}}{c}\right)^{1/2}$$

For the Crab nebula this radius is about 10% of the nebular radius. There must clearly be some sort of a shock transition at this radius R_s. Let us consider the relativistic wind first. At the shock the energy of streaming motion will get randomized so that beyond the shock the outward velocity will be less than $c/\sqrt{3}$. At the boundary of the nebular cavity the expansion velocity of the relativistic particles must match that of the walls of the cavity itself. From the simple estimate one had for the shock radius one can conclude that the outward velocity of the relativistic gas accumulated in the cavity will decrease smoothly and roughly as $1/r^2$.

The presence of the walls of a cavity also has a significant influence on the magnetic field of the nebula. Since the magnetic field of the pulsar is anchored at the boundaries of a cavity each time the pulsar turns around its open field lines get wound up once. In the case of an isolated pulsar this spiral will be expanding with an outward velocity nearly equal to the velocity of light, and its field strength will decrease as $1/r$. But as we just saw, beyond the shock radius the outward velocity of the wind decreases. In fact, in the simplified picture that we considered it decreases as $1/r^2$. Therefore the toroidal field will increase radially. Observations do not support this simple picture of field increasing radially, but the scenario due to Rees and Gunn, has all the essential ingredients for trying to understand the nature of the magnetic field in the nebula.

Around the same time that Rees and Gunn were trying to understand the origin of the wound up field of the Crab nebula, Pacini and Salvati (1973) made a pioneering attempt to understand the secular evolution of the magnetic field, particle content, and luminosity of an expanding nebula produced and maintained by a central pulsar. This study is particularly important because it provides the necessary framework for understanding the properties of a pulsar-produced nebula in terms of the basic characteristics of the pulsar itself, viz., its rotation period and magnetic field. Later, while discussing the progenitors of pulsars we shall briefly discuss the intriguing question of the poor association between pulsars and supernova remnants. To anticipate this discussion, the basic problem is the following. One expects all young and active pulsars to produce a synchrotron nebulosity around them (in analogy with the Crab pulsar). The radiation from such nebulae is essentially isotropic, and therefore its detection should not depend upon any particular viewing geometry, as is the case of pulsars whose radiation is highly beamed. Therefore one would expect to see all the nebulae produced by young pulsars in the Galaxy regardless of whether one detects the pulsar itself or not. It is in this context that one will be interested to know how typical the Crab nebula is.

The relativistic particles and the magnetic fields responsible for the nebular emission are injected by the central pulsar at the expense of its rotational energy. The rate of loss of rotational energy of the pulsar evolves with time as

$$L_{PSR}(t) = \frac{L_0}{(1 + t/\tau_0)^\alpha}$$

where L_0 is the initial energy loss rate, and τ_0 the initial spindown timescale. α is an exponent which is connected with the breaking index n as $\alpha = (n+1)/(n-1)$. For pure magnetic dipole radiation the breaking index n should be equal to 3. Pacini and Salvati assumed that a certain fraction ϵ_p of the energy released is injected in the form of relativistic particles and a fraction of the energy released ϵ_m goes into the magnetic field. The injected energy spectrum of particles is assumed to be a relatively flat power law

$$J(E,t) = K(t)E^{-\gamma}, \quad E < E_{max}$$

The evolution of the magnetic energy content is obtained by using essentially the first law of thermodynamics.

$$\frac{dW_B}{dt} + \frac{1}{3}\frac{W_B}{V}\frac{dV}{dt} = \epsilon_m \, L_{PSR}(t)$$

where $V(t)$ is the volume of the cavity, and W_B is the magnetic energy content, i.e., $V \times B^2/8\pi$. The solution of the above equation tells us about the secular decrease in the magnetic energy of the nebula. Once the magnetic field is known the evolution of the energy of an individual particle can be obtained from

$$\frac{1}{ER} - \frac{1}{E_i R_i} = c_1 \int_{t_i}^{t} B^2(t)/R(t)dt$$

where E_i is the initial energy of the particle injected at time t_i. The left hand side of the equation represents adiabatic or expansion losses, whereas the right hand side represents synchrotron radiation losses. Based upon these simple assumptions Pacini and Salvati were able to show that the evolution of a nebula such as a Crab nebula consists of three major phases: (i) The initial phase lasts for a very short time after the supernova explosion during which the expansion of the cavity is negligible. If v is the expansion velocity of the cavity, R_0 its initial radius, then this phase refers to times $t \ll R_0/v$. (ii) The second phase lasts till the initial spindown timescale τ_0 of the central pulsar. For the Crab pulsar this is ~ 300 years. During this phase the pulsar output can be considered to be roughly constant at its initial value L_0. Most of the magnetic flux of the nebula is generated during this phase. As the nebula expands the nebular luminosity after an initial increase slightly decreases with time. (iii) At times $t > \tau_0$ the pulsar output decreases significantly. The adiabatic losses also become important, and as a consequence the luminosity of the nebula decreases rapidly with time. To illustrate how the luminosity of the nebula at any given time depends on the parameters of the pulsar, the secular evolution of the nebular luminosity is shown in Fig. 19 for two cases. In the panel on the left the three figures correspond to three different values of magnetic field strength, while in the right panel all three pulsars have same magnetic field but different initial periods.

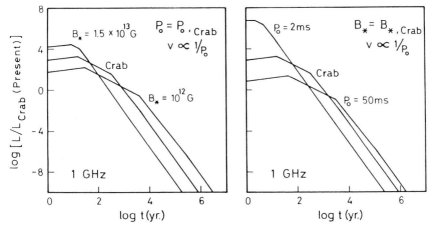

Fig. 19. The secular evolution of the radio luminosity of pulsar-produced synchrotron nebulae such as the Crab nebula. The three phases mentioned in the text are clearly seen. In the figure on the left the luminosity evolution of three nebulae, all powered by central pulsars with the *same* initial period (equal to the estimated initial period of the Crab pulsar), but with *different* magnetic fields are shown. In the right panel the magnetic fields of the central pulsars are the same, but the initial periods are different.

3 The Physics of Neutron Stars

We shall now turn to a discussion of some of the properties of neutron stars. For a detailed account we refer the reader to the excellent book by Shapiro and Teukolsky (1983). Let us briefly recall some of the relevant parameters we have already encountered: $M \sim M_\odot$, $R \sim 10$ km, Mean density $\sim 10^{14}$ g cm^{-3},

Moment of inertia $\sim 10^{45}$ g cm^2, Magnetic field $\sim 10^{12}$ G, Rotation period $\sim$ millisecond to hundreds of seconds, Surface temperature $\sim 10^6$ K.

3.1 Internal Structure

We now would like to explore its internal structure in somewhat more detail. It is an oversimplification to think of a neutron star being made up of essentially neutrons (although as we shall see this is certainly true in the core) and supported against gravity by the degeneracy pressure of the neutrons. To have an overall picture of the cross-section of the neutron star one may approach the problem from the following first principles point of view. In the formation of a neutron star nuclear processes take place sufficiently rapidly that matter in most of the star, except perhaps the outermost layers, can be thought of to be in *complete nuclear equilibrium*. In other words, strong, electromagnetic and weak interactions adjust the nuclear composition at each point to the thermodynamically most favourable one. During the process of core collapse and neutronisation of matter, each time a proton is converted into a neutron, a neutrino is emitted which escapes from the star carrying away energy and entropy. Within a few years after its birth in a supernova explosion the neutron star would have cooled to a fairly low temperature – low compared to characteristic excitation energies in nuclei which are $\sim$ MeV (10^{10} K). Therefore, to a first approximation one may regard the neutron star matter as being in its lowest possible energy state at each point, subject only to local charge neutrality and conservation of baryons.

What is the absolute ground state of matter? Let us imagine that one specifies the number of baryons per unit volume and also overall charge neutrality. For matter at zero pressure the lowest energy state consists of (i) the nucleus arranged into Fe^{56} nuclei, the most tightly bound nucleus, (ii) the nuclei arranged in a lattice to minimize the configurational energy, (iii) electrons being in a ferromagnetic state. Therefore the absolute ground state of matter at zero pressure is a ferromagnetic iron lattice.

As one goes deeper into the star, and as the density increases due to gravitational pressure, the characteristics of the matter changes. By the time density reaches 10^4 g cm^{-3}, atoms are completely ionized. When the density is $\sim 10^7$ g cm^{-3}, the electron gas is degenerate and fully relativistic.

Around the same density (or equivalently electron Fermi energies greater than ~ 1 MeV) Fe^{56} is no longer the lowest energy state because of the onset of inverse beta decay. Electrons at the top of the Fermi sea combine with the protons in the nucleus to form neutrons and neutrinos which escape thus lowering the energy of the system. Under these conditions, Ni^{62} is the most stable nucleus. This continues to be so till one reaches the density $\sim 2.7 \times 10^8$ g cm^{-3} when Ni^{64} becomes preferred. Thus, through this electron capture or inverse beta decay process, the stable nuclei become more and more neutron-rich as the density increases. The sequence of equilibrium nuclei up to a density of 4.3×10^{11} g/cm^3 is as follows:

Fe^{56}, Ni^{62}, Ni^{64}

Se^{84}, Ge^{82}, Zn^{80}, Ni^{78}, Fe^{76}

Mo^{124}, Zr^{122}, Sr^{120}, Kr^{118}

The nuclei mentioned above have closed shell configurations, and have low energies. The nuclei in the second line have 50 neutrons, and those on the third line have 82 neutrons. Across a phase transition from one nucleus to another the pressure remains continuous. Since the pressure is mainly due to the relativistic electrons the density of protons must be continuous. However, the density of neutrons must obviously increase at each transition. Thus, each phase transition is accompanied by a density increase

$$\frac{\Delta \rho}{\rho} \sim -\frac{\Delta(Z/A)}{(Z/A)}$$

For example, in the Fe^{56} to Ni^{62} transition the density increases by $\sim 3\%$.

In the sequence of neutron-rich nuclei mentioned above, Kr^{118} is so neutron-rich that the last neutron is just barely bound to the nucleus. Above this density viz. $\sim 4.3 \times 10^{11}$ g cm^{-3} the neutrons begin to "leak out" of the nuclei and form a degenerate liquid very much like the degenerate electron gas in metals. This regime is known as the "neutron drip regime". This phenomenon continues until one reaches a density $\sim 2.5 \times 10^{14}$ g cm^{-3} when nuclei begin to touch one another, and merge into a continuous fluid of neutrons, protons and electrons, and all of them will be degenerate. The relative numbers of neutrons and protons can be easily determined from the condition of equilibrium under beta decay

$$n \to p + e^- + \bar{\nu}_e, \quad e^- + p \to n + \nu_e$$

This implies that the chemical potential of the neutron must be equal to the sum of the chemical potentials of the protons and electrons $\mu_n = \mu_p + \mu_e$. A very simple exercise shows that the proton concentration will be $\sim 5\%$ and there will be an equal number of electrons as required by overall charge neutrality.

The very neutron-rich nuclei, as well as the neutrons in the neutron sea are stabilised against beta decay by the very high degeneracy of the electrons. Consider the beta decay of a neutron. This process will be inhibited as long as the energy of the electrons produced as a decay product is less than the Fermi energy of the degenerate electron gas. This has an important consequence for the minimum mass of neutron stars which we will discuss later.

Let us at this stage summarize our overview of the structure of a neutron star regarded as the ground state of matter (see Fig. 20). One expects a solid crust whose outer layers are made up of Fe^{56} nuclei. The stability of the pulses from radio pulsars certainly requires that the emitting region must be anchored to a rigid solid. As we go down the crust one encounters more and more exotic neutron-rich nuclei till we eventually reach the critical "neutron drip density". Above this density one has a solid of nuclei and a degenerate liquid of neutrons which have leaked out of the nucleus. Above a density $\sim 2.5 \times 10^{14}$ g cm^{-3}

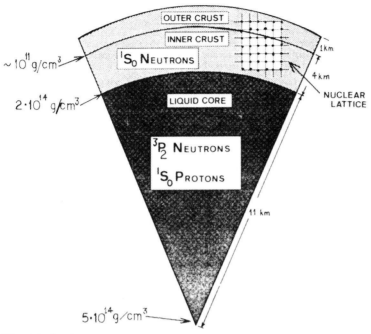

Fig. 20. A schematic cross-section of a $1.4M_\odot$ neutron star based upon a *stiff* equation of state (from Sauls 1989).

we essentially have a degenerate liquid of neutrons, with a small admixture of protons and electrons.

3.2 The Equation of State

What one would really like to do is to construct models for neutron stars such as Chandrasekhar did for white dwarfs. But this requires a detailed knowledge of the equation of state of neutron star matter viz. the dependence of pressure on density $P = P(\rho)$. We shall briefly outline the general principles that underlie the determination of equation of state at densities beyond white dwarf densities. The actual details are very complicated and there is a whole civilization of people who devote their entire scientific career to calculating newer and better equations of state! There are many outstanding Reviews on this topic; we refer in particular to Pethick and Ravenhall (1991) and Baym (1991), and references cited there.

The calculation of equation of state may be broadly divided into two regimes: (a) densities below the neutron drip, and (b) densities above the neutron drip. Let us first consider the lower density regime, i.e., $\rho < 4 \times 10^{11}$ g cm^{-3}.

Equation of State Below Neutron Drip Density

The problems of determining the equilibrium nucleus at any given density is simply that of minimizing the total energy per unit volume at fixed baryon density and demanding overall charge neutrality. One may formally write the energy density of a mixture of nuclei and free electrons as

$$\varepsilon = n_N M(A, Z) + \varepsilon_e(n_e)$$

where $M(A, Z)$ is the energy of the nucleus (A, Z), including the rest mass of the nucleons. n_N is the number density of nuclei and n_e is the electron density. The quantity $M(A, Z)$ is not known experimentally for the kind of very neutron-rich nuclei that one expects to have in the inner crust, and therefore must be deduced theoretically. This is usually done by means of a *semi-empirical mass formula*. Historically the first equation of state calculated using this approach was by Harrison and Wheeler (1958). The semi-empirical mass formula is based on the "liquid drop" model of the nucleus. The number of nucleons A in a given nucleus is determined by balancing the nuclear surface energy which favours large nuclei ($A^{-1/3}$ per nucleon) and the nuclear Coulomb energy which favours small nuclei ($A^{2/3} (Z/A)^2$ per nucleon). Once one has calculated the energy density of the nuclei + electrons it is a straightforward matter to calculate the pressure, and thus the equation of state. In the Harrison-Wheeler treatment the nucleon charge and the atomic mass were regarded as continuous variables in the spirit of the semi-empirical mass formula. Nuclei, however, have discrete values of A and Z, and the shell structure within the nuclei plays an important role in determining the binding energy of the nuclei. These factors were taken into account by Salpeter in a famous paper in 1961, and he calculated the nuclear composition and the equation of state up to the neutron drip regime. A decade later Baym, Pethick and Sutherland (1971a) improved upon Salpeter's treatment by including the lattice energy also into the calculation. The lattice energy viz., the net Coulomb interaction energy of all the nuclei and the electrons is given by

$$E_{lat} = -1.82 \frac{Z^2 e^2}{a}$$

per nucleus for a body centered cubic lattice with a lattice constant a. The inclusion of lattice energy has a quantitative effect and shifts the balance between the surface energy and the Coulomb repulsion slightly towards larger nuclei. This happens simply because at densities approaching the neutron drip density the lattice energy reduces the repulsive Coulomb nuclear energy by $\sim 15\%$, thus favouring slightly larger nuclei.

The equation of state thus calculated is shown in Fig. 21. For comparison we have also shown the equation of state for an ideal electron gas with a nuclear composition of pure Fe^{56}, as well as that of the ideal (n, p, e^-) gas.

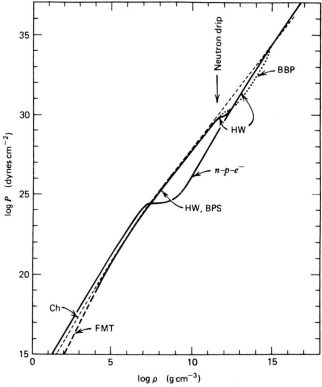

Fig. 21. Several equations of state below neutron drip. HW refers to the Harrison-Wheeler, and BPS to the Baym, Pethick, Sutherland equations of state. The ideal degenerate electron gas model $\left(\mu_e = \frac{56}{26}\right)$ is labelled Ch (for Chandrasekhar). The curve labelled $n - p - e^-$ corresponds to an ideal (n, p, e^-) gas. As expected this deviates from the Chandrasekhar equation of state only for $\rho \gtrsim 2 \times 10^7$ g cm^{-3} where inverse β-decay becomes important (from Shapiro and Teukolsky 1983).

Equation of State Above the Neutron Drip

Let us now consider the properties of condensed matter above the neutron drip density 4.3×10^{11} g cm^{-3}. Let us subdivide this regime into two regions. (1) an intermediate density regime from neutron drip to the nuclear density $\rho_0 = 2.8 \times 10^{14}$ g cm^{-3}, the density at which the nuclei begin to dissolve into a fluid core, and (2) the high density range above nuclear density.

The pioneering effort in the neutron drip to nuclear density regime is due to Baym, Bethe and Pethick (1971b). The essential new ingredient in their approach is that whereas they also used the semi-empirical mass formula, they incorporated in it many results from detailed manybody calculations. The basic point is the following: Above the neutron drip density one has free neutrons present outside the nuclei. This has two effects: (a) the neutrons exert pressure on the nuclei, and (b) when the free neutron density becomes comparable to

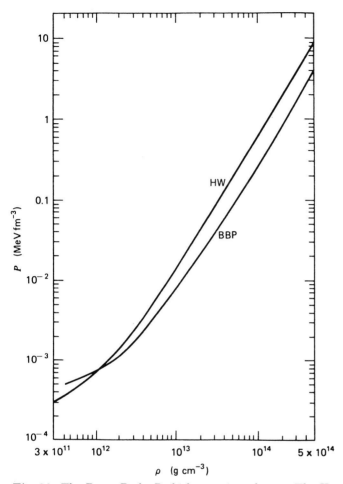

Fig. 22. The Baym-Bethe-Pethick equation of state. The Harrison-Wheeler equation of state is also shown for comparison (from Baym et al. 1971b).

the density inside the nuclei (and this happens when the nuclei themselves are very neutron-rich) the matter inside and the matter outside the nuclei are very similar, and this greatly reduces the nuclear surface energy. The equation of state calculated by Baym, Bethe and Pethic is shown in Fig. 22. Also shown for comparison is the extrapolation of the Harrison-Wheeler equation of state to densities above the neutron drip density.

The Liquid Interior

The calculation of the equation of state from nuclear density to $\sim 10^{15}$ g cm^{-3} requires: (i) determining the nuclear potential for the nucleon interaction, and (ii) techniques for solving the manybody problem. The nucleon-nucleon interaction potential must be deduced from pp and pn scattering experiments at energies below ~ 300 MeV. The manybody Schrödinger equation has to be solved to find the energy density as a function of baryon density and the most popular techniques use some form of variational principle or other. There are two important checks: The deduced 2-body potential is used to compute the saturation energy and the density ρ_0 of symmetric nuclear matter, as well as the binding energies of light nuclei with $A < 4$. The progress made during the last decade may be summarized as follows. The 2-body potentials fail to produce sufficient binding of the light nuclei, and the equilibrium density is too high. To get the binding energies of the light nuclei right one has been forced to take into account 3-body forces acting between nucleons.

For illustration, the energy per nucleon calculated with the best available nuclear potentials is shown in Fig. 23. The curves marked UV14 and AV14 were obtained using the 2-body potentials obtained by two different groups (Paris-Urbana and the Argonne groups, respectively), and UVII and TNI include 3-body interactions. One notices that the inclusion of the 3-body interactions has a distinct effect of "stiffening" the equation of state compared to that calculated with 2-body forces alone. The pressure P is directly calculated from E by the relation $P = \rho^2 \delta E/\delta \rho$. Please note that there may be considerable uncertainty in the equation of state because of the treatment of the 3-body interactions, particularly at higher densities.

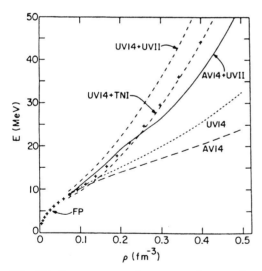

Fig. 23. Energy per nucleon *versus* baryon density with and without 3-body forces in pure neutron matter. UV14 and AV14 use only 2-body forces, while UVII and TNI represent inclusion of 3-body forces (in two different models) (from Wiringa, Fiks, Fabrocini 1988).

3.3 Stability of Stars

At densities greater than twice the nuclear density there are further uncertainties. We shall return to them a little later. Armed with these equations of state we are in a position to construct models of neutron stars. But before that it is worth asking ourselves the following question: The central density of a white dwarf with Chandrasekhar mass is only 10^9 g cm^{-3}. For a neutron star one is talking about central densities well in excess of 10^{14} g cm^{-3}. Why are there no stable stars with central densities between these two limits? This is a very basic question which is worth understanding. As a prelude to this we would now like to digress a little and discuss the stability of stars. Here, by stability we mean *dynamical* stability, or in other words the response of a star to radial perturbations. In particular, we wish to highlight the role of the adiabatic index in determining the stability of stars. Let us recall that the adiabatic index which is the ratio of the specific heat at constant pressure to specific heat at constant volume is formally defined as follows;

$$\gamma_{ad} = \frac{\delta \ln P}{\delta \ln \rho}\bigg|_s$$

It follows from the theory of stability of self-gravitating systems that in order for a star to be stable against radial oscillation this adiabatic index should be greater than 4/3. Strictly speaking, one would refine this statement: *For stability the adiabatic index averaged with respect to pressure over the entire star should be greater than 4/3*, where pressure-averaging is defined as follows:

$$\bar{\gamma} \equiv \int_0^R \gamma \, P \, r^2 \, dr \bigg/ \int_0^R P \, r^2 \, dr$$

To convince ourselves that this is indeed reasonable, let us consider a simple case where we compress the star homologously and adiabatically. First one must define what one means by homologous contraction. In comparing two masses M and M' with radii R and R' one often considers homologous points at which the relative radii are equal, i.e.

$$\frac{r}{R} = \frac{r'}{R'}$$

We say that two configurations are *homologous* if the homologous mass shells $m/M = m'/M'$ are located at homologous points. Let us now consider the homologous changes to a star. Let us consider a sphere of radius $r = r(m)$ in a star in hydrostatic equilibrium. The pressure on this surface is the weight of the layers above it per unit area.

$$P = \int_m^M \frac{GM}{r^2} \frac{1}{4\pi r^2} \, dm$$

Now let us compress this star adiabatically and homologously. The right hand side of this equation will vary as

$$\left(\frac{R'}{R}\right)^{-4}$$

while the left hand side will vary as

$$\left(\frac{\rho'}{\rho}\right)^{\gamma_{ad}} = \left(\frac{R'}{R}\right)^{-3\gamma_{ad}}$$

In writing these we have used both adiabaticity and homology. Therefore, if the adiabatic index is greater than 4/3 the *pressure increases more with contraction than the weight*, and the star will expand again. In other words, it will be stable. If the adiabatic index is less than 4/3 the opposite will be true, and the star will be unstable. $\gamma = 4/3$ is a neutral equilibrium case because the compression again leads to hydrostatic equilibrium. At this stage let us recall that for an ideal monotomic gas the adiabatic index is equal to 5/3, while for an ideal relativistic gas it is 4/3.

In the discussion so far we have considered only Newtonian gravity. The inclusion of general relativity makes an important correction to this stability criterion (Chandrasekhar 1964). Allowing for the effects of general relativity one finds that the critical value of the adiabatic index is given by

$$\gamma_{crit} = \frac{4}{3} + \kappa \frac{GM}{c^2}$$

where κ is a number of order unity that depends on the structure of star. The sense in which general relativity modifies the Newtonian result is as expected; since gravitational effects are stronger in general relativity it tends to destabilize configurations.

Having calculated the equation of state one can proceed to calculate the adiabatic index. Figure 24 shows a plot of the calculated adiabatic index as a function of mass density. At low densities γ is greater than 4/3, and as inverse beta-decay becomes more and more important γ decreases and hovers very close to 4/3. Just above the neutron drip density γ drops well below 4/3 signifying an instability. The reason being that as the mass density increases the main effect is an increase in the number of free neutrons but these free neutrons when they first appear have very low momenta and contribute very little to the pressure. As the density of free neutrons begins to grow they begin to contribute significantly to pressure, the matter becomes stiffer and γ begins to rise. It becomes greater than 4/3 again when the degeneracy pressure of the neutrons becomes of the order of the electron pressure and this occurs around the density $\sim 7 \times 10^{12}$ g cm^{-3}. This plot clearly tells us why there are only two classes of dense cold stars — the white dwarfs, and neutron stars — with nothing in between. Since most of the matter in a star is at densities close to the central density, when $\gamma(\rho_0) < \frac{4}{3}$ the star will not be stable (in Newtonian physics). Stars with central densities below the neutron drip will be stable — these are the white dwarfs. Stars with central densities such that γ is greater than 4/3 are the neutron stars. We shall presently see that the branch of stable neutron stars actually starts at a central density $\sim 10^{14}$ g cm^{-3}. Thus at a very basic level one can say that it is the fact that nuclei begin to become very neutron-rich, and eventually neutrons begin to drip out of the nuclei that is responsible for the existence of two independent branches of cold stars.

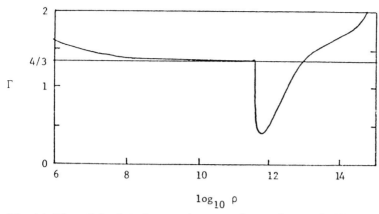

Fig. 24. The adiabatic index as a function of mass density. In Newtonian physics this index has to be greater than 4/3 for stability. Inclusion of general relativistic effects increases the critical value to above 4/3.

3.4 Neutron Star Models

Finally we are ready to construct models of neutron stars. Historically the first persons to address this question were Oppenheinmer and Volkoff (1939). Their main motivation was the following: As we have already discussed, by 1932 Chandrasekhar had shown that in models where gravitational pressure is balanced by the pressure of cold degenerate Fermi gas there exist no stable equilibrium configurations for mass greater than $\sim 1.4 M_\odot$. By 1936 Gamow had argued that in sufficiently massive stars neutron cores will form when all thermo-nuclear sources of energy had been exhausted. Oppenheimer and Volkoff wanted to know whether the picture of Gamow would be right for arbitrarily heavy stars, or to put it in their own words, "to investigate whether there is some upper limit to the possible size of such a neutron core". The "Chandrasekhar mass limit" for a neutron star is $\sim 5.73 M_\odot$. Let us recall that such a mass limit was obtained by equating the gravitational pressure to the relativistic degeneracy pressure of the neutrons. Oppenheimer and Volkoff realized that whereas such an approach was rigorously valid for determining the maximum mass of a white dwarf, there is an improvement that needs to be made if one wants to consider neutron stars. The Chandrasekhar mass limit was obtained on the basis of Newtonian gravitational theory. If one is considering very massive stars and very high densities, effects of general relativity would have to be included.

In order to calculate the radius of a star for a given mass one has to integrate the equation of hydrostatic equilibrium with an assumed equation of state. When effects of general relativity are taken into account the equation of hydrostatic equilibrium reads as follows.

$$\frac{\delta P}{\delta r} = \frac{-G\left[m(r) + 4\pi r^3 P(r)/c^2\right]\left[\rho(r) + P(r)/c^2\right]}{r^2\left[1 - \frac{2\,Gm(r)}{rc^2}\right]}$$

This equation is known as the Tolman-Oppenheimer-Volkoff equation. If one lets $c \to \infty$ then it reduces to the well known Newtonian equation of hydrostatic equilibrium. The way one goes about calculating general relativistic models is very similar to the way one constructs white dwarf models.

1. Pick a value of the central density ρ_c. The equation of state gives the value of the central pressure.
2. The equation is integrated out from $r = 0$ using the initial conditions at the centre. Each time a new value of the pressure is obtained the equation of state gives the corresponding density.
3. The value $r = R$ at which $P = 0$ is the radius of the star, and the mass contained within this is the total mass of the star.

The equilibrium configurations obtained by Oppenheimer and Volkoff are shown in Fig. 25. From this they obtained a maximum mass for a neutron star $\sim 0.7 M_\odot$, with a radius 9.6 km, and a central density of $\sim 5 \times 10^{15}$ g cm^{-3}. This value for the maximum mass is considerably smaller than the Chandrasekhar mass limit for a neutron star which, as we said, is $\sim 5.73 M_\odot$. This arises due to two reasons:

1. At the maximum mass of the sequence the neutrons are only mildly relativistic but not ultra-relativistic as in the case of a white dwarf with a limiting mass.
2. In the case of white dwarfs one did not make a distinction between the rest mass and the gravitational mass. In the case of the neutron star one must make this distinction.

As for the mass-radius relation for neutron stars, this is easy to obtain for low density neutron stars. Using the theory of polytropes one can derive the following relations (Shapiro and Teukolsky 1983):

$$R = 14.64 \left(\frac{\rho_c}{10^{15} \text{ g cm}^{-3}} \right)^{-1/6} \text{ km}$$

$$M = \left(\frac{15.2 \text{ km}}{R} \right)^3 M_\odot$$

As expected, although there is a maximum mass for neutron stars there is no corresponding minimum mass in the Oppenheimer-Volkoff theory: $M \to 0$, $R \to \infty$ as $\rho_c \to 0$. (this is the same in the Chandrasekhar theory of white dwarfs.) In reality, there will be a minimum mass for neutron stars because such a star will be unstable against beta decay. You will recall our earlier statement that neutrons in a neutron star are stable because of the very high Fermi energy of the electrons. The only way the neutrons can decay is if the electrons had sufficient energy to climb to the top of the Fermi sea. Since the Fermi energy decreases with density it is not difficult to see that at a critical lower mass, with a consequent lowering of density, beta decay can suddenly set in and the neutron star will be unstable.

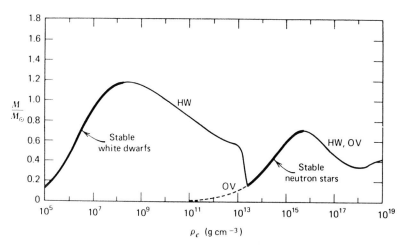

Fig. 25. A plot of gravitational mass *versus* central density (from Shapiro and Teukolsky 1983). The curves obtained using the Oppenheimer-Volkoff (OV) and the Harrison-Wheeler (HW) equations of state are shown. This figure clearly shows why stable white dwarfs and stable neutron stars form two distinct sequences. It may also be seen that there is no minimum mass for neutron stars in the Oppenheimer-Volkoff theory.

Oppenheimer and Volkoff assumed that the matter consists of neutrons, and they used the ideal Fermi gas equation of state as given by Chandrasekhar. One can try to improve upon it by considering an equilibrium mixture of non-interacting neutrons, protons and electrons. The neutron to proton ratio reaches a maximum at $\sim 10^{12}$ g cm^{-3} and then decrease to 8 as ρ increases beyond the nuclear density. Models of neutron stars constructed with this slightly modified equation of state do not differ from the Oppenheimer-Volkoff result, as it should be expected. For example,

$$M_{max} \sim 0.72 M_\odot, \quad R \sim 8.8 \text{ km}, \quad \rho_c \sim 5.8 \times 10^{15} \text{ g cm}^{-3}$$

One can now try to join the Oppenheimer-Volkoff equation of state to the Harrison-Wheeler equation of state, which in turn joins smoothly to the white dwarf equation of state. The resulting mass vs central densities curve is shown in Fig. 25. This shows that one must expect a minimum mass for a neutron star

$$M_{min} \sim 0.18 M_\odot, \quad R \sim 300 \text{ km}, \quad \rho_c \sim 2.6 \times 10^{13} \text{ g cm}^{-3}$$

More Recent Models

Now we turn to more recent attempts. While discussing the modern equations of state we remarked that there is considerable uncertainty in the 3-body interactions, particularly at higher densities. Despite these uncertainties most of these modern equations of state seem to yield similar models for neutron stars.

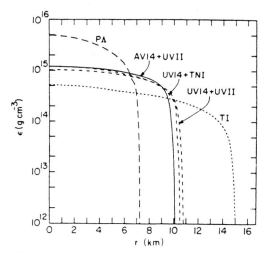

Fig. 26. Density profiles of a $1.4M_\odot$ neutron star calculated with various neutron matter equations of state. All the models with modern potentials (see Fig. 23 for the relevant legends) have a radius ≈ 10.5 km, and a central density $\sim 6\rho_0$. The crust is only ~ 1 km thick (from Wiringa, Fiks, Fabrocini 1988).

For example, Fig. 26 shows the mass density profile of a $\sim 1.4M_\odot$ neutron star. All the models with modern potentials have a radius 10.4 km to 11.2 km, and a central density $\sim 6\rho_0$. The crust in these models is only ~ 1 km thick.

Figure 27 shows the gravitational mass *vs* central densities for a sequence of stars calculated with the same equation of state. One sees that the maximum mass for neutron stars is $\sim 2M_\odot$. The mass radius relations derived from these equations of state is shown in Fig. 28. The main message from this is that the radii of the stars calculated with modern potentials vary very little with mass. At this stage it may be worthwhile to summarize once again the internal structure of neutron stars as deduced from modern equations of state (refer to Fig. 20):

1. The density near the surface is $\sim 10^6$ g cm^{-3}. In this region one expects both the temperature and the magnetic field to have an important effect on the equation of state.
2. The outer crust: This covers the density region 10^6 to 4.3×10^{11} g cm^{-3} and this is a solid region in which a Coulomb lattice of heavy neutron-rich nuclei exists in beta equilibrium with a relativistic electron gas.
3. The inner crust spans the density region $\sim 4.3 \times 10^{11}$ g cm^{-3} to $\sim 2.5 \times 10^{14}$ g cm^{-3}. This consists of a lattice and a degenerate neutron liquid which we shall call "crustal neutron fluid".
4. Above $\sim 2.5 \times 10^{14}$ g cm^{-3} one essentially has a neutron fluid with an admixture of degenerate protons and electrons. As we shall presently argue, the neutron fluid (in the crust, as well as the core) and the proton fluid will be in a superfluid state.
5. One may wonder whether the neutron fluid extends all the to the centre of

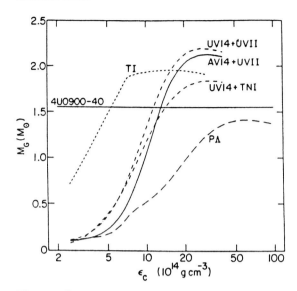

Fig. 27. Gravitationl mass *versus* central density. The maximum mass for the modern equations of state is $\sim 2M_\odot$ (from Wiringa, Fiks and Fabrocini 1988).

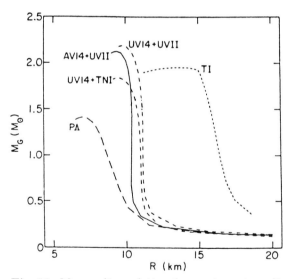

Fig. 28. Mass-radius relation for neutron stars (from Wiringa, Fiks and Fabrocini 1988).

the star or whether there is a transition from a neutron liquid or a solid to some sort of exotic matter such as "quark" matter in the very core of the star. We shall make a few comments about this a little later.

Maximum and Minimum Mass of Neutron Stars

The minimum mass of a stable neutron star may be deduced by setting the mean value of the adiabatic index equal to the critical value $\sim 4/3$. To recall, for stability against radial oscillations the adiabatic index should be greater than $4/3$. We discussed the dependence of the adiabatic index on density for the Baym, Bethe and Pethic equations of state. We saw that γ drops below $4/3$ at the neutron drip density $\sim 4.3 \times 10^{11}$ g cm^{-3}, and does not rise above $4/3$ till $\sim 7 \times 10^{12}$ g cm^{-3}. One can use this equation of state to deduce what is the mass for which the *average adiabatic index* is equal to $4/3$. The result is that the minimum mass is $\sim 0.1 M_\odot$, its central density $= 1.5 \times 10^{14}$ g/cm^3, and the radius $= 164$ km.

It is interesting that the central density for a neutron star with a minimum allowed mass is $\sim 10^{14}$ g cm^{-3} which is very much larger than $\sim 7 \times 10^{12}$ g cm^{-3} at which γ becomes greather than $4/3$. The reason for this is rather straightforward. While discussing the stability of a star against gravitational collapse one is interested in the mean value of γ. Secondly, this mean value must be substantially greater than $4/3$ for stability in general relativity.

To summarize, one can say that since equation of state is relatively well understood from densities less than $\sim 10^{14}$ g/cm^3, one can regard the above results for the minimum mass and the radius of a neutron star as relatively well established.

Unfortunately, a similar statement cannot be made concerning the maximum mass of neutron stars. This is primarily because of the uncertainties in the equation of state above $\sim 2.5 \times 10^{14}$ g cm^{-3}. We saw, however, that there is a reasonable convergence of the deduced maximum mass from the more recent equations of state all of which yield a maximum mass $\sim 2 \, M_\odot$.

Given these uncertainties one can ask the following question: Can one appeal to some fundamental principle which will set an exact upper limit to the mass of the neutron star? The answer is "yes". There are two fundamental requirements that we can put:

1. The equation of state should satisfy the microscopic stability condition viz.

$$\frac{dP}{d\rho} \geq 0$$

If this condition were violated matter would spontaneously collapse.
2. We require that the causality condition is satisfied, viz.

$$\frac{dP}{d\rho} \leq c^2$$

i.e., the velocity of sound is less than the velocity of light.

About twenty years ago, Rhoades and Ruffini (1974) attempted to determine the maximum mass of a neutron star subject to the above two conditions, and assuming further that general relativity is the correct theory of gravity and that one knows the equation of state below a certain matching density ρ_0. They performed a variational calculation from which they concluded that the maximum mass is obtained for the equation of state given by

$$P = P_0 + (\rho - \rho_0) c^2, \quad \rho \geq \rho_0$$

For ρ_0 they chose $\sim 4.6 \times 10^{14}$ g cm^{-3} and matched the above equation of state to the Harrison-Wheeler equation of state below this density. This is not a very crucial assumption because only a few per cent of the mass is in regions of density less than ρ_0. *From this analysis they concluded that the maximum mass for a neutron star cannot exceed $\sim 3.2 M_\odot$.* At this stage one may wonder whether rotation of a neutron star might increase this limiting mass. It does, but only marginally. Unlike in the case of the white dwarfs where rotation can increase the maximum mass substantially, in the case of the neutron star people have convinced themselves that rotation cannot increase the mass by more than $\sim 20\%$.

3.5 The Observed Masses of Neutron Stars

After this lengthy theoretical discussion full of uncertainties let us look at neutron stars in nature !

X-Ray Binaries

The masses of X-ray emitting stars in a number of massive binaries have been deduced from observations. The basis of this method is, of course, Kepler's laws. Let the star 1 emit a pulse or have a spectral feature. This will be Doppler shifted with the amplitude

$$v_1 = \frac{2\pi}{P} a_1 \sin i$$

where P is the orbital period, and i the angle of inclination of the line of sight to the orbit normal. Now from Kepler's laws one knows that

$$f(M_1, M_2, i) \equiv \frac{(M_2 \sin i)^3}{(M_1 + M_2)^2} = \frac{P v_1^3}{2\pi G}$$

the quantity f on the left hand side is traditionally known as the "mass function", and depends only on the observable quantities viz., the period and the measured doppler modulation. Since one does not have *a priori* information about the angle of inclination i one cannot proceed further without further assumptions. In the case of half-a-dozen X-ray binaries it has been possible to measure the mass functions for both the optical star, as well as the X-ray emitting neutron star. The masses of neutron stars so obtained are shown in Fig. 29. The error bars signify the uncertainties in the mass determination.

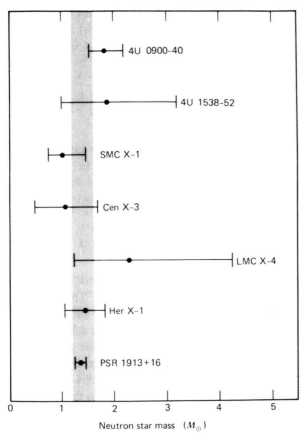

Fig. 29. Derived masses of neutron stars in binary systems. Except for PSR 1913+16 which is a double neutron star system, the other six are binary X-ray pulsars (from Rappaport and Joss 1983).

Binary Pulsars

In the case of the Hulse-Taylor pulsar and a few other binary pulsars one has been able to precisely determine the masses of both the stars, thanks to general relativistic effects, which have been measured with great precision. For example, the advance of the periastron and second order doppler shift. The masses of pulsars derived in this fashion are shown in Fig. 30. A few comments are in order regarding these deduced masses. First of all, none of these deduced masses are in conflict with any of the theoretical arguments we have given so far either in terms of minimum mass or maximum mass. Second, the masses deduced for the Hulse-Taylor system, as well as the others, are a spectacular confirmation for stellar evolution theories according to which neutron stars should have masses very close to the Chandrasekhar mass limit for white dwarfs. Third, a mass $\sim 1.4 M_\odot$ definitely rules out very soft equations of state.

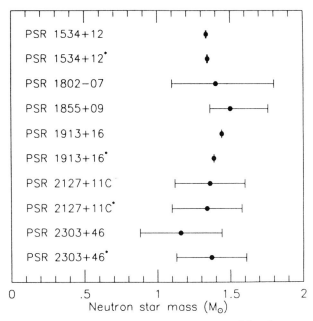

Fig. 30. Masses of 10 neutron stars measured by observing relativistic effects in binary pulsar orbits. Asterisks after pulsar names denote comparisons to the observed pulsars (from Taylor 1995).

3.6 Exotic States of Matter

In the discussion so far we have considered only three constitutent particles in the interior of a neutron star – neutrons, protons and electrons – all obeying Fermi-Dirac statistics. It is reasonable to ask whether mesons such as π mesons and K-mesons can also exist in condensed matter. If they can, then one might expect some interesting astrophysical consequences. π mesons can be either electrically neutral, or positively or negatively charged. The "rest mass" of a π meson is ~ 139.6 MeV. Therefore one can naively expect the following: As we have already remarked, since the neutrons, protons and electrons in the core are in beta equilibrium $\mu_n = \mu_p + \mu_e$. Thus when $\mu_n - \mu_p = \mu_e$ exceeds the rest mass of the π meson it will be energetically favourable for a neutron at the top of the neutron Fermi sea to decay to a proton and a π meson: $n \rightarrow p + \pi^-$. In neutron star matter in beta equilibrium one would expect this to happen at densities slightly higher than nuclear density. In this discussion one has ignored the interaction of the π meson with the background medium. When one takes this into account one expects the π meson to appear at a density around twice the nuclear density.

If π mesons do occur one would expect them to form a *Bose-Einstein condensate* at sufficiently low temperatures because they obey Bose statistics and not Fermi statistics. A fundamental property of an ideal gas of Bose particles is that below a certain critical temperature there will be macroscopic occupation of the zero-momentum state. Let T_0 be this critical temperature. The number of particles in the zero-momentum state is given by the following expression:

$$N(0) = N\left[1 - \left(\frac{T}{T_0}\right)^{3/2}\right]$$

It follows that at very low temperatures essentially most of the particles will be in the zero-momentum state. When this happens one says that there is Bose condensation. A fundamental consequence of such a Bose condensation into zero-momentum state is that it will greatly "soften" the equation of state. The reason being that the particles in zero-momentum state do not contribute to pressure. Therefore, possible existence of pions in the cores of neutron stars is of more than academic interest.

The formation of K-mesons is also likely in a neutron star. The estimated critical density above which they may occur is around three times the nuclear density. If these strange mesons appear then they, too, would form a condensate. An important consequence of pion condensation or Kaon-condensation is that they will have fairly large neutrino luminosities, which will have an important consequence for the cooling of neutron stars as we shall presently see.

Quark Matter

We saw that at density $\sim 2.5 \times 10^{14}$ g cm^{-3} nuclei merge together and dissolve into a continuous fluid of neutrons and protons. But according to the modern point of view, neutrons and protons are not fundamental particles, but they are composed of even more fundamental entities called "quarks". Therefore it is not unreasonable to expect that when one "squeezes" the neutrons and protons into each other, beyond a certain density the more appropriate description will be in terms of the more fundamental quarks. In the technical jargon, one would say that the quarks are now "de-confined". The meaning of this is simply the following: Free quarks do not exist because the forces between quarks increases with distance. By the same token at very short distances or very high temperatures or energies the forces between quarks may be neglected and they may be regarded as essentially "free". Inside, a heavy particle like a proton or a neutron the quarks and the gluons which are responsible for the interaction between the quarks are "confined". But at very high densities and temperatures one may regard the matter as consisting of a Fermi sea of quarks. It is interesting to ask whether such a quark-gluon plasma exists near the very centre of a neutron star. This is a very complex issue, and the answer is far from certain. According to the best available estimates, at very low baryon density the phase transition from baryonic matter to a quark gluon plasma will occur at a temperature $\sim 10^{12}$ K. Alternatively, at very low temperatures such a phase transition will occur at a density somewhere between 5 to 10 times the nuclear density. From our earlier discussion one can say that such densities are probably not reached even in the very centre of neutron stars with mass $\sim 1.4 M_\odot$ if one were to believe in the modern equations of state. But obviously one should be a bit modest and regard this matter as unsettled rather than rule out the possibility of a "quark soup"

Superfluidity in Neutron Stars

We next turn to a discussion of superfluidity in neutron stars. As we know, Baade and Zwicky invented neutron stars in 1934 to explain the origin of cosmic rays and the supernova phenomena. These exotic objects were eventually discovered in 1967. In the intervening three decades while the majority of physicsts and astronomers took no note of the Baade and Zwicky's suggestion, the physics of such highly condensed stars and their astronomical consequences were studied mainly in Russia — by Landau's school and by a group of young astrophysicists working with Zeldovich (of which Novikov was a key member).

An understanding of the microscopic origin of superconductivity eluded physicists for four decades. The problem was finally cracked in 1957 by Bardeen, Cooper and Schriefer for which they were awarded the Nobel Prize. This was a very great triumph for theoretical physics, and the BCS theory is till today the most successful of all manybody theories. Landau and his pupils, like many other schools around the world, were also in the pursuit of the theory of superconductivity. Soon after the news of the BCS theory came to Moscow, A.B. Migdal argued that superfluid state should be expected inside neutron stars (Migdal 1959). Using the newly discovered BCS theory he showed that the high density neutron fluid should be in a superfluid state. Following this Ginzburg and Kirzhnits (1964) showed that protons in the interior will also be in a superfluid state; being charged they will be superconducting. It is worth explicitly pointing out that these remarkable papers were published before the discovery of pulsars. Ten years later when the first glitch was observed in the Vela pulsar, superfluidity of the interiors of neutron stars was invoked to understand the long relaxation time after the glitch. Indeed, till today this continues to be the strongest argument for the existence of superfluid interiors.

An essential concept in understanding the phenomenon of superfluidity is *Bose condensation*. As we have already discussed, below a certain critical temperature a large fraction of bosons will occupy the zero-momentum state. In liquid helium at least superfluidity is closely related to Bose-Einstein condensation. The condensate is described by an *order parameter* of the form

$$\Psi(\mathbf{R}) = |\Psi(\mathbf{R})|\, e^{i\theta(\mathbf{R})}$$

The most basic property of the condensate is that the amplitude of this order parameter is *phase coherent* over the entire superfluid. Thus, if the condensate phase is known at any given point, then one can predict the phase at macroscopic distance away through

$$\theta(\mathbf{R}') = \theta(\mathbf{R}) + \frac{2m}{\hbar}\mathbf{V}_s \cdot (\mathbf{R}' - \mathbf{R})$$

where $\mathbf{V}_s$ is the local velocity of the condensate, i.e., the superfluid velocity. Why the collection of helium nuclei can in principle Bose-condense and become a superfluid can be easily understood in terms of the fact that they obey Bose-Einstein statistics. Although neutrons and protons obey Fermi statistiscs, the helium nuclei consisting of two neutrons and two protons obey Bose statistics. But in the interior of a neutron star we are talking about a fluid consisting of neutrons and protons. Why should these Fermi particles Bose-condense and become superfluids?

Let us digress for a moment and ask why electrons in metals become superconducting for after all they, too, obey Fermi statistics? The essential ingredient is the formation of the so called "Cooper pairs" of electrons. Since the force between electrons is a repulsive one would not expect them to form bound pairs. But at sufficiently low temperatures they in fact form bound pairs because the net interaction between two electrons becomes attractive. This attraction between two electrons is mediated by phonons or lattice vibrations. This is a rather subtle phenomenon but one may gain a qualitative understanding of it in the following fashion: Imagine a particular electron going through the lattice. Let us freeze the motions of this electron for a moment. Because of this excess local negative charge the neighbouring positive ions in the lattice will get pulled in a little bit, or in other words the lattice will locally "deform". Another electron can tend to get attracted towards this region because this lattice deformation will appear as a net excess of positive charge to the electron. This is a pedestrian way of understanding how a lattice can mediate to create net attractive interaction between the electrons. Actually speaking, this occurs only for the electrons on the surface of the Fermi sphere with equal and opposite momenta. In addition, in the case of most terrestrial superconductors spins of the electrons are paired in an antiparallel fashion. The formation of Cooper pairs is essential for the formation of the condensate and the phenomenon of superconductivity. In the case of the neutrons and protons that we are considering, the necessary ingredients for the formation of a condensate of Cooper pairs is an attractive interaction between two neutrons or two protons on the Fermi surface with zero total momentum. The existence of a Fermi sea guarantees the formation of a bound Cooper pair no matter how weak the interaction is as long as it is attractive. The strength of the interaction, of course, has an important effect on the temperature at which superconductivity or superfluidity occurs. In the case of conventional superconductors the net attractive interaction between the two electrons of the Cooper pair is only $\sim$ milli-electron volts. This is the reason why superconductivity is essentially a very low temperature phenomenon. At high temperatures the Cooper pairs get broken up. Incidentally, the big challenge of the recently discovered high temperature superconductors is to find a mechanism which will give a net attractive energy between electrons which is comparable to room temperatures.

In neutron star matter the origin of attraction is the natural nucleon-nucleon interaction
$$V_{nn} = V_{central}(|r|) + V_{S0}(|r|)\mathbf{S} \cdot \mathbf{l}$$
where the central part of the potential is attractive at long range and repulsive

at short distances. The second term on the right hand side is the spin-orbit interaction. Many authors have calculated the transition temperature for the neutron and proton fluids using the BCS theory. Typical values of transition temperatures range from $\sim 0.1\,\text{MeV}$ to $1\,\text{MeV}$, i.e., $\sim 10^9\,\text{K}$ to $\sim 10^{10}\,\text{K}$. *It is very important to appreciate that these are very high temperatures compared to the ambient temperatures for even the youngest neutron stars.* For example, various estimates suggest that the interior temperature of even the young Crab pulsar cannot be more than $\sim 10^8\,\text{K}$. Therefore, *a neutron star is a very cold object compared to the transition temperatures that are estimated.* And it is for this reason that one must take the occurrence of superfluidity and superconductivity seriously.

At this stage we wish to mention in passing some subtle aspects of the superfluid states in neutron stars. The conventional laboratory superconductor is what one would call "s-wave, spin-singlet superconductor". In other words, the total spin of the pair is zero (spin-singlet) and the orbital angular momentum of the pair is also zero (s-wave). This essentially means that the orbital motion of the pairs is isotropic and the spins of the two fermions are paired in a magnetically inert way.

Not all terrestrial superfluids and superconductors are like He^4. The most outstanding counterexample are the various superfluid phases of liquid He^3. It is known that He^3 is described by a spin triplet $(S =| \bar{s}_1 + \bar{s}_2 |= \hbar)$ p-wave $(l = \hbar)$ order parateter. Now it turns out that whereas the crustal neutron fluid and the proton fluid in the core will indeed form a singlet s-wave superfluid, this is not so for the neutron fluid in the *core*. The neutron fluid in the core will form a $^3P_2$ superfluid in which the spins of the fermions are parallel. Thus the core superfluid is like liquid He^3. This may be seen in Fig. 31 in which we have plotted the derived transition temperatures as a function of density for various phases of superfluid in the neutrons and protons. As may be seen, the core superfluid, (with a density of $\sim 4 \times 10^{14}$ g cm^{-3}) would prefer to form $^3P_2$ superfluid. But notice that protons in the interior will form a $^1S_0$ supercondcutor. This is simply because there are eight times fewer protons than neutrons in the core. Density-wise the core proton fluid is similar to the crustal neutron fluid. It is gratifying to note that this prediction viz. that there are likely to be two kinds of neutron superfluids in neutron stars was pointed out by Ruderman and his collaborators even before superfluidity of He^3 was discovered on Earth (Hoffberg et al. 1970)! To summarize, the neutron superfluid in the crust is similar to liquid He^4, the proton superconductor in the core is similar to most laboratory superconductors, but the neutron superfluid in the core is a more exotic superfluid similar to liquid He^3. The ferromagnetic alignment of the spins of the two neutrons has many important consequences, particularly for the coupling of the superfluid core to the rotating crust of the star. The occurrence of superfluidity and superconductivity in the interior is far from an academic issue. As we shall see later, they have profound astrophysical consequences. Just to mention two examples, superfluidity of the neutron liquid is intimately connected to the phenomenon of glitch as well as the slow post-glitch recovery, and superconductivity of the protons has a profound implication for the evolution of the magnetic fields of neutron stars.

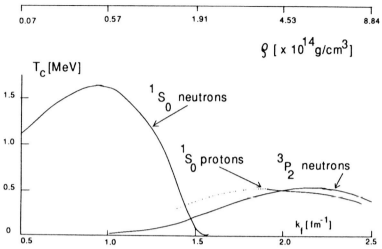

Fig. 31. Superfluid transition temperatures as a function of density. The calculation for neutron matter is from Hoffberg et al. (1970) and for proton matter from Chao et al. (1972). As may be seen, the cross-over from s-wave pairing to p-wave pairing for neutrons occurs at a density $\sim 2 \times 10^{14}$ g cm^{-3} i.e., roughly the density at the interface between the crust and the core. (From Sauls 1989)

The Nature of the Rotating Superfluid

At this stage we want to bring in an essential complication into the scenario described above. And this arises due to the rotation of the neutron star. If one regards the crust as a vessel, then the superfluids that we have so far talked about are in a rotating vessel. One is therefore forced to ask "what is the nature of the thermodynamic state of the superfluid in a rotating vessel"? This is of course a well understood and fascinating question.

It is well known that a "superflow" is a purely *potential flow*; the condensate cannot support circulation because

$$\boldsymbol{\nabla} \times \mathbf{V}_s = 0$$

The reason why this is so can be appreciated rather simply. For the sake of simplicity let us consider a $^1S_0$ superfluid. As we remarked earlier, the superfluid velocity, or the velocity of the condensate, is given by the gradient of the phase of the order parameter

$$\mathbf{V}_s = \frac{\hbar}{2m} \boldsymbol{\nabla} \theta(\mathbf{R})$$

It is because of this that the "Curl" of the superfluid velocity field is zero, and hence our conclusion that a superfluid cannot support circulation. The resolution of this famous question was first given by Onsager (1949) and independently by Feynman (1955) in one of their most famous papers. They pointed out that a superfluid in a rotating vessel will be "perforated" with quantized vortices with each vortex having a quantum of circulation equal to $h/2m$ where m is the bare mass of the Fermion. The number of vortices is obtained readily from the following simple equation

$$\oint \mathbf{V}_s \cdot d\mathbf{l} = (\Omega R)(2\pi R) = N_v \frac{h}{2m}$$

Here, Ω is the angular velocity of rotation, R is the radius of the vessel, and N_v is the number of Onsagar-Feynman quantized vortices. Figure 32 shows a sketch of the distribution of such vortices parallel to the rotation axis, and also the superfluid velocity along the line through the centre of the star. One notices that the superfluid velocity deviates from the classical rigid body value of (ΩR) only very close to the centre of individual vortices where the field of that particular vortex dominates over the average velocity field of all the other vortices. Thus, as Ginzburg and Kirzhnits noted in their pre-pulsar paper, on a macroscopic scale the neutron superfluid threaded by these vortices of required number will "mimic" rigid body rotation, and thus have a classical moment of inertia. Since it is of great historical interest, in Fig. 33 we show an actual photograph of quantized vortices developing in a rotating beaker with superfluid helium. As we start from the left and go to the right the angular velocity of the beaker is being increased. The white patches are the quantized vortices. One sees that as the angular velocity of the beaker is increased the number of vortices increases precisely as Onsager and Feynman had predicted. Further, notice that these vortices arrange themselves into a lattice with specific symmetry in order to minimise the configurational energy. One notices how quickly the triangular lattice forms. Needless to say, this process is reversed if we slow down the beaker viz. the number density of vortices will decrease proportionately through a radial outward movement of the vortices and their eventual destruction at the walls of the beaker. (We refer to Sauls 1989 for a pedagogic review of superfluidity in neutron stars.)

3.7 Cooling of Neutron Stars

Measurement of the surface temperatures of neutron stars can tell us a lot about the interior hadronic matter and neutron star structure. The fact that the current upper limits to the surface temperature of the Crab pulsar are $\sim 10^6$ K puts important constraints on the thermal history of a neutron star. When neutron stars form in supernova explosions they are expected to have very high interior temperatures in excess of $\sim 10^{11}$ K. How does this temperature drop to a few million degrees in a thousand years? The main cooling mechanism in a hot neutron star is neutrino emission. One expects a newly born neutron star to have cooled to a temperature $\sim 10^9$ K in a few days. In the subsequent discussions we shall not address this question since this period is so short compared to the subsequent cooling phase.

Since cooling by neutrino emission is the most dominant mechanism let us discuss this briefly. At very high temperatures $T > 10^9$ K the dominant mode

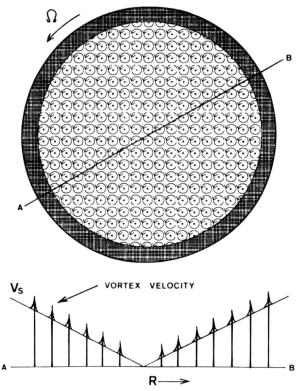

Fig. 32. A schematic representation of the *vortex state* in the neutron superfluid (from Sauls 1989). The radial dimension of the vortex core is ~ 100 fermi, and the mean spacing between them is $\sim 4 \times 10^{-3}$ cm for a pulsar with a rotation period ~ 100 ms. The lower figure shows how the average velocity field due to all the vortices mimics rigid body rotation.

of energy loss is from the so called URCA reactions

$$n \to p + e^- + \bar{\nu}_e, \qquad e^- + p \to n + \nu_e$$

This process is quite important if the matter is not degenerate. However, by the time the neutron star has cooled to $\sim 10^9$ K the nucleons become highly degenerate and these reactions are greatly suppressed. This can be seen as follows: We have already seen that in beta equilibrium $\mu_n = \mu_p + \mu_e$ where the three quantities refer to chemical potentials of the neutrons, protons and electrons. Only neutrons that are capable of decaying are those which line within kT of the Fermi surface. Hence, by energy conservation the final proton and electron must also lie within kT of their Fermi surface. The escaping neutrino must also have energy $\sim kT$. Because the Fermi momentum of the electrons and protons are small compared with the Fermi momentum of the neutron, the final protons and electrons, as well as neutrinos must have small momenta. But the initial neutrons which decay must have a momentum comparable to the Fermi

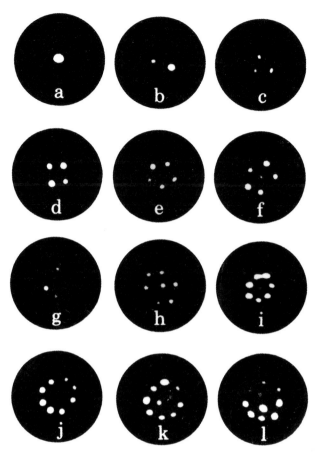

Fig. 33. Onsager-Feynman quantized vortices in a rotating beaker containing superfluid helium (from Yarmchuk et al. 1979).

momentum of the neutron which is very large. Therefore, *such a beta decay of the neutron cannot conserve momentum if it conserves energy*. In order for this process to work there must be a momentum sink which can be provided by a "bystander" which absorbs the excess momentum. This was first pointed out by Chiu and Salpeter in 1964, and they suggested a modified URCA reaction as shown below:

$$n + n \to n + p + e^- + \bar{\nu}_e, \qquad n + p + e^- \to n + n + \nu_e$$

The resulting neutrino luminosity from this modified process has been calculated by a number of people, and a typical result is shown below (see Shapiro and Teukolsky 1983 for a detailed discussion):

$$L_\nu = (5.3 \times 10^{39} \text{ erg/s}) \frac{M}{M_\odot} \left(\frac{\rho_{nuc}}{\rho}\right)^{1/3} T_9^8$$

The Salam-Weinberg theory of weak interaction allows for futher neutrino processes such as shown below:

$$n + n \rightarrow n + n + \nu + \bar{\nu}, \qquad n + p \rightarrow n + p + \nu + \bar{\nu}$$

The luminosity from these processes also varies as T^8 but when one takes the pre-factors into account these are far less important than the modified URCA process. The above discussion does not include the possibility that neutrons and protons may be in a superfluid state.

The temperature of a neutron star can be calculated as a function of time. The thermal energy of the star resides mainly in the degenerate nucleons. Neglecting interaction between nucleons the heat capacity of these particles is given by

$$C_v = N c_v = \left.\frac{dU}{dT}\right|_V$$

where c_v is the specific heat per particle. The cooling equation is

$$\frac{dU}{dt} = C_v \frac{dT}{dt} = -(L_\nu + L_\gamma)$$

where L_ν is the total neutrino luminosity, and L_γ is the photon luminosity. Assuming a black body emission from the surface at an effective surface temperature T_e we have

$$L_\gamma = 4\pi R^2 \sigma T_e^4 = 7 \times 10^{36} \text{ erg/s} \left(\frac{R}{10 \text{ km}}\right)^2 T_{e,7}^4$$

One can now insert the appropriate photon and neutron luminosity into this equation, and integration gives us the time for a star to cool from an initial interior temperature T_i to a final T_f. The schematic cooling curve vs. time for various processes are shown in the Fig. 34. Each line assumes only a single process. One sees that photon cooling becomes important only when the interior temperature has dropped to $\sim 10^8$ K. These cooling curves were calculated assuming that the neutrons and protons were normal and not superfluid. Superfluidity has two effects on the cooling curve. First, it modifies the specific heat. As superfluidity sets in the specific heat jumps discontinuously to a value greater than in the normal state and then falls off exponentially at low temperatures. This tends to decrease the cooling rate immediately below the transition temperature and increase it at much lower temperatures. A second effect of superfluidity is to suppress neutrino processes in the interior. So far we have been talking about the internal temperature. It is generally believed that the surface temperature will be about two orders of magnitude smaller than the interior temperatures. We have so far also neglected the role of magnetic field. The effect of the magnetic field may be quite significant near the surface because it can reduce the photon opacity. Thus, for a given internal temperature the decrease in the opacity will increase the surface temperature and therefore increase the photon luminosity from the surface. The figure showing the cooling rate also shows the effect of a

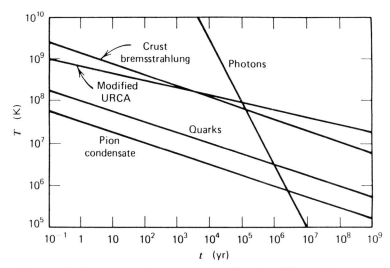

Fig. 34. Cooling curves for neutron star interior. The various curves correspond to various processes acting alone (from Baym and Pethick 1979).

possible condensate of pions in the interior. As we already remarked, pion condensation will increase the neutrino luminosity although we did not elaborate on this. The same applies if there is quark matter in the interior.

These calculations are of immense value in the context of modern X-ray observations of neutron stars and the interiors of nearby supernova remnants. Soon after the Einstein observatory was launched there was a sensitive search to look for black body X-ray emission from neutron stars in the centres of nearly 50 supernova remnants, in particular the half-a-dozen or so historical remnants. Sadly, these efforts did not meet with much success. The immediate conclusion that theorists drew from this is that neutron stars must cool much faster than in standard theories and one must therefore invoke exotic states of matter in their interior such as pion condensation. But this conclusion may be premature. In particular, recent ROSAT observations seem to suggest that there is no need to panic and invoke exotic states!

With this we conclude our overview of the properties of neutron stars.

4 The Progenitors of Neutron Stars

Having discussed the physics of neutron stars we now turn to their progenitors. This is a pertinent question in a course devoted to stellar remnants. Let us first address this question from a theoretical point of view.

We saw earlier that Chandrasekhar's theory of ideal (non-relativistic) white dwarfs predicted that all stars will ultimately find peace as white dwarfs. The exact theory, however, predicted that there is an upper limit to the mass of white dwarfs viz., $1.4 M_\odot$. In our historical discussion one was left speculating about the fate of stars more massive than $1.4 M_\odot$.

To pursue this question further one must appeal to modern theories of stellar evolution. Since the question of the progenitors of neutron stars is rather important and topical, we shall digress a little and recapitulate the main conclusions of modern theories. (For further details we refer to the excellent treatise by Kippenhahn and Weigert 1990.)

An essential complication that could not have been anticipated in the 1930s is the fate of the helium core formed due to central hydrogen burning. It turns out that as the mass of the helium core grows to $\sim 0.45 M_\odot$ the central temperature becomes high enough for helium burning to set in. Unfortunately, in low mass stars the helium core is degenerate. This can have a catastrophic consequence for reasons which are not difficult to understand. Let us do a thought experiment in which we increase the temperature of the core. Let us first consider a case when the core is non-degenerate. The increase in temperature will result in an increased energy production, which in turn will cause further heating. Fortunately this does not occur because if the core is non-degenerate then the pressure responds to an increase in the temperature: consequently the core will expand. This expansion will result in a cooling of the core and a decrease in the energy production. This is the safety valve that operates in the Sun, for example. Unfortunately, there is no such safety valve if the core is degenerate. Here an increase in the temperature of the core *will* result in an increased energy production; the degeneracy pressure of the electrons is insensitive to the increase in the temperature, and therefore the core will not expand. This will result in a dramatic increase in the burning rate resulting in a run-away burning or in other words a helium bomb! Fortunately this does not happen when helium ignites in the degenerate cores of low mass stars. There is a flash, but before it gets out of hand electron degeneracy is lifted and the safety valve begins to operate once again. After a flash helium burns in a quiescent fashion building up a carbon-oxygen core which, in turn, will become degenerate. As the degenerate carbon core contracts it will get hotter and hotter due to secondary effects. It turns out that if the carbon core has a mass $< 1.39 M_\odot$ it will never get hot enough for carbon to ignite. Therefore one can say with considerable confidence that all stars with mass $< 1.4 M_\odot$ will find peace as white dwarfs. The life history of such low mass stars is shown in Fig. 35 where the central temperature is plotted against the central density. The slanting line delineates the region of degeneracy from the classical region. What is the fate of more massive stars? This may also be seen in the figure. To be specific, let us consider the evolution of a $7 M_\odot$ star. Unlike in the case of a $1 M_\odot$ or $2 M_\odot$ star the helium core remains non-degenerate when it ignites. Therefore there is no associated helium flash. But soon the carbon-oxygen core does become degenerate. If the star is sufficiently massive then the degenerate carbon-oxygen core will grow in mass to the critical value of $1.39 M_\odot$ when carbon *will* ignite. Since the core is degenerate this will result in a run-away thermo-nuclear reaction. Unfortunately, unlike in the case of helium flash the star may not be able to save itself in this case. The reason may be traced to the fact that carbon burning is extremely temperature-sensitive with the reaction rate proportional to T^{140}. Therefore, before degeneracy is lifted and the safety valve becomes operative the star may blow itself apart. Some years ago this was taken

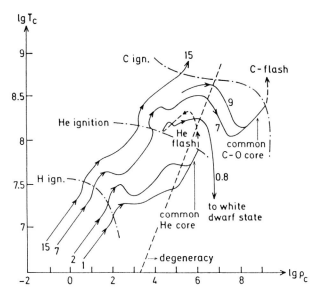

Fig. 35. Evolution of the central temperature T_c (in K) and central density ρ_c (in gm cm^{-3}) for stars of different masses (from Kippenhahn and Weigert 1990). The hydrogen, helium and carbon ignition zones are indicated by the dashed-dotted lines. The sloping dashed lines demarcates the regions of electron degeneracy and non-degeneracy.

to be a certainty, and was proposed as a model for Type I supernovae. Although the physical arguments underlying this conclusion may be sound — there are some loopholes in this also — the following possibility was overlooked. It may be that in stars of such intermediate mass the carbon-oxygen core will never grow to $1.4 M_\odot$. This can happen, for example, due to sudden and dramatic mass loss . The observational clue came from the discovery of white dwarfs in some of the relatively young open clusters. While the presence of white dwarfs in open clusters is not surprising by itself what was surprising was that in these clusters the main sequence turn-off mass was quite high — in some cases as large as $5 M_\odot$ or $6 M_\odot$. This implies that the progenitors of the white dwarfs must have been more massive than the turn-off mass. Thus for reasons which are not fully understood, stars more massive than, say, $6 M_\odot$ are able to lose sufficient amount of matter and eventually find peace as white dwarfs. This scenario is schematically shown in Fig. 36. The solid curves show the decrease in the stellar mass due to mass loss. Also shown is the curve of growth of the degenerate core as a function of times (dashed line). When the degenerate carbon core reaches the critical mass of $1 M_\odot$ it will ignite. The question is whether the star will be able to lose mass rapidly enough to prevent the core from ever reaching this critical mass. We see from this figure that the carbon core will not ignite in stars with initial mass less than a certain critical value M_i (min). Observation of white dwarfs in open clusters seems to suggest that this minimum mass may be as high as 7 or $9 M_\odot$. Let us now turn to stars more massive than, say, $9 M_\odot$ and ask what is their

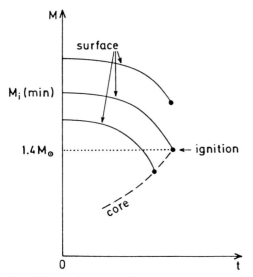

Fig. 36. A schematic diagram showing the growth of the mass of the core, and the decrease in stellar mass due to mass loss. The carbon core will ignite when its mass reaches about $1.4M_\odot$. All stars with initial masses less than M_i (min) will avoid this catastrophe (from Kippenhahn and Weigert 1990).

eventual fate. Their life history is shown in Fig. 35. In these stars the core never becomes degenerate, and the burning is controlled at all stages. Why is this so? We touched upon this in our introductory chapter, but let us recall the arguments. Let us ask for the criteria for gas to be non-degenerate. Under laboratory situations one would say that the gas is non-degenerate if kT is very much greater than the Fermi energy. But in a stellar context one has to take into account the role of radiation pressure as well. But the fundamental criterion remains the same viz., the gas may be considered as non-degenerate if the pressure of the electrons calculated from Boyle's law exceeds the pressure calculated according to Fermi-Dirac statistics. We showed earlier that if the radiation pressure is more than 9.2% of the total pressure then the gas cannot become degenerate (Chandrasekhar 1932). This is an exact result. Let us know return to the life history of a massive star. The reason why degeneracy never set in in the cores of such stars is that radiation pressure is always greater than 9.2% of the total pressure. In such stars successive stages of burning proceed in an uneventful manner all the way till the silicon burning results in the formation of the iron-group elements. Degeneracy quickly sets in in the inert iron core, but we have fortunately reached the final act of the drama. As more silicon is converted to iron, the mass of the degenerate core quickly grows to $1.4M_\odot$. It then collapses, and finds peace as a neutron star. This may be the fate of stars in the mass range $10 - 40M_\odot$. Stars even more massive than this may end up as black holes. Since the mass function of stars is fairly steep, the majority of the progenitors of neutron stars must have masses close to the lower limit for the formation of neutron stars.

4.1 Pulsar Birth Rate

We shall now appeal to observations to constrain the above mentioned theoretical expectations. A quantity of considerable interest is the birth rate of neutron stars. In principle this would be the sum of the birth rate of radio pulsars and the birth rate of X-ray emitting neutron stars in binary systems. But it turns out that the birth rate of radio pulsars is much higher than that of the X-ray sources, and therefore the former will be a good measure of the birth rate of neutron stars.

Let us remind ourselves once again about the observed population of pulsars. Figure 11 shows about 560 pulsars with their periods and derived magnetic fields (which are mostly in the range $\sim 10^{12} - 10^{13}$ G). You will notice that there are a handful of very low field pulsars with fields in the range $\sim 10^8 - 10^9$ G. These are the millisecond pulsars. For the present we shall not be concerned with them since they have a very low birth rate of the order of 1 in 10^6 yr. There are many ways of estimating the birth rate of pulsars from the observed population. For example, if one is able to derive an average "age" or "lifetime" for the pulsars then given their total number in the Galaxy one can estimate a birth rate. But this is not a very reliable method because it is not easy to estimate an *average lifetime*. The lifetime of a pulsar is, of course, the time it takes to cross the "death line". But this time will depend on the magnetic field of the pulsar since the period derivative of the pulsar is proportional to B^2. The derived lifetime will also depend on the specific assumptions made about the possible decay of the magnetic field. But more importantly, such an oversimplified method does not allow for the possibility that the birth rate of pulsars with different field strength may not be the same.

4.2 Pulsar Current

A more reliable way of calculating the birth rate is by calculating the *current of pulsars*. This is defined in analogy with the electrical current in a wire. Let us consider a period bin of width ΔP centred at P. Pulsars move into the bin from the left and move out of the bin to the right. In addition, some pulsars could in principle be born in this bin with an initial period P, or even die while traversing this bin. It is possible to judiciously choose the period bin such that the birth or death of pulsars within the bin are unimportant. In such a situation the current of pulsars across such a bin should be conserved. And how does one define the current? It is essentially the number of pulsars in the bin multiplied by their "velocity" across the bin. Since pulsars within a period bin may have different magnetic field strength (or period derivatives), a better way to formally define the current is as follows:

$$J(P) = \frac{1}{\Delta P} \sum_i \dot{P}_i$$

where the summation is performed over all pulsars in the bin under consideration. *The maximum value attained by the current so defined is nothing but the birth rate of pulsars.* Not quite! We must remember that we have not allowed for the fact that the observed population of pulsars is only a small fraction of the true galactic population. To calculate the true galactic birth rate one has to make two corrections:

1. Since the radiation from pulsars is "beamed" one can "see" only a certain fraction of pulsars even in principle. The *duty cycle* of pulsars gives us a handle on the opening angle of the radiation cone, which in turn determines the beaming factor f. In the pulsar community this factor is taken to be 0.2, or in other words one out of five pulsars will be beamed towards us. One can get more sophisticated and build a model in which this beaming factor changes as the pulsar ages etc.

2. The second correction has to do with different kinds of *selection effects* such as (i) low flux of pulsars, (ii) pulse smearing due to the dispersion in the interstellar medium, (iii) pulse broadening due to interstellar scattering etc. All these factors affect the detectability of pulsars. Over the years a great deal of thought has gone into how one might correct for such selection effects (Narayan 1987). Basically what one has to do is the following. Consider a pulsar with a period P, period derivative $\dot{P}$ (with a consequent luminosity L), at a certain distance and in a particular direction in the Galaxy. One can now ask whether a particular survey for pulsars would have detected it. By dividing the Galaxy into "cells", one can estimate the detection probability for pulsars with a given P and $\dot{P}$ by a Monte Carlo technique. The reciprocal of this detection probability will give us the desired *scaling factor* $S(P, \dot{P})$ which will enable us to construct a true Galactic population of pulsars from the observed population.

One is now in a position to define the true current of pulsars:

$$\bar{J}(P) = \frac{1}{\Delta P} \sum_i \dot{P}_i S\left(P_i, \dot{P}_i\right) \frac{1}{f}$$

This represents the galactic birth rate of pulsars. The average pulsar current calculated this way is shown in Fig. 37. One sees that the current continues to rise till a period ~ 0.6 s. This clearly shows that not all pulsars are born spinning very rapidly like the Crab pulsar, as conventional wisdom would have predicted. After this the current remains roughly constant till $P \sim 2$ s. Beyond this the current drops as death of pulsars becomes a significant factor. *The maximum value of the current yields us a birth rate of pulsars of 1 in 75 years.* It is worth repeating again that the birth rate is quite sensitive to several things, perhaps most importantly the distance scale to pulsars. The rate quoted above was calculated using the latest distance model due to Taylor and Cordes (1993). The same analysis with the "previous-best" distance model gives a birth rate of 1 in 40 years.

Before proceeding further, let us try to compare the birth rate of pulsars with some related rates, viz., the supernova rate and the birth rate of supernova remnants.

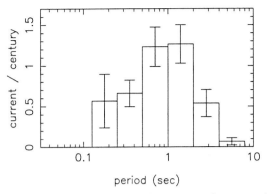

Fig. 37. The current distribution as a function of period. As may be seen, the current reaches its maximum value around a period of 0.5 s, and begins to decline at around 2 s. The maximum value of the current corresponds to a pulsar birth rate of about 1 in 75 years.

The most reliable estimate of the Galactic supernova rate is from the oriental historical records. Based on the nine recorded supernovae in the last two millenia Clark and Stephenson (1977) have estimated the frequency of supernovae in the Galaxy to be once in ~ 30 years. Another method is to estimate the rate of supernovae in external galaxies with morphology similar to ours. This is a tricky procedure, but the latest estimates of extragalactic supernova rate translates to a Galactic rate of 1 in ~ 100 years. Before comparing these with the birth rate of neutron stars one must remember that not all supernovae are triggered by the birth of a neutron star. There is consensus that neutron stars should be associated with Type II supernova. There is an emerging consensus that Type Ib supernovae may also be associated with massive binaries. If this is true then these, too, would leave behind neutron stars. Therefore the relevant ratio is Type II : Type Ib : Type Ia. If one goes by the ratio quoted in the literature, it appears that Type Ia supernovae may be a minority, and one would not be too wrong in comparing the total supernova rate with the pulsar birth rate. If one does that, then given the intrinsic uncertainties one may conclude that the supernova rate and the pulsar birth rate are not inconsistent; although the supernova rate is somewhat higher.

Turning to supernova remnants (SNRs), an estimate of their birth rate is even more tricky. Everything hinges on the assumed model for estimating the life time of SNRs. In the standard model all SNRs are assumed to be in the Sedov or self-similar phase of expansion. If one accepts this for a moment, then it yields a SNR birth rate ~ 1 in 120 years (Clark and Caswell 1976). But this assumption needs to be questioned. The standard model is justifiable only in the case of those remnants that are expanding in a relatively dense interstellar medium with a number density ~ 1 atom/cc. However, if the ambient density is much lower then the free expansion phase will last longer, and one would be grossly *overestimating* the age of a given remnant. During the past decade there is growing observational evidence (ultraviolet and soft X-ray observations) for a much more rarefied component of the interstellar medium with a relatively large filling factor. Theoretical ideas of Spitzer (1978) and McKee and Ostriker (1977) are also consistent with this. If one allows for the possibility that a certain fraction of SNRs may be expanding in a rarer component of the interstellar medium, then the estimated SNR birth rate will be much higher than that suggested by the standard model, and reasonably consistent with the supernova rate, and the pulsar birth rate (Bhattacharya 1987). So all may be well.

While it is certainly worthwhile to check if the birth rates of SNRs and pulsars match, it is equally important to ask whether neutron stars are physically associated with the known SNRs. Since SNRs are relatively short lived objects compared to pulsars it is reasonable to expect such an association. The most famous association is of course the pulsar in the Crab Nebula (Fig. 38). In fact, soon after pulsars were discovered and it became clear that they must be neutron stars, one went and looked for one in the Crab Nebula – and found it. The discovery of a pulsar in the Vela SNR soon followed. But during the next two decades although the catalogue of SNRs increased in size, the number of known associations remained essentially two. In the recent past one or two more associations have been added to the list. But the absence of neutron stars near the centres of most SNRs is a cause for worry. Earlier while faced with the absence of point X-ray sources in SNRs one invoked rapid cooling of the neutron stars, or that the majority of SNRs may not have left behind any stellar remnant. It is harder to explain the absence of radio pulsars in SNRs. According to some, this, too, could easily be understood in terms of the beaming of pulsar radiation, as well as the various selection effects discussed earlier which make it hard to detect pulsars. But this is not convincing either for the following reason. It is well understood that the synchrotron radiation from the Crab nebula is due to the relativistic wind from the central pulsar. If all young pulsars are spinning rapidly as the Crab pulsar then one would expect them to produce a surrounding synchrotron nebula which should be seen regardless of the viewing geometry. Thus all SNRs with active central pulsars should have a hybrid radio morphology: A shell emission arising from the relativistic particles accelerated in the shoked ejecta, plus a central emission due to the pulsar produced nebula. Unfortunately this is not observed – most SNRs have a shell morphology with hollow interiors (Fig. 39). Some years ago it was suggested that this can only be understood if the majority of pulsars are born with long initial periods (Srinivasan et al. 1984). Since the mechanical luminosity of the pulsar, which is ultimately converted into the electromagnetic luminosity of the nebula, is determined by its period and magnetic field strength, it follows that pulsars born with long initial periods will produce low luminosity synchrotron nebulae. This certainly suggested a reasonable way out of the dilemma of the poor association between pulsars and SNRs. Earlier while discussing pulsar current we saw that the current continues to build up till a period 0.5 seconds, implying a wide distribution of initial periods. This is consistent with the above conjecture. Another way out of this problem is if the pulsars had migrated out of the SNR! This possibility was never considered earlier because this would require the space velocity of

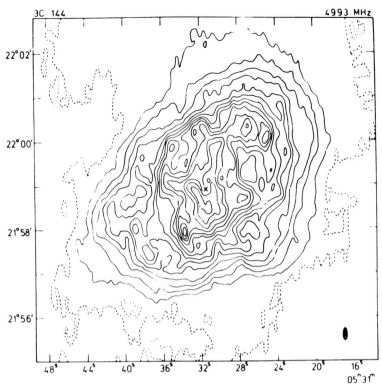

Fig. 38. The Crab Nebula at 2.7 GHz. The contours refer to total intensity. The + sign indicates the position of the pulsar which supplies both the particles and the magnetic field which produce this centrally concentrated synchrotron nebula (from Wilson 1972).

the pulsar to be large compared to the average expansion velocity of the SNR, and this appeared contrary to the observed velocity distribution of pulsars. The thinking has changed dramatically during the past couple of years or so. One has now measured space velocities of a few pulsars as large as $800 - 1000$ km s^{-1}. It is not yet clear if these represent the extreme tail of the distribution, or whether pulsars are intrinsically high velocity objects. If the latter is true then it offers another way to reconcile the poor association between pulsars and SNRs. After this digression let us return to our discussion of the progenitors of pulsars. Given a reasonable estimate for the pulsar birth rate one should now try to pin down the mass range within which stars end their lives as neutron stars. To do this one must compare the birth rate of pulsars to the death rate of stars.

Let $b(M)$ and $d(M)$ be the birth rate and death rate, respectively, of stars with mass greater than M (in units of number/pc^2/yr). For sufficiently massive stars ($M \gtrsim 3 M_\odot$) the two rates can be assumed to be equal. The birth rate of stars can be inferred from the counts and their theoretical lifetime on the main sequence. For illustration we give below the rate derived by Tinsley (1977)

$$\log M = -0.47 \log d(M) - 4.3$$

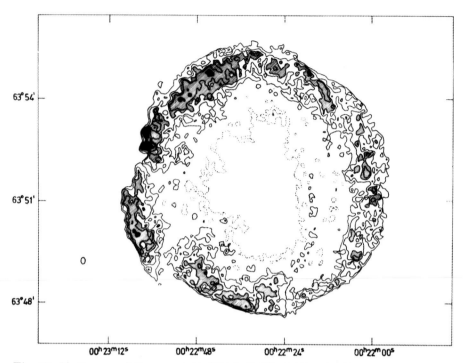

Fig. 39. Radio map of the remnant of Tycho's SN of AD 1572 (from Duin and Strom 1975, *the Westerbork Synthesis Radio Telescope is operated by the Foundation for Research in Astronomy with the financial support of the Netherlands Organisation for Scientific Research NWO*). The spherical shape and hollow interior are typical of young shell-type remnants.

The derivation of the birth rate of stars also involves considerable uncertainty. The spatial distribution of O and B stars is patchy. Also, star counts are done as a function of spectral types which have to be converted to mass etc. But there is one other complication in comparing the birth rate of pulsars with the death rate of stars. The birth rate of pulsars is a *galactic* rate, while star counts yield a *local* rate. Therefore one has to convert one of them suitably. One way to do this is to assume a radius for the Galaxy and convert the Galactic rate to a local rate *(per square parsec)*. This is straightforward to do if the distribution of pulsars was uniform not only in the azimuthal coordinate, but also as a function of the galacto-centric distance. But this would not be a good assumption. There is observational evidence that there are more pulsars, SNRs, HII regions, giant molecula clouds etc. around 5 kpc from the galactic centre than in the local neighbourhood. Therefore in going from a galactic rate to a local rate one must assume a larger *effective radius* for the Galaxy than its actual dimension.

Figure 40 shows the integrated death rate of stars using the initial mass function due to Miller and Scalo (Wheeler et al. 1980). One sees that a galactic birth rate of 1 in 80 years would require that all stars above $\sim 12-15 M_\odot$ must

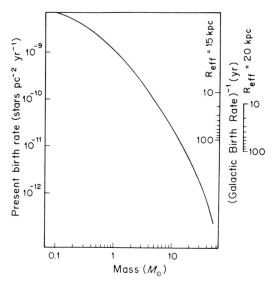

Fig. 40. Integrated stellar birth rate as a function of mass. This has been adapted from Wheeler et al. (1980). On the right hand side we have shown the galactic pulsar birth rate for two assumed effective radii for the Galaxy.

leave behind neutron stars. Pundits of stellar evolution would find this quite comfortable. One the other hand, if the birth rate is 1 in 40 years then all stars above $\sim 7-8 M_\odot$ must produce neutron stars. This may be inconsistent with the conventional wisdom. Some years ago Blaauw (1985) independently reached the conclusion that the massive O-B stars in associations cannot by themselves account for the local population of pulsars, and that lower mass *field stars* must make the major contribution to the pulsar birth rate. If this conclusion is correct then Blaauw argued that

> *Pulsars are, on a galactic scale, tracers of regions of past spiral structure rather than of active spiral structures.*

4.3 Do Pulsars Trace Spiral Arms?

Let us elaborate on this beautiful point. If the progenitors of the majority of pulsars were massive stars which delineate spiral arms then one would expect the majority of pulsars also to be located close to the leading edge of spiral arms. But if the majority of pulsars have progenitors less massive than, say, $10 M_\odot$, then since the lifetime of such stars may be as long as 50 – 60 Myr one would expect these stars to explode at substantial distances from the leading edge of the spiral arms. This is because of the relative motion between the spiral density wave and the matter in the Galaxy. If this was the case then one would not expect any strong correlation between the *present* distribution of pulsars and the present location of the spiral arms. One would, instead, find a correlation with past spiral structure.

Does the matter distribution lead the spiral pattern or lag behind it? This depends on the galacto-centric distance. Since the Galaxy is rotating differentially, and the spiral pattern rigidly, inside the *co-rotation radius* R_c the spiral pattern will lag behind the matter, and the converse will be true outside the co-rotation radius. From a detailed dynamical modelling of the gas distribution and their motions in our Galaxy it appears that the co-rotation radius is approximately 12 – 15 kpc (Burton, 1971). Since the majority of pulsars are inside the solar circle their circular velocities should be larger than that of the spiral pattern at the corresponding radius and consequently the present distribution of pulsars should be ahead of the present spiral pattern.

Recently an attempt was made to test this remarkable conjecture by Blaauw (Ramachandran and Deshpande, 1994), viz. one tried to look for a correlation between the present distribution of pulsars and the location of the spiral arms in the past. An essential ingredient in this analysis is the distance estimate to the known pulsars. Like in the birth rate analysis the recent electron density distribution due to Taylor and Cordes (1993) was used. The distribution of pulsars derived from this model and projected on to the plane of the Galaxy is shown in Fig. 41. The *dots* indicate the pulsars and the contours show the electron density distribution. As may be seen, the observed density of pulsars is systematically higher in the solar neighbourhood as might be expected from various selection effects. To be able to systematically correct for such a bias the sample of pulsars was restricted to those which, in principle, should have been detected by any one of the major eight surveys. The next step is to construct the true galactic distribution of pulsars from the observed distribution. Once again this involves the computation of scale factors. While deriving the current of pulsars we calculated the scale factor as a function of P and $\dot{P}$. In the present context one wants to ask a slightly different question, viz., given any location in the Galaxy where a pulsar has in fact been detected one wants to calculate the *probability* of detecting a pulsar at that location were it to have a different period or different magnetic field. This will enable one to calculate the scale factor as a function of position in the Galaxy. The procedure adopted was the following. In Fig. 42 we have shown the *true* number distribution of pulsars in the B-P plane. This distribution is derived from the observed B-P distribution by computing the scale factor in various "(B, P) bins" as described in Section 4.2. *The true distribution so derived is equivalent to a probability distribution for the occurrence of the observed periods and magnetic fields.* Thus, given a particular location in the Galaxy and given this probability distribution one can calculate the detection probability or the fraction of pulsars that are likely to be discovered at that particular location. From this one can derive the scale factors and the true galactic distribution of pulsars (Ramachandran and Deshpande, 1994). As mentioned earlier there is a relative azimuthal angular motion between the spiral pattern and the matter in the Galaxy. Assuming a flat rotation curve for the Galaxy the relative angular rotation is given by

$$\beta(t, R) = V_{rot} \left(\frac{1}{R} - \frac{1}{R_c} \right) t \qquad (1)$$

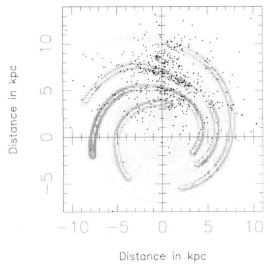

Fig. 41. The electron density distribution derived by Taylor and Cordes (1993) is shown as a contour diagram. The *dots* indicate the location of the pulsars estimated from this model and projected on to the plane of the Galaxy.

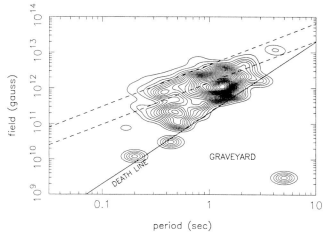

Fig. 42. The *true* number distribution of pulsars. The contours have been smoothed with a function shown in the bottom right hand corner (Deshpande et al. 1995).

where V_{rot} is taken to be 225 km/s as recommended by the IAU (see Kerr and Lynden-Bell 1986). Using this relation the pulsar distribution can be "rotated" as it were with respect to the spiral arms as delineated by the electron density distribution, and one can look for a correlation between the two at some past epoch. The expected correlation (Blaauw, 1985) can, in principle, be smeared due to three effects: (1) the spread in the birth places of the progenitors, (2) the motion of the progenitors between their birth and death, and (3) motion of the pulsars after their birth. The first two effects may not be significant, but the smearing due to the space velocities of the pulsars acquired at their birth could be important. If one finds a correlation despite this then one can turn it around to set limits on the space velocities of pulsars.

Before giving the results of the analysis it is worth recalling two assumptions that have been made in this analysis: (1) the "arm component" of the electron density in the Taylor-Cordes model adequately describes the mass distribution in the spiral structure. This is a reasonable assumption since the electron density model is based on the observations of giant HII regions. (2) In order to define the circular velocity of the spiral pattern a value of 14 kpc has been assumed for the co-rotation radius. As mentioned before, this value is consistent with the detailed modelling of neutral hydrogen in the Galaxy.

The conclusion arrived at by Ramachandran and Deshpande (1994) is shown in Fig. 43. Surprisingly there are two significant features, one corresponding to 60 Myr ago and the other to the present epoch. The correlation feature at the present epoch, viz., the correlation between the distribution of pulsars and the *present* spiral structure is most likely an artefact of the apparent clustering of pulsars in the "arm regions". This can happen, for example, if the spiral arm component of the electron density distribution is over-estimated relative to the smooth component. This can also happen if the location of the spiral arms in the model is in error. In the former case one would expect the effect to be more pronounced in the inner Galaxy. This is indeed the case. Pulsars which contribute the correlation at the present epoch are mostly between 5 to 7 kpc from the galactic centre. Let us now turn to the other feature in Fig. 43 namely

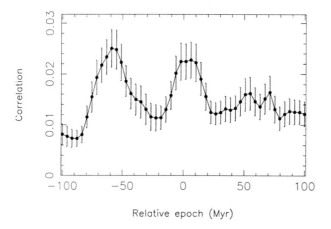

Fig. 43. A plot of the correlation between the mass distribution in the spiral arms and the present pulsar distribution as a function of different past epochs. A co-rotation radius of 14 kpc has been assumed in this analysis. The error bars indicate 1σ deviation on either side. There are two strong features in this plot. It is argued in the text that the picture corresponding to the present epoch is most likely an artefact. The feature at −60 Myr has a very high statistical significance of 99.95% (Ramachandran and Deshpande 1994).

the one which corresponds to a correlation between the pulsar distribution and the spiral arms some 60 Myr ago. The following tests were done to test the significance of this feature. The longitudes of the pulsars were scrambled and the whole analysis was repeated. Similarly, the distances to the pulsars were varied randomly by about 30% and the analysis was repeated. Based on many tens of thousands of simulations it was concluded that the correlation maximum at -60 Myr has a significance level of 99.95%. This should be compared with a significance level of about 93% for the other feature, which supports our earlier conjecture that it must be an artefact.

In our opinion the above analysis lends strong support to the prescient remarks of Blaauw that "pulsars are, on a galactic scale, tracers of regions of past spiral structure rather than of active spiral structure". Since the average lifetime of the pulsars in the sample is 10 ± 2 Myr, the above analysis leads one to the conclusion that the average lifetime of the progenitors of the pulsars must be about 50 Myr. This is roughly the lifetime of stars with masses $\sim 7 M_\odot$. This would suggest that a pulsar birth rate of 1 in 80 years derived earlier might be an under-estimate. The strong correlation found between the present distribution of pulsars and the location of spiral arms in the past argues against pulsars being high velocity objects. We feel that it may be hasty to conclude that the majority of pulsars are very high velocity objects.

To conclude this discussion of the progenitors of neutron stars, let us summarize.

1. The birthrate of pulsars is ~ 1 in 80 years.
2. The supernova rate is somewhat higher ~ 1 in 30 years.
3. The supernova remnant birthrate tell us more about the inhomogenities in the interstellar medium; they do not give a reliable estimate of the supernova rate itself.
4. There is a highly statistically significant correlation between the present distribution of pulsars and the location of the spiral arms some ~ 60 Myr ago.
5. This would suggest that the majority of the progenitors of neutron stars might have masses in the range $\sim 7 - 10 M_\odot$.

5 Binary Pulsars

So far our discussion has centred around solitary neutron stars. We now turn to a discussion of pulsars in binary systems. The first binary pulsar PSR 1913+16 was discovered by Hulse and Taylor in 1975. Although this was one among a population of nearly 150 pulsars, it was, in many ways, the most remarkable. The binary characteristics of this pulsar are summarized below:

Period (P)	:	59.02 ms
Period derivative $(\dot{P})$	:	8.63×10^{-18} s/s
Magnetic field (B)	:	2.3×10^{10} G
Distance (d)	:	7.13 kpc
Orbital period (P_{orb})	:	$7^h 45^m$
Eccentricity (e)	:	0.617

It was clear even early on that the companion must be a stellar mass object, and most likely another neutron star. Given these extraordinary characteristics, this system was soon to become the most important and also the most successful laboratory for verifying various predictions of General Relativity and other theories of gravity. The most spectacular result of timing this pulsar was, of course, the clear demonstration of the decrease in the orbital period due to the emission of gravitational radiation. It was therefore a matter of great pride and satisfaction to the physics and astronomical community that Taylor and Hulse were awarded the 1993 Nobel Prize for Physics. Since all the beautiful results that have emerged from a study of this pulsar have been based on timing the arrival time of the pulses, in the following we shall focus primarily on the characteristics of the pulsar itself.

The most striking feature of PSR 1913+16 is the anomalous combination of short rotational period (59 ms) and low magnetic field ($\sim 3 \times 10^{10}$ G). This is seen very clearly in Fig. 44. Most of the pulsars have magnetic fields in the range $\sim 10^{12} - 10^{13}$ G and periods $\sim$ second. The exceptions are the Crab and Vela pulsars which are spinning quite rapidly ($P = 33$ ms and 89 ms, respectively). But this is understood in terms of their relative youth. This would lead us to believe that the Hulse-Taylor pulsar with a period of 59 ms must be a very young pulsar. But then why is its magnetic field more than two orders of magnitude smaller than that of a typical pulsar? At this point one could, of course, say that this is an odd case of a neutron star born with a low field. It may well be so. But there is a second "coincidence" that should make us feel uncomfortable, viz., how come the one odd pulsar is also the only one in a binary system!

> "*Any evidence,* said Miss Marple to herself, *is always worth noticing. You can throw it away later if it is only a coincidence."* – Agatha Christie

So one is forced to ask whether there is any causal connection between the characteristics of this pulsar and the fact that it is in a binary?

Concerning the low field of this pulsar there is a more straightforward hypothesis one can advance viz. that this is a very old pulsar and its field has decayed substantially over the course of time. The idea that the magnetic fields of neutron stars may decay had been in the literature. Gunn and Ostriker (1969) had invoked ohmic decay of the magnetic field to reconcile the discrepancy between the dynamic age and the characteristic age of some pulsars. If the low field of PSR 1913+16 is to be understood in terms of decay, then the short rotation period posed a problem: one would expect an old pulsar to have a long period. Indeed, one would have expected it to cross the death line and cease to function as a pulsar. But the binary pulsar is spinning rapidly. Therefore, whether one thought of the pulsar as young (as the short period would suggest) or old (as the

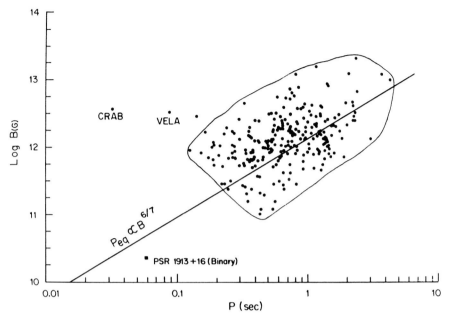

Fig. 44. Most pulsars lie within the 'island' shown schematically. The Hulse-Taylor binary pulsar with its anomalous combination of short period (59 ms) and low field ($\sim 3 \times 10^{10}$ G) stands out of the population. Also shown is the *spin up line* or *critical equilibrium period line*.

low field suggests) one is forced to conclude that the binary history may have something to do with all this.

Even prior to its discovery there were suggestions by Pringle and Rees (1972) and Davidson and Ostriker (1973) that accretion from the companion could spin up a neutron star. Faced with the absence of binary pulsars(!) Bisnovatyi-Kogan and Komberg (1974) had suggested that magnetic fields of neutron stars in binaries might be buried during accretion; such low field neutron stars would obviously not function as pulsars.

Soon after the Hulse-Taylor pulsar was discovered Bisnovatyi-Kogan and Komberg (1976) and Smarr and Blandford (1976) argued that it could have had such a history. Let us now proceed to unravel this mystery.

The first step to understand is how such a binary consisting of two neutron stars could have formed. This is schematically shown in Fig. 45.

1. Let us consider the evolution of a binary where masses of stars are $\sim 25 M_\odot$ and $\sim 10 M_\odot$, respectively. The primary star, being more massive, evolves more rapidly and about $\sim 5 \times 10^6$ years after the birth of the system it overflows its Roche Lobe and transfers its entire hydrogen rich envelope to its companion.

2. The system now consists of a helium star of $\sim 8.5 M_\odot$ and a main sequence mass of $\sim 26.5 M_\odot$.

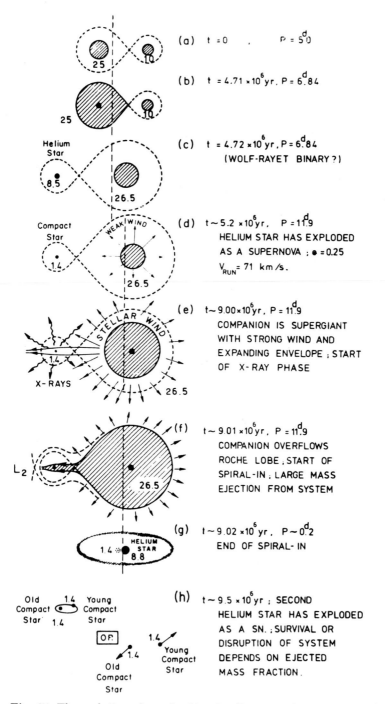

Fig. 45. The evolution of massive binaries (from van den Heuvel 1983).

3. Helium stars are short lived because they are over luminous. Detailed computations show that within $\sim 5 \times 10^5$ years after the mass exchange the helium star will terminate its helium burning and carbon burning, and the degenerate core will collapse to form a neutron star. The accompanying supernova explosion is not expected to disrupt the binary since the mass ejected will be less than half the total mass of the system.

4. Then there is a long "quiet phase": the companion will still be in the early stages of hydrogen burning, and the neutron star will not be seen either as a rotation powered radio pulsar or as an accreting X-ray binary.

5. During the brief interval of time $\sim 10^4$ years when the massive companion is an early supergiant the neutron star will be seen as an X-ray source. As soon as the supergiant begins to fill its Roche Lobe the X-ray source will be extinguished. Since the neutron star cannot accept all the transferred mass much of it will leave the system. The system will shrink very rapidly due to the very large loss of angular momentum. The spiralling-in of the neutron star will terminate when only the helium core of the companion is left. Soon the helium star will explode. This time the binary will most likely disrupt and one will have two run away neutron stars. In the rare case when the binary is not disrupted, the system will resemble PSR 1913+16. The fact that we don't see two pulsars should not be surprising given the narrow beams of pulsars. Also, both the neutron stars may still not be functioning as pulsars. (There are many outstanding Reviews on binary evolution by van den Heuvel. For example, see van den Heuvel 1983, 1985.)

Now one has to ask the following question: is the pulsar we do see the first-born or the second-born neutron star. If it is the remnant of the second supernova then it must have been born with the presently observed field, and we are back to the old dilemma. *Therefore one has to conclude that PSR 1913+16 must be the first-born neutron star in the system. Its magnetic field is low because it has presumably decayed in the time interval between the first and the second supernova explosions.* It now remains to understand why an old neutron star is spinning fast.

5.1 Spinning Up a Star!

To understand this one must examine the rotational history of a neutron star in a binary system. Before proceeding further let us introduce three important lengths in the problem.

1. *Light cylinder radius*: This is the radius at which an object in a synchronous orbit around the neutron star will be moving with the velocity of light.

2. *The Alfvén radius R_A*: This is the radius at which the pressure of the magnetic field of the star becomes comparable to the ram pressure of the accreting matter. This defines the magnetosphere of the star. Inside this radius the energy density

of the magnetic field dominates, and consequently the motion of the accreting matter whose infall has been arrested will now be dictated by the field. The Alfvén radius is determined by the surface magnetic field strength B and the accretion rate $\dot{M}_a$:

$$R_A = \left(\frac{B^4 R^{12}}{8GM\dot{M}_a^2}\right)^{1/7}$$

3. *The Corotation radius R_c*: This is the radius of the synchronous orbit $R_c = (GM/\Omega^2)^{1/3}$.

The rotational evolution of a neutron star in an accreting binary can be broadly described as follows (see Fig. 46).

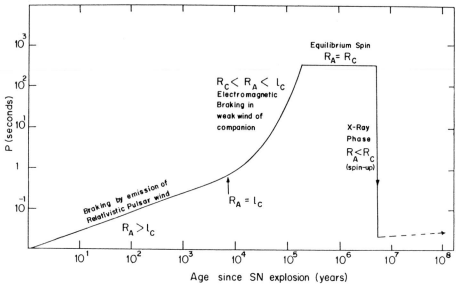

Fig. 46. The rotational history of a neutron star in a binary system. The various phases described in the text are labelled.

Phase I: When the first-born neutron star is young and spinning rapidly the pressure of the pulsar radiation — low frequency dipole radiation and relativistic wind — will be sufficient to prevent the stellar wind plasma from the companion from entering the light cylinder. In this phase the neutron star will slow down like any normal pulsar. As the pulsar slows down and its activity declines, the stellar wind will be able to penetrate the light cylinder or the induction zone of the pulsar. From this point onwards the plasma can and will influence the rotation of the pulsar.

Phase II: When the infall of the matter is arrested by the magnetic pressure at the Alfvén radius, the matter will be forced to corotate at this radius with the magnetic field which is rigidly anchored to the star. In this early phase when the stellar wind is weak and the pulsar is spinning fast, the Alfvén radius will be larger than the corotation radius. Consequently the matter cannot corotate with the star and will be expelled from the magnetosphere. But since the plasma had entered the induction zone it will exert a strong electromagnetic torque on the magnetosphere. This will rapidly brake the neutron star. Theoretical estimates suggest that even with a small mass loss rate by the companion $\sim 10^{-9} M_\odot/\text{yr}$ the characteristic braking timescale will be $\sim 10^4$ yr. Consequently pulsar activity will soon stop, and the neutron star will be braked to periods ~ 100 s within a million years or so.

Phase III: As the pulsar slows down, the corotation radius will eventually become larger than the Alfvén radius. Accretion of matter onto the neutron star is now possible and it will become an X-ray source. The fact that several known X-ray pulsars have periods $\sim$ hundreds of seconds bears testimony to this. However, since the matter from the companion brings with it specific angular momentum accretion will have the consequence of spinning up the neutron star. This will now bring in the corotation radius. This spin-up will terminate as soon as the corotation radius becomes, once again, smaller than the Alfvén radius. Accretion will once again be inhibited, and the neutron star will experience a braking torque. Consequently, the rotation period of the neutron star will tend to slightly oscillate around the *equilibrium value* P_{eq}, at which $R_A = R_C$.

$$P_{eq} = \left(\frac{1}{2^{17/14} \pi G^{5/17}} \right) M^{-5/7} R^{18/7} B^{6/7} \dot{M}_a^{-3/7}$$

It can be easily seen that this equilibrium period corresponds to the Keplerian period at the inner edge of the accretion disc (Davidson and Ostriker 1973; Illarianov and Sunyaev 1975; van den Heuvel 1977). The important thing to notice is that this period is uniquely determined by the magnetic field and the accretion rate.

Phase IV: *The critical equilibrium period*: When the companion evolves and the stellar wind increases, the magnetosphere will get compressed dramatically. The neutron star will become a strong X-ray source, and will be rapidly spun-up. Such a rapid spin-up with a characteristic timescale $\sim 10^3 - 10^4$ yr is actually observed in some X-ray pulsars. Soon the companion will explode, the debris will clear, and the now rapidly spinning first-born neutron star will begin its second lease of life as a radio pulsar. It would have been resurrected from the graveyard of pulsars, and will have an anomalous combination of short period and low field.

We now wish to ask the following questions: to what period will the neutron star be spun up at the end of the X-ray phase? Can an accreting neutron star be spun up to arbitrarily short periods? The answer is "no", because there is an upper limit to the rate at which a neutron star can accrete matter. This upper limit, $\dot{M}_{Edd}$, corresponds to an X-ray luminosity equal to the Eddington luminosity limit (Srinivasan and van den Heuvel 1982).

$$L_X = \eta \dot{M}_a c^2 \leq L_{Edd}$$

$$L_{Edd} = \frac{4\pi c G m_p M}{\sigma_T}$$

where L_X is the X-ray luminosity, η an efficiency factor.

This concept originally introduced in the context of the theory of stars represents the maximum luminosity of a star in radiative equilibrium. The critical accretion rate $\dot{M}_{Edd} \approx 7 \times 10^{17}$ g s^{-1}. Since the equilibrium period of a neutron star is determined by its magnetic field at the time of accretion and by the accretion rate, *the minimum period to which a neutron star can be spun up will be uniquely determined by the magnetic field:*

$$P_{eq} = 1.9 \text{ ms} \left(\frac{B}{10^9 \text{ G}}\right)^{6/7} \left(\frac{\dot{M}_a}{\dot{M}_{Edd}}\right)^{-3/7}$$

The above equation defines the spin-up line for a neutron star accreting at the Eddington rate (see Fig. 44). The proximity of the Hulse-Taylor pulsar to this spin-up line gives strong support to this scenario.

Let us summarize. We set out to understand the anomalous characteristics of the Hulse-Taylor pulsar, viz., the peculiar combination of low magnetic field and short period. If one rejected the possibility that the pulsar was born with this anomalous combination of P and B then one was forced to seek a causal relationship between these characteristics and its binary history. The scenario outlined above and which is now widely accepted may be summarized as follows (Fig. 47):

1. PSR 1913+16 is the first-born neutron star in the binary.
2. Its magnetic field is low because it has decayed in the time interval between its birth and the X-ray phase.
3. During the heavy mass transfer phase it was spun up to a short period.
4. On general grounds one expects the final period to be determined solely by its magnetic field strength during accretion.

A key assumption that led to a unique prediction for the rotation period after spin up was that the accretion rate was limited by the Eddington limit. It must be borne in mind that the concept of Eddington luminosity limit may be transcended in the case of an accretion neutron star. There are several reasons why the luminosity may exceed this limiting value.

(i) The effective opacity may be less than that provided by Thompson scattering. This may happen, for example, in the accretion column above the magnetic poles where the scattering cross-section is $< \sigma_{Th}$ for photons whose frequency is below the cyclotron frequency.
(ii) Strictly speaking the Eddington limit is applicable only for spherical accretion. In columnar accretion, unless the scattering cross section is highly anisotropic, radiation can escape from the sides of the column.
(iii) The Eddington limit can be violated in unsteady processes.

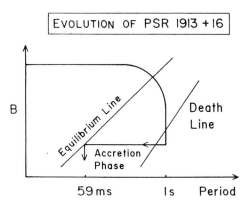

Fig. 47. PSR 1913+16 as a *recycled* pulsar. It is argued that this pulsar is the first-born pulsar in the binary whose field had decayed between its birth and the onset of mass accretion during which it was spun up to an *equilibrium period* determined by its magnetic field.

Having made this cautionary remarks, we would like to emphasize that from an observational point of view there is strong evidence for a luminosity cut-off in the distribution of X-ray sources, and this cut-off is $\sim 10^{38}$ erg s^{-1} consistent with the Eddington limit for a $1 M_\odot$ neutron star!

5.2 The Population of Binary Pulsars

So far we have discussed the first and the most famous binary pulsar. The population of binary pulsars has grown significantly during the last two decades. The present situation is shown in Fig. 48. There are about a dozen binary pulsars with massive companions. One of the recently discovered ones is very similar to the Hulse-Taylor system; the companion is most likely another neutron star. Two binary pulsars have optical companions. One of them is a Be star, and the other in the SMC is a B star. The Be star binary has a highly eccentric orbit with an orbital period ~ 1200 days. One sees from the figure that with the exception of this binary the rest are all located to the *right* of the spin-up line, lending strong support to the recycling scenario outlined here.

5.3 Are Many Pulsars Processed in Binaries?

So far we discussed pulsars which *are* in binary systems. Although there are many such systems now, they are still a minority compared to the population of solitary pulsars. But it is quite conceivable that a fraction, perhaps even a significant fraction, of solitary pulsars were born in binary systems. After all, as we saw earlier, most binaries will disrupt during the second supernova explosion – if there is one – releasing two pulsars. The second-born pulsar will be indistinguishable from pulsars born of solitary stars. But as we discussed in great detail, the first-born pulsar, the recycled one, will have distinct characteristics.

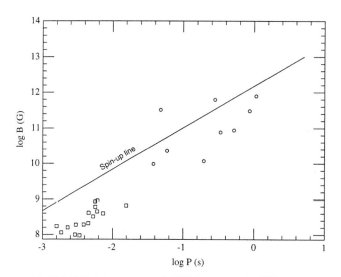

Fig. 48. The population of binary pulsars.

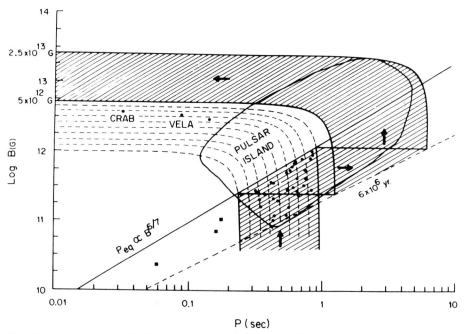

Fig. 49. It is conceivable that a fraction of solitary pulsars in the main population are such recycled pulsars. They are expected to be located to the *right* of the equilibrium period line. Two possible evolutionary scenarios are shown for the low field pulsars located in the bottom right hand corner of the island. The dashed lines show evolutionary tracks with rapid field decay, and the hatched track shows their evolution (traced backwards) in the recycling scenario (from Radhakrishnan and Srinivasan 1981).

They will have a magnetic field-period combination that would locate them perhaps close to, at any rate to the right of the spin-up line.

Does the population of solitary pulsars contain many recycled pulsars? Radhakrishnan and Srinivasan (1981) were the first to ask this question some fifteen years back. If the magnetic field of the neutron star had decayed significantly before the onset of mass transfer then the recycled pulsar would be spun up to relatively short periods and the recycled pulsar will stand out of the main island of pulsars. If on the other hand the magnetic field of the first-born neutron star had not decayed significantly for some reason or other, then after being spun up the pulsar would be deposited close to spin-up line, but this time *inside* the island of solitary pulsars (see Fig. 49). As we shall see later when we discuss the decay of the magnetic fields, this is what one would expect in massive binaries as opposed to intermediate mass or low mass binaries. In this scenario one would expect the population of solitary pulsars to the right of the spin-up line to be an admixture of pulsars born from single stars or second-born in binaries, and recycled pulsars. Is there any evidence for this? If so, can we estimate the fraction that are recycled? Let us consider the former question first. Earlier while discussing the birth rate of pulsars we had introduced the concept of the current of pulsars. To recall, the current of pulsars along the period axis is essentially the number of pulsars in a period bin of width ΔP multiplied by their "velocity" or period derivatives.

$$\bar{J}(P) = \frac{1}{\Delta P} \sum_i \dot{P}_i S\left(P_i, \dot{P}_i\right) \frac{1}{f}$$

If all pulsars were born with relatively short periods then the current will quickly build up and remain constant till deaths diminish the current. On the other hand, if there is an *injection* of recycled pulsars into the island then one expects to see a step in the current. There is some evidence for this in the data (Fig. 37). But how do we know this step in the current is not due to solitary pulsars born with long initial periods? If the step in the current is due to recycled pulsars making their appearance then one would expect them to have that particular combination of periods and fields as to locate them close to the spin-up line. There is some evidence for this as may be seen in Fig. 50. If one looks at the current in various magnetic field "windows" then one finds evidence for such an injection only in the range $\sim 10^{12} - 10^{12.5}$ G. This would locate this injection very close to the spin-up line, although corresponding to an accretion rate somewhat in excess of the standard Eddington value. Concerning the fraction of solitary pulsars that come from binaries, Deshpande et al. (1995) conclude that as much as 10% of solitary pulsars may have had a binary history.

5.4 Pulsar Velocities

We shall now turn to a very topical subject and which has an intimate connection with neutron stars in binaries, viz., the space velocities of pulsars. That pulsars should have substantial velocities was anticipated in a remarkable paper by Blaauw nearly ten years before their discovery. In an attempt to explain

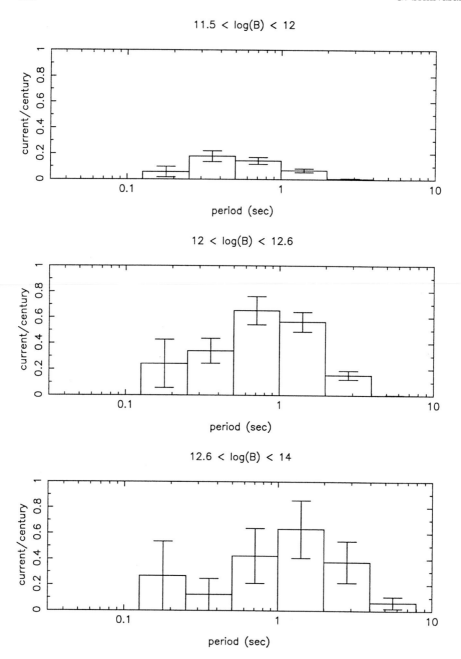

Fig. 50. The current distribution shown in Fig. 37 has been binned into three magnetic field ranges. As may be seen, it is only in the central panel which corresponds to $12 < \log B < 12.6$ that one sees a possible *step* in the current at a period of about 0.5 s.

the high velocities of the "run away" stars, Zwicky (1957) and Blaauw (1961) conjectured that their velocities may be attributed to the disruption of a binary system in which the more massive star exploded as a Type II supernova. Since a Type II supernova was expected to leave behind a "stellar remnant" Blaauw argued that the "remnant" should also have a velocity with respect to the centre of mass of the pre-supernova binary. Thus, when the runaway star also eventually explodes as a Type II supernova there will be two high velocity stellar remnants. As a test of this scenario for the origin of high velocities of run-away stars, Blaauw suggested that it might be worthwhile looking for such high velocity stellar remnants of supernovae!

Soon after puslars were discovered it became clear that in addition to angular velocity they also have substantial linear velocities. There were two pieces of evidence for this. First, as was shown by Trimble (1968), the optical pulsar in the Crab nebula has a measured transverse velocity 100 km s^{-1}. Secondly, Gunn and Ostriker (1970) noticed that the long period pulsars are significantly farther away from the galactic plane compared to the short period ones. Based on this they hypothesized that pulsars may be born with short periods and fairly close to the plane of the Galaxy; as they age and their periods lengthen they migrate from their birth place due to velocities presumably acquired at birth. From a statistical analysis of the small number of pulsars known at that time they concluded that pulsars must have space velocities $\gtrsim 100$ km s^{-1}.

The proper motions of nearly 100 pulsars have now been measured. Essentially, two techniques have been used: (i) Interferometric measurements, and (ii) Scintillation measurements. For 26 pulsars there are very accurate measurements of proper motion. The drifts of pulsar scintillation pattern and arrival times over several years provides an alternative but less-accurate method for determining proper motions. Based on this, Cordes has estimated the transverse speeds of about 70 pulsars. (For a recent discussion of pulsar velocities we refer to Lyne 1995.)

5.5 Migration from the Plane

About ten years ago, Lyne and his collaborators concluded that the proper motion data was consistent with all pulsars farther than 100 parsec from the plane moving *away* from the plane. This was consistent with the conjecture by Gunn and Ostriker mentioned above. This situation has changed considerably during the last ten years or so. Figure 51 shows the present position in galactic longitude and z-distances for all pulsars with latitude less than 30 degrees. (For these pulsars the determination of z-velocity is relatively uncontaminated by the unknown radial component of the velocity.) The location of the pulsar is indicated by a dot, the length of the tadpole indicates approximate distances the pulsar has travelled in the last million years. The plot clearly shows that a significant number of pulsars are moving *towards* the plane. The most straightforward interpretation of this would be that these pulsars are oscillating in the galactic potential, and we are seeing them coming towards the galactic plane after, perhaps, a quarter cycle of oscillation about the galactic plane. But there might be more to this.

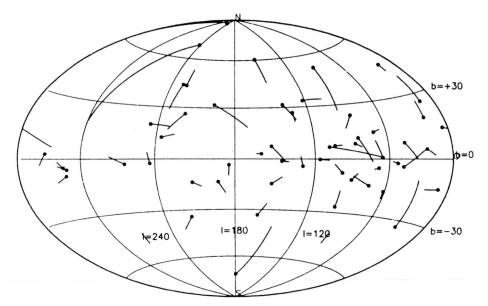

Fig. 51. A galactic distribution of pulsars and their velocity vectors. The *tails* represent the approximate paths travelled during the last million years (After Harrison, Lyne and Anderson 1993).

5.6 The Origin of Pulsar Velocities

We have already mentioned a pre-pulsar conjecture by Blaauw that the disruption of binaries provides a natural explanation for space velocities of pulsars. In 1970 Shklovskii suggested that if the mass ejection in the supernova explosion was not perfectly symmetric then one would expect fairly large velocities for pulsars. In fact, he estimated that pulsar velocities may be as large as 1000 km s^{-1}, and indeed there could be a galactic "halo" of pulsars. But this idea fell out of favour because the observed velocities of pulsars were only 100 km s^{-1}. This would require remarkable fine-tuning of the asymmetry in the supernova explosion in order to produce such low velocities. Or, to put it differently, the asymmetry should be less than $\sim 1\%$. In an alternative mechanism due to Harrison and Tademaru (1975), a neutron star is propelled like a rocket due to an asymmetry in the emission of the magnetic dipole radiation. For such an effect to be present, the magnetic dipole should be both displaced from the centre of the star, and be skewed with respect to the rotation axis. In this mechanism the velocity of the pulsar will be parallel to the rotation axis. Unfortunately, observations ruled out this mechanism, too. There is a simple way to test this hypothesis viz., by checking if the direction of motion and the rotation axis coincided (Morris et al. 1976). The projection of the rotation axis on the plane of the sky can be inferred from observations by appealing to the polar cap model which we discussed earlier. Quite simply, the direction of the linear polarization at the centre of the pulse coincides with the projection of the rotation axis on the plane of the sky. The proper motion is also a projection of the space velocity on the plane of the sky. If the velocity vector and the rotation axis coincide as required by the 'rocket' mechanism, then, so should their projections. Unfortunately, as Anderson and Lyne (1983) concluded, this is not the case.

Therefore, the most popular mechanism during the last decade for the origin of pulsar velocities was the binary origin. But the basic difficulty was that if binary disruption was the sole mechanism then all pulsars would have to be born in binaries. Earlier we concluded that only $\lesssim 10\%$ of pulsars might have come from binaries. Therefore one had to look for additional scenarios.

Around this time asymmetric supernova explosions were revived, but ironically, while trying to understand a *binary* pulsar system! From a detailed study of the possible progenitors of the binary companion of the Hulse-Taylor pulsar, Burrows and Woosley (1986) concluded that the sytem would have become "unbound" during the second supernova explosion unless the explosion was slightly asymmetric. Because the masses of the two stars in this system have now been measured very accurately, thanks to general relativistic effects, they were able to quantitatively conclude that there must have been an asymmetry at $\sim 3\%$ level. In other words, were it not for a "conspiracy" due to the asymmetry in the second supernova the binary would have been disrupted. Quite independently, from a detailed Monte-Carlo simulation Dewey and Cordes concluded that unless supernova "kicks" are invoked one will be overproducing the population of binary pulsars. But all this was circumstantial evidence. Now with measured velocities of ~ 1000 km s^{-1} such as for PSR 2224+65 one can safely conclude that supernovae do give "kicks" to neutron stars at birth. A velocity of ~ 1000 km s^{-1}, if confirmed, cannot possibly be explained in terms of an orbital velocity in a binary system. For it would imply an orbital period ~ 1 minute.

One may conclude this discussion of the origin of pulsar velocities in the following manner. For neutron stars which were born from solitary stars, their space velocities must be traced to an asymmetry in the explosive event in which they were born. For pulsars born in binary systems there are two possibilities. As we have already discussed, the sytem is unlikely to be disrupted in the first explosion. When the system is disrupted in the second explosion, the velocity of the first born recycled pulsar will essentially be its orbital velocity at the time of the disruption of the binary. As for the second born pulsar, its space velocity will be a vector sum of the orbital velocity of its progenitor and the supernova kick that it acquired at its birth. As to the nature of the distribution of velocities at birth, it is still not possible to give a definitive answer. The distribution of measured velocities certainly has a "tail" at high velocities, contrary to what the observation of a decade ago revealed (Lyne and Lorrimer 1994). But whether pulsars are truly high velocity objects, or whether the high velocity "tail" is an artefact of observational *selection effects* against low velocity objects, is too early to conclude.

6 Millisecond Pulsars

The discovery of millisecond pulsars was one of the most spectacular discoveries in astronomy in recent years. When the first of these, PSR 1937+21, with a period of 1.5 milliseconds was discovered in 1982 (Backer at al. 1982) it created a great deal of excitement. For one thing, a neutron star spinning 642 times a second – as newspapers across the world headlined it – was spinning very close to the theoretical limits. Secondly, with such a rapid spin rate this pulsar would be a powerful source of gravitational waves even with a modest equatorial elipticity.

The first thought that occurred to one was that it was a newly born pulsar. If its magnetic field was of the order of the canonical value of $\sim 10^{12}$ G then its age would be a mere 3 – 4 years. Since the estimated distance to this pulsar was only of the order of 1 kpc one wondered why one had not seen the supernova explosion in which this pulsar was born. In the IAU circular announcing the discovery of this pulsar the authors had suggested that the period derivative of this pulsar may be as large as $\sim 3 \times 10^{-14}$ s s^{-1}. This yielded a characteristic age $P/2\dot{P} \sim 700$ years, comparable to the age of the Crab pulsar. But somehow this did not seem right. For example, why had the supernova not been seen by the historical observers? At any rate, even if the supernova went unnoticed, there should be a pronounced SNR surrounding the pulsar. But there wasn't any. Of course, there is always the possibility that the neutron star was born in an event in which not much mass was ejected; in that case on would not expect a pronounced SNR. But there was a second paradox that was harder to circumvent. We know that the Crab pulsar produces and maintains the Crab Nebula – it supplies the relativistic particles and the magnetic field. The total luminosity of the Crab Nebula nicely matches the total energy loss rate of the pulsar. If the reported slowing down rate of the millisecond pulsar was taken seriously then there should have been a synchrotron nebula surrounding it which is many thousand times more luminous than the Crab Nebula. However there is no pronounced activity.

The Case of the Dog That Did Not Bark at Night Those of you who are fans of Sherlock Holmes will remember the story of a horse named Silver Blaze which was stolen from its stable. As always, the local inspector had called in Sherlock Holmes from London. After his investigations as Sherlock Holmes was about to return the inspector asked "Mr.Holmes, is there anything else that you would like to draw my attention to?". Holmes replied: "Yes, the curious incidence of the dog that night". The inspector exclaimed "the dog did not do anything that night!". And Holmes replied "*that* is the curious incidence!". Here was a pulsar spinning thirty times more rapidly than the Crab pulsar, and there was no nebulosity surrounding it! That was the vital clue. This implied that the measured value of $\dot{P}$ must be grossly in error, and that it had to be at least five orders of magnitude smaller than the estimates (Radhakrishnan and Srinivasan 1982). Converting the measured period and the upper limit to the period derivative to a magnetic field strength, they were led to conclude that the magnetic field of this pulsar could not be very much more than $\sim 10^8$ G, i.e., 4 to 5 orders of magnitude smaller than the magnetic field of the typical pulsar. This extremely anomalous combination of ultrashort period and extremely small magnetic field immediately reminded them of the Hulse-Taylor pulsar. Although this pulsar was solitary, they advanced the hypothesis that it must be a "recycled" pulsar, but somehow the binary must have got disrupted (Radhakrishnan and Srinivasan 1982).

Let us briefly recall what we discussed in the previous chapter about recycled pulsars (see Fig. 47). The basic idea is that the magnetic field of the first born pulsar in a binary system could in principle decay substantially during the time between its birth and the onset of mass transfer from the companion. During the X-ray phase this old neutron star whose period might be many hundreds of seconds at that stage will be spun up to a 'critical equilibrium period', which is nothing but the Keplerian period at the boundary between the magnetosphere and the accretion disc. Given enough time, an equilibrium state will be achieved in which the corotation speed at the magnetospheric boundary equals the Keplerian speed at that distance. As we saw earlier, this period is determined solely by the magnetic field strength of the star and is indicated as the spin-up line in the figure. If the conjecture that this solitary millisecond pulsar must be a recycled one was correct, then there was a clear prediction to be made about its magnetic field. As may be seen from Fig. 52 the predicted value was $\sim 5 \times 10^8$ G. This scenario was also independently advanced by others (Alpar et al. 1982, Fabian et al. 1983). A few months later when the period derivative of this pulsar was accurately measured the derived mangetic field turned out to be $\sim 4.7 \times 10^8$ G, in excellent agreement with the recycling scenario.

But there was the embarrassment that the pulsar has no companion. This led to many ingenious attempts to get rid of the companion after the pulsar had been spun up. These included (i) coalescence of two neutron stars, (ii) coalescence of a neutron star and a relatively massive white dwarf, (iii) coalescence with a low mass degenerate companion, and so on. Even as these attempts were in progress the second millisecond pulsar PSR 1953+29 with a period of 6.1 milliseconds was discovered. This one *was* in a binary, with a circular orbit and a low mass companion. As we shall see a little later these are exactly the characteristics one would expect in the recycling scenario. The population of millisecond pulsars has grown considerably during the last decade. There are now nearly two dozen millisecond pulsars in the galactic disc, and interestingly, 90% are in binaries. This should be contrasted with a binary fraction of the order of 5% for the entire population of radio pulsars. The fraction of binaries among the millisecond pulsars in globular clusters is somewhat smaller and is of the order of 50%, but this is what one would expect in globular clusters where "ionization" of binaries is a distinct possibility. In any case, even this 50% is a much larger fraction than for the general population of disc pulsars. With the exception of those in globular clusters the orbits of the binaries are exceedingly circular. In almost all the cases the companion is a very low mass star with the most probable mass being $\lesssim 0.3 M_\odot$. Although the present evidence for millisecond pulsars being spun up is reasonably compelling, the solitary nature of

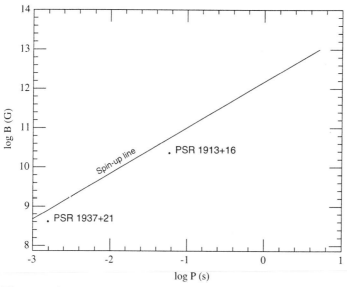

Fig. 52. The proximity of the millisecond pulsar 1937+21 to the spin-up line strongly supports the hypothesis that although at present solitary it might be a recycled pulsar like PSR 1913+16.

the original 1.5 millisecond pulsars prompted alternative scenarios in which they were born with ultrashort periods and very low magnetic fields. While this cannot be ruled out, the preponderance of such pulsars in binary systems remains to be explained. Some years ago, Pacini (1983) countered this objection by advancing an argument that one expects a low mass binary to survive the formation of a neutron star only if the accompanying supernova "fizzles out". One way to do this is to extract the gravitational energy released over a very long period of time, rather than suddenly as in the conventional core collapse scenario. This is achievable if the magnetic field is very low and if the core of the progenitor was spinning sufficiently close to the break-up limit. This will naturally result in an ultrashort period pulsar. While this elegant argument explains why the binary will not be disrupted, it still leaves unanswered the question as to why most low field pulsars occur in binary systems. Even if one were to come up with a plausible argument for this it should be remarked that the *evolved nature* of the companions of the millisecond pulsars strongly suggests that there must have been a prolonged phase of mass transfer in these systems so that even if these pulsars were born with low fields *their presently observed periods* are most likely due to spin up during the X-ray phase.

Returning to the solitary millisecond pulsar for a moment, one of the remarkable suggestions that had been made by Ruderman and his collaborators is that the millisecond pulsar may have vapourized the companion (Ruderman et al. 1989). The discovery of the 1.6 millisecond eclipsing pulsar PSR 1957+20 provides a spectacular example of this process (see Fig. 53). From the very long

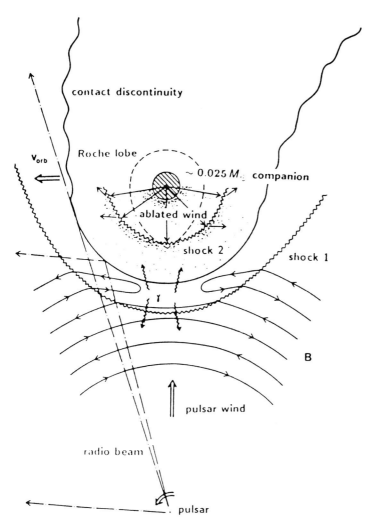

Fig. 53. The eclipsing millisecond pulsar PSR 1957+20 may be ablating its companion that spun it up and resurrected it from the grave yard (from Phinney et al. 1988).

duration of the eclipse one has been able to conclude that the companion star with mass $\sim 0.01 M_\odot$ is being ablated by the pulsar radiation. Timing data shows that the orbital period of this binary is decreasing on a very short time scale ~ 30 Myr due to heavy loss of mass and associated angular momentum from the system. This suggests that the companion will be blown away in a relatively short time scale $\sim 10^8$ yr leaving behind a solitary millisecond pulsar. Thus it may be that pulsars spun up to ultrashort periods in very close binaries will turn out to be 'ungrateful' and blow away their companions which resurrected them from the graveyard!

6.1 The Moral of Millisecond Pulsars

The Progenitors of Millisecond Pulsars

If millisecond pulsars are indeed recycled then one can draw several interesting conclusions concerning their progenitors. For example, to spin up a neutron star to periods as short as a few milliseconds it must accrete a very large amount of mass ($\sim 0.1 M_\odot$). Even at the Eddington accretion rate ($\dot{M}_{Edd} \sim 10^{-8} M_\odot$ yr^{-1}) it would take $\sim 10^7$ yr to do this. The duration of the mass transfer phase in massive binaries would be very much less than this. Therefore the progenitors of millisecond pulsars must be low mass X-ray binaries in which the companion star has a mass $\lesssim 1 M_\odot$. In these systems as mass transfer from the less massive companion proceeds, the orbit tends to expand thereby keeping the mass transfer stable over long periods. The slow evolution of the low mass donor star ensures a prolonged mass transfer phase. According to the standard model, LMXBs can be broadly divided into three classes according to their initial orbital period P_{b_0}.

Wide Systems ($P_{b_0} \lesssim 1 - 2$ Days) In these systems the secondary star will fill the Roche lobe when it is a subgiant or a giant. As the mass transfer proceeds the orbit will widen. However, due to the nuclear evolution the secondary will expand further and re-establish contact with the Roche lobe. The mass transfer will thus be sustained till the entire envelope is transferred and only the degenerate core of the donor star remains. At this point the binary will detach and mass transfer will cease. The end product of this evolution will be a white dwarf with mass $\sim 0.2 - 0.4 M_\odot$. The orbit will now be very wide and circular with an orbital period in the range of tens of days to many years. As Joss and Rappaport (1983) showed, the 6 millisecond pulsar with an orbital period of 117 days could be the end product of a low mass binary system with an initial orbital period of 12 days. The duration of the mass transfer phase will depend upon how wide the orbit is initially. This may be easily seen as follows. If the original binary is rather wide then the secondary will make contact with its Roche lobe later in its evolutionary phase. Since the rate of expansion of a star increases progressively on the giant branch one would expect the duration of the mass transfer phase to be correspondingly shorter in the case of a wide binary than in binaries with an initially tight orbit. For a $1 M_\odot$ companion star, the average mass transfer rate and the duration of the X-ray phase can be shown to be as follows:

$$<\dot{M}> \approx 8 \times 10^{-10} \left(\frac{P_{b_0}}{1d}\right) M_\odot \text{ yr}^{-1}$$

$$\tau_X \approx 10^9 \left(\frac{1d}{P_{b_0}}\right) \text{ yr}$$

Close Systems ($P_{b_0} \lesssim 12$ Hours) These systems will be brought into contact due to the loss of orbital angular momentum arising out of gravitational radiation and magnetic breaking – the secondary will still be on the main sequence. Magnetic breaking works as follows. The secondary star is expected to be in synchronous rotation with the orbit due to strong tidal interaction. Because of the stellar wind from the system there will be a strong spin-down torque on the secondary via the secondary's magnetic field. As the secondary spins down tidal effects will attempt to bring it back into synchronous rotation by spinning it up at the cost of the orbital angular momentum. After a while the binary will be brought into contact due to the shrinking of the orbit and mass transfer will begin. This will have two consequences: (i) due to mass loss the main sequence star will shrink, and (ii) the orbit will expand because the mass donor is the less massive star. But continued angular momentum loss will ensure that the system will remain intact.

The discovery of PSR 1953+29 with an orbital period of 117 days and PSR 1855+09 with an orbital period of 12.3 days have vindicated the hypothesis that low mass X-ray binaries must be the progenitors of millisecond pulsars.

The Population of Millisecond Pulsars

The first attempts to estimate the total number of millisecond pulsars in the Galaxy was made soon after the discovery in 1986 of the **third** millisecond pulsar PSR 1855+09 (Bhattacharya and Srinivasan 1986). The most extraordinary fact was that all the three millisecond pulsars were within ~ 25 parsec of the galactic plane. Even if one grants that for some unknown reason millisecond pulsars do not acquire significant velocities at birth, given that they are very old objects their proximity to the plane was very hard to understand. To recall, normal pulsars have a scale height ~ 350 parsec. It should be emphasized that the observed scale height of pulsars is much larger than the scale height of massive stars which is only ~ 60 parsec. According to present estimates the total number of LMXBs in the Galaxy must be ~ 125. The twenty or so for which reasonably accurate distances are known, suggest that LMXBs must have a scale height ~ 300 parsec, consistent with their belonging to the old disc population. To repeat once again, it is astonishing that all the three millisecond pulsars known at that time were within ~ 25 parsec of the Galactic plane. The only satisfactory resolution to this dilemma is that this must be due to *selection effects* i.e., it must be simply due to the fact that intensive searches for millisecond pulsars have so far been made only in a limited region of the sky. Adopting a scale height of ~ 300 parsec for millisecond pulsars (comparable to that of their progenitors, the LMXBs) a simple scaling suggests that there should be over thousand millisecond pulsars in the Galaxy. Simultaneously and independently van den Heuvel and his collaborators reached a similar conclusion based on the ratio of the number of millisecond pulsars to normal pulsars (van den Heuvel et al. 1986). A little later, from a deep survey of a 300 square degree area of the sky Stokes et al. (1986) concluded that binary millisecond pulsars might constitute nearly 10% of the total pulsar population, i.e., their total number could be as large as 10,000.

A much more detailed analysis of the statistics of millisecond pulsars was subsequently undertaken by Kulkarni and Narayan (1988). They took into account the various selection effects and concluded that the total number of active millisecond pulsars in the Galaxy could be as large as $\sim 10^5 - 10^6$.

Residual Fields of Neutron Stars

Even if the total number of millisecond pulsars in the Galaxy is only a few thousand there is an apparent problem. As mentioned before, the population of LMXBs is unlikely to be more than ~ 100. The resolution of this puzzle is straightforward. If the basic premise viz., that millisecond pulsars evolve from LMXBs is correct then the *lifetime* of these pulsars must be 20 – 30 times longer than the X-ray phase of their progenitors. Assuming that the lifetime of LMXBs is $\sim 10^8$ year one is forced to conclude that millisecond pulsars must live for more than $\sim 10^9$ years. At least in one case there is direct observational evidence to support this. The optical white dwarf companion of PSR 1855+09 has been detected and the standard cooling theory for white dwarf suggests that it must be $\gtrsim 10^9$ years old, implying that the pulsar must be even older than that. This long lifetime of millisecond pulsars is to be contrasted with the lifetime of $\sim 10^7$ years for normal pulsars.

This longevity of millisecond pulsars has a profound implication for the secular decay of the magnetic fields of neutron stars (Fig. 54). While discussing the recycling scenario we assumed that somehow the magnetic field of the neutron star decays from its original value $\sim 10^{12}$ G to $\sim 10^8$ G before the onset of mass transfer. Only then can one explain the observed millisecond period as a critical equilibrium period. But there is no reason to assume that the magnetic field will stop decaying after this. Therefore one would expect the recycled pulsar to have a finite lifetime corresponding to the time it would need to cross the death line for a second time. But this is inconsistent with the lifetime of millisecond pulsars in excess of $\sim 10^9$ years inferred both from observations, as well as from theoretical arguments. This can only be reconciled if magnetic fields of millisecond pulsars do not decay after they have been spun up. In other words, the observed magnetic fields in millisecond pulsars are in some sense *asymptotic fields*. The fact that the previously known two millisecond pulsars at that time had fields $\sim 5 \times 10^8$ G led Bhattacharya and Srinivasan (1986) to suggest that most millisecond pulsars will have magnetic fields of this order. In particular they predicted that the third millisecond pulsar must also have a similar magnetic field. Subsequent observations confirmed this. The present situation is that the majority of millisecond pulsars in fact have magnetic fields of this order (see Fig. 48). The above discussion is summarized in Fig. 55. The apparent discrepancy between the number of LMXBs in the Galaxy and the expected large number of millisecond pulsars in the Galaxy cannot be understood unless millisecond pulsars live essentially for ever. This in turn requires that their magnetic fields do not decay significantly after their spin up and that their presently observed field represents some sort of an asymptotic value. The recycling scenario also predicts that the majority of ultrafast pulsars will be in binaries with circular orbits and low mass evolved stars as companions. Finally, since the period derivative of a pulsar is inversely proportional to its period one would expect a piling up of pulsars towards longer periods. Given their incredibly small slowing down rate and the age of the Galaxy one would expect this piling up to occur in a period range $\sim 6 - 10$ milliseconds. Again, this is consistent with the present data.

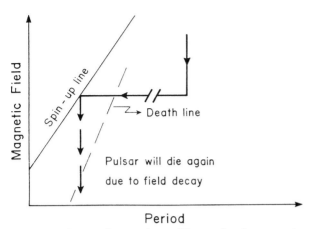

Fig. 54. The conclusion that millisecond pulsars must essentially live for ever implies that their magnetic fields cannot decay indefinitely.

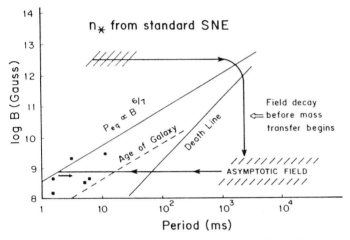

Fig. 55. Spinning up a star to millisecond period. In this scenario advanced by Bhattacharya and Srinivasan (1986), the magnetic field of a neutron star with a low mass companion decays till it reaches an *asymptotic value*. It is then spun up to an equilibrium period $\sim$ a few milliseconds. They predicted that $B \sim 5 \times 10^8$ G may be the value of the asymptotic field. As may be seen in Fig. 48, the overwhelming majority of millisecond pulsars have fields of this order.

Constraints on the Equation of State

According to the recycling theory the period of a spun up pulsar will be determined by its magnetic field strength. Nevertheless, the extraordinary fact that two of the fastest millisecond pulsars PSR 1937+21 and PSR 1957+20 have almost identical periods of 1.6 ms (within 3% of each other) is very difficult to understand just in terms of the standard scenario. In order for two spun up pulsars to have identical spin periods the accretion rate, magnetic field, and the mass and radius would have to be fine-tuned to a remarkable degree of precision. For example, the derived magnetic fields of these two pulsars differ by a factor of about 3! This led Friedman et al. (1988) to suggest that 1.5 milliseconds may be a limiting period determined by gravitational effects. It is now generally accepted that because of instability to non-axisymmetric perturbations driven by gravitational radiation, the limiting period of a neutron star is likely to be slightly longer than predicted by Newtonian gravity. This limiting period, like its Newtonian counterpart, depends upon the equation of state of the neutron star matter. Thus, if the period of the two fastest pulsars is interpreted as the limiting period determined by general relativistic effects then one can derive some constraints on the equation of state of the interior of these pulsars. The conclusion that Friedman et al. arrived at is that the equation of state of neutron star matter must be quite stiff.

6.2 A New Population of Gamma-Ray Sources?

Pulsars have a pride of place in gamma-ray astronomy. This is as it should be. For, after all, the Crab and Vela pulsars are the best studied gamma-ray sources. There have been suggestions in the literature that some of the unidentified COS B gamma-ray sources may indeed be pulsars. There have also been estimates of the pulsar contribution to the gamma-ray luminosity of the Galaxy. In such discussions attention is usually focussed on "young pulsars", of which the Crab and Vela are the prototypes.

In the preceding discussion we saw compelling reasons for expecting a very large population of millisecond pulsars in the Galaxy. Although their magnetic fields are four orders of magnitude smaller than that of the Crab and Vela pulsars, the fastest among them have an energy loss rate comparable to that of the Vela pulsar. This is, of course, by virtue of their ultrarapid spin rate. Therefore on very general grounds one would expect gamma-ray emission from millisecond pulsars also.

What can we expect from them? The details, of course, will depend upon the assumed model for the pulsar magnetosphere and the gamma-ray emission mechanism. In one of the models, high energy photons are produced by primary particles accelerated in the polar cap region, and the radiation mechanism is curvature radiation as the particles stream along the open field lines. In high field pulsars, radiation above $\sim$ GeV will be strongly attenuated because of pair production in the magnetosphere. As Usov (1983) has pointed out, since millisecond pulsars have rather low magnetic fields, primary gamma-rays with energies up to $\sim 10^{11}$ eV will be able to escape the magnetosphere and the expected flux above $\sim 10^{11}$ eV may be comparable to that from the Crab pulsar in the same energy range. In an alternative model due to Ruderman and his collaborators (Cheng, Ho and Ruderman 1986), gamma-rays are produced by a combination of synchrotron radiation and inverse Compton effects in the outer regions of the magnetosphere. Their model predicts a gamma-ray spectrum and luminosity similar to that from the Vela pulsar, i.e., an intensity spectrum roughly proportional to 1/E in the MeV to GeV range, with about 1% of the total spin down power emitted as gamma-rays. It may well be that in the coming decade several millisecond pulsars may be identified as discrete gamma-ray sources.

But it is equally interesting to estimate the collective contribution of millisecond pulsars to the gamma-ray background of the Galaxy (Bhattacharya and Srinivasan 1991). Their conclusion is summarized in Fig. 56. The 'dots' show the intensity observed by the SAS-2 satellite at different galactic latitudes averaged over $\pm 60°$ in galactic longitude around $l = 0°$. The expected contribution of the population of millisecond pulsars numbering $\sim 5 \times 10^5$ is also shown for comparison. The 'dash' and 'solid' lines correspond to the two models mentioned earlier viz., the curvature radiation model, and the synchrotron- cum-inverse Compton model, respectively. There are two points worth making. First, the contribution to the gamma- ray background may exceed that of the standard pulsar population by a large factor. Second, the effective "width" of the latitude profile of the contribution due to millisecond pulsars is expected to be much larger than that due to normal pulsars. This is in spite of the fact that normal pulsars have a scale height ~ 400 parsec, larger than the assumed scale height of ~ 300 parsec for the millisecond pulsars. The reason for this is that although normal pulsars have a large scale height, the younger short period ones among them are expected to be rather close to the galactic plane. Since these short period pulsars make the most dominant contribution to gamma-ray production, the latitude profile is rather narrow. In the case of the millisecond pulsars, even the "young ones" have a large scale height ~ 300 parsec since this is the scale height of their progenitors viz., the LMXBs. Further, even if a millisecond pulsar is born near the plane and migrates many hundreds of parsecs away from the plane it would still not have slowed down appreciably because of its very low magnetic field. Therefore, the latitude profile of the gamma-ray intensity from the millisecond pulsar should be at least as wide as that of the progenitors. Further observations by the gamma-ray observatory and further gamma-ray missions will clarify this situation.

7 Magnetic Field Evolution

We shall now turn to a discussion of the evolution of the magnetic fields of neutron stars. The magnetic field plays an important role in the life history of a neutron star. The vast majority of neutron stars discovered so far are radio pulsars, and very strong magnetic fields are crucial for pulsar activity. The rate at which they slow down and how long they function as pulsars are also determined by the magnetic field. Again, it is the strong magnetic fields which collimate the accreting matter in binary systems resulting in the X-ray intensity modulated at the spin period. We saw how the magnetic field of an accreting neutron star determines the period to which it will be spun up. It would be a fair description to say that the magnetic field of a neutron star plays the same role in their evolution as the mass of a star does in stellar evolution. Indeed, the magnetic field-period diagram is the analog of the H-R diagram in stellar physics.

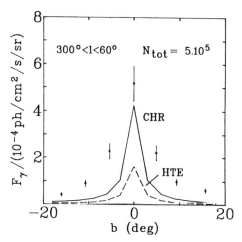

Fig. 56. The estimated contribution of millisecond pulsars to the diffuse gamma-ray background is compared with the intensities observed by SAS-2 at different galactic latitudes (Fichtel, Simpson and Thomas 1978). Results for two different magnetospheric models (Cheng, Ho and Ruderman 1986; Harding, Tademaru and Esposito 1978) are shown. Both the data and the theoretical estimates are averaged over $\pm 60°$ in galactic latitude around $l = 0°$ (from Bhattacharya and Srinivasan 1991).

7.1 The Nature of the Field

What is the nature of this magnetic field? From the sweep of the position angle of the linearly polarized radiation from pulsars one has been able to deduce that the magnetic field has a predominantly dipole structure at distances ranging from a few stellar radii to a few hundred radii. Regarding the origin of this field one is less certain. There are essentially two scenarios: (i) The observed fields of neutron stars are fossil fields, and (ii) the magnetic field is in fact built up after the formation of the neutron star either through a battery mechanism or a dynamo process.

Fossil Field

Even before pulsars were discovered, Woltjer (1964) and Ginzburg and Kizhnits (1964) had independently conjectured that since magnetic flux is likely to be conserved during the collapse of the cores of the progenitor stars the stellar remnants could have field strength as large as $\sim 10^{14}$ G. Although the observed fields of neutron stars are consistent with this hypothesis the scenario cannot be conclusively proved since one knows very little about the field strengths in the cores of massive stars.

Field Generation After Birth

Woodward (1978) was the first to explore the possibility that the magnetic fields of neutron stars may be generated after their birth. The next major step was taken by Blandford and his collaborators (Blandford et al. 1983) following on an earlier idea due to Urpin and Yakovlev (1980). The basis of this idea is a thermoelectric battery mechanism. Although these represent interesting ideas we shall not pursue them in this lecture because in our opinion there is no compelling observational evidence to suggest that the magnetic fields of neutron stars are generated after their birth. Indeed, it appears that one may have to contrive to do so.

Therefore, in what follows we shall assume that the observed fields of neutron stars are in fact fossil fields amplified during the core collapse. If this is indeed the case, then one expects the magnetic flux to permeate the entire star more or less uniformly. As we saw, the interior of a neutron star consists of a fluid core which is essentially a neutron fluid with an admixture of $\sim 1\%$ protons and electrons. In the classical picture the exterior magnetic field should be understood in terms of current loops in the interior. For example, the predominantly poloidal magnetic field must be due to toroidal current loops in the interior.

As we saw earlier, such a classical picture is unlikely to be obtained in the interior of a neutron star. We argued that the neutrons will be in a superfluid state and the protons in a superconducting state. At first sight it would appear that the onset of superconductivity of the protons might have a drastic effect on the magnetic property of the star. For example, when the interior cools and superconductivity sets in, one might expect the magnetic field to be expelled due to the Meissner effect. This is very unlikely to happen in the case of a neutron star. The only way the magnetic flux can be expelled is if the electrons are expelled from the core, but this cannot happen because it would result in enormous electric fields being produced. Flux expulsion can in principle occur due to ohmic diffusion, but experts who have looked at this problem very carefully concluded long ago that the timescale for such expulsion is $> 10^8$ yr. Therefore, unlike in a laboratory situation, proton superconductivity will set in retaining the magnetic field although this will cost some energy.

The question one now has to ask is 'what is the nature of the magnetic field trapped in a superconductor?'. If it is a so-called Type I superconductor like lead or mercury then it will accommodate the field by forming sandwiches with alternate layers of superconducting material which will be field-free and normal material containing the field. But it turns out that the proton matter will not be a Type I superconductor but a Type II superconductor. Alloys which become superconducting are of this type, including the recently discovered high temperature superconductors. Whether a material becomes Type I or Type II depends upon the relative magnitude of two important lengths in the problem: The coherence length ξ, and the London penetration depth λ. The coherence length is a measure of distance within which the properties of a superconductor cannot change appreciably. The London penetration depth is the screening length, i.e., the strength of an externally applied magnetic field would fall to $1/e$ of its surface

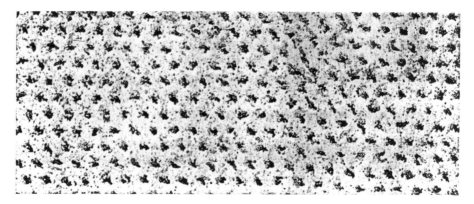

Fig. 57. Abrikosov flux lattice in a superconductor. Each fluxoid has a quantum of flux $\phi = hc/2e = 2 \times 10^{-7}$ G cm^2.

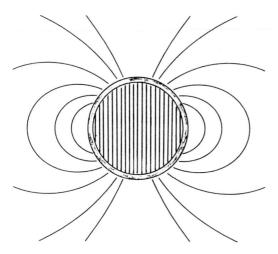

Fig. 58. A schematic drawing which shows the nature of the core field in a neutron star. Since the proton fluid is believed to be a Type II superconductor, the flux is confined to an array of quantized fluxoids. In a neutron star with a surface field of 10^{12} G the number of fluxoids will be $\sim 10^{31}$.

value at a depth λ inside the superconductor. If $\xi > \sqrt{2}\lambda$ the superconductor will be in a Type I state, and will exhibit perfect diamagnetism. On the other hand, if $\xi < \sqrt{2}\lambda$ the superconductor will exhibit a Type II behaviour. Such a superconductor will allow the flux to penetrate, not homogeneously but confined to quantized flux tubes with each flux out or fluxoids carrying a quantum of magnetic flux $\phi = hc/2e = 2 \times 10^{-7}$ G cm^2. These quantized flux tubes or Abrikosov vortices are the analog of the Onsagar-Feynman vortices in a rotating superfluid which we introduced earlier. The number of these fluxoids threading the superconductor will obviously be determined by the flux. Figure 57 shows

Neutron Stars

such a lattice of Abrikosov vortices. According to all the estimates made so far the coherence length of the proton superconductor in the core of a neutron star will be much less than the London penetration depth, and hence the proton will exhibit a Type II behaviour. To summarize, in this quantum picture the fossil field is entrained in an array of fluxoids that thread the superconducting interior as shown in Fig. 58.

7.2 Field Decay

Early Evidence for Field Decay from Radio Pulsars

Very soon after pulsars were discovered, and Gunn and Ostriker (1969) pointed out that their magnetic fields may be decaying rather rapidly over a timescale $\sim 10^6$ yr. What they had noticed in the statistics of the dozen or so pulsars known at that time was that in some cases the dynamic age of the pulsar (inferred from their distance from the galactic plane and an assumed average velocity of pulsars) was much smaller than the characteristic age $P/2\dot{P}$. They argued that this discrepancy can easily be understood if the magnetic fields decay rather rapidly. This idea became quickly accepted and became an essential ingredient in most evolutionary scenarios for radio pulsars. The case for rapid field decay seemed to be strengthened when the measurements of the proper motion of pulsars became available. For those pulsars for which a proper motion is measured one can compute their kinetic age t_k defined as the ratio of the z-distances from the plane and the z-component of their velocity. As data accumulated it seemed to confirm the earlier conjecture by Ostriker and Gunn. Unfortunately, what was not critically looked at was that there is an inherent error in the estimate of the kinetic ages of pulsars. This arises due to the fact that the radial component of the velocities are unknown, and also the fact that there is considerable uncertainty in the birth place of the pulsars in question. In view of this, no strong conclusion should have been drawn from the apparent discrepancy between the kinetic ages and the spin-down ages. Nevertheless, field decay in radio pulsars became an accepted paradigm.

The Present Situation

A recent and more careful analysis of the enlarged population of pulsars now numbering more than 500 suggests exactly the opposite, viz., that there is no evidence for field decay in the population of radio pulsars. This result has been arrived at independently from two entirely different approaches to pulsars statistics. One of these methods is based upon an analysis of pulsar *current* (Srinivasan 1989), and the other on population synthesis (Bhattacharya et al. 1992). We shall not elaborate on these methods here, but merely state that the conclusion of these studies is that if the magnetic fields of neutron stars decay, then they probably do over a timescale ~ 100 Myr. The bottom line here is that since this is much longer than the characteristic lifetime of pulsars one cannot infer anything about the decay of magnetic fields over much longer timescale. One can merely say that there is no evidence for field decay during their lifetime as pulsars. This statement, however, does not preclude the possibility of field decay over much longer timescales. To test this one has to look at evidence of field decay among 'dead' pulsars. During the last couple of years the existence of strong magnetic fields $> 10^{12}$ G has been inferred in some of the γ-ray burst sources through the discovery of cyclotron absorption lines in their X-ray spectrum. If these particular gamma-ray bursters are very old solitary neutron stars then one may be forced to conclude that the magnetic fields of solitary neutron stars do not decay significantly even over timescales comparable to the age of the Galaxy. But this issue remains controversial at present.

Theoretical Expectations

One can now have a sigh of relief, because rapid decay of the magnetic fields as originally suggested by Ostriker and Gunn is not consistent with one's theoretical expectations. In a beautiful paper published soon after the discovery of pulsars, Baym, Pethick and Pines (1969) questioned the conclusion arrived at by Ostriker and Gunn. Their arguments can be summarized as follows:

1. If the electrons and protons in the interior are not superconducting, the timescale for the magnetic field decay by ohmic dissipation will be governed by the electrical conductivity of the interior. The main mechanism contributing to the resistivity is the scattering of electrons by protons; electrons are scattered little by the neutrons. Calculations show that the conductivity $\sigma \sim 1.5 \times 10^{29}$ s^{-1}. This very large conductivity is the consequence of the extreme degeneracy of protons. For a neutron star of 10 km radius the flux decay time τ_d is $\tau_d \sim 4 \pi \sigma R^2 / c^2 \sim 10^{13}$ yr in the interior of the star, much larger than the age of the Universe. In estimating the electrical conductivity in the interior, Ostriker and Gunn had ignored the fact that the protons were extremely degenerate. This resulted in a much smaller conductivity, and therefore a much smaller decay timescale.

2. As we have already discussed, if the proton fluid was in a superconducting state then the flux will be confined to an array of quantized fluxoids. Flux can be expelled from such a superconducting region only by the migration of the quantized fluxoids to the boundary of the superconducting region. The characteristic time for a spontaneous expulsion of flux from a Type II superconductor is again very much larger than the age of the Universe.

Thus, the careful analysis of Baym, Pethick and Pines showed that regardless of whether the interior of a star is normal or superconducting, there are strong theoretical reasons for rejecting the idea of spontaneous field decay in neutron stars. As we saw, this is entirely consistent with the recent studies of the population of solitary radio pulsars.

The Low Fields of Binary Pulsars

Nevertheless, there is strong circumstantial evidence that the magnetic fields of some neutron stars do decay. Invariably, these happen to be neutron stars with a binary history. Let us recall some phenomenological evidence that lead us to conclude this. For example, from the fact that accreting neutron stars in massive X-ray binaries pulsate one concludes that they must be endowed with strong magnetic fields: Accretion is collimated by the magnetic field towards the polar cap resulting in the modulation of the X-ray intensity with a period equal to the rotation period of the neutron star. Neutron stars in massive binaries are obviously relatively young objects and like the majority of pulsars they are endowed with strong magnetic fields. In contrast, neutron stars in low mass X-ray binaries do not pulsate, and this is commonly understood in terms of their magnetic fields being several orders of magnitude smaller than their counterparts with massive companions. Since neutron stars in LMXBs are very old objects we conclude that very old neutron stars have low magnetic fields. The most straightforward conclusion one can draw from this is that magnetic fields of neutron stars seem to decay with time *provided they are in a binary system*!

As we saw, this conclusion is enormously strengthened by the observed characteristics of binary radio pulsars such as the Hulse-Taylor pulsar and also the millisecond pulsars. As already discussed, the alternative point of view viz. that the low field pulsars were born with low fields is not a very attractive proposition for two reasons: (1) While about $\sim 3\%$ of all known pulsars in the galactic disc are in binary systems, pulsars in binaries account for $\sim 70\%$ of known disc pulsars with $B < 10^{10.5}$ G. This huge preponderance of neutron stars with a binary history among the low field pulsars points to an intimate connection between field decay and their binary history. If these pulsars were merely born with such low fields it is hard to see why hardly any solitary pulsar has very low field strength. (2) Almost all the low field pulsars in the galactic disc are located exclusively to the *right* of the spin-up line. This is precisely what one would expect if they were spun up. If pulsars were born with such low magnetic field strength then one would expect to see a fraction of them to the left of the spin-up line. Therefore one is forced to the conclusion that whereas magnetic fields of solitary pulsars do not decay significantly, the fields of neutron stars born and processed in binaries do decay, and very significantly.

Asymptotic Fields

1. While discussing the millisecond pulsars we concluded that they must live essentially for ever. Although during their first life they might have functioned only for a few million years, in their reincarnation they live for ever. This can only be understood if the field decay does not continue after they have been resurrected from the graveyard. From the fact that the overwhelming majority of millisecond pulsars have magnetic fields $\sim 10^8 - 10^9$ G, one may conclude that this value of the magnetic field represents some sort of an asymptotic field of neutron stars which have descended from low mass binaries.

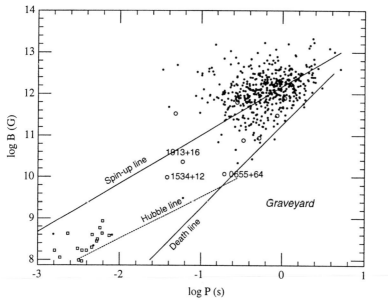

Fig. 59. A B-P plot showing the present population of pulsars. The dots are solitary pulsars, circles represent binary pulsars with massive companions (neutron stars or white dwarfs), and squares are millisecond pulsars with low mass white dwarf companions. All binary pulsars to the right of the spin-up line are believed to be *recycled* pulsars with their present fields being an asymptotic limit. Binary pulsars with massive companions have residual fields which are an order of magnitude more than the residual fields of millisecond pulsars.

2. This remarkable property that the decay of the magnetic field stops after the neutron star has been spun up is not peculiar to the millisecond pulsars. This seems to be the case even for neutron stars from massive binary systems (see Fig. 59). Consider the binary pulsar PSR 0655+64 with a massive white dwarf companion. The estimated cooling age of the rather low surface brightness of the white dwarf is in excess of $\sim 10^9$ yr. Since one expects the pulsar to be the first-born member of the binary, it must be even older than the white dwarf. Nevertheless this pulsar is endowed with a magnetic field $\sim 10^{10}$ G. This suggests that the presently observed field must be the residual or asymptotic field (this was first pointed out by Kulkarni 1986). But there is an interesting point to be noticed here. The population of binary pulsars may be divided into two subgroups – those with massive companions (either another neutron star or a massive white dwarf), or very low mass white dwarf companions with mass $\sim 0.2 M_\odot$. Binary pulsars with massive companions have residual fields which are an order of magnitude more than the residual fields of millisecond pulsars. To summarize the phenomenological evidence about field decay:

(i) There is no evidence for field decay in the population of solitary pulsars. Indeed, there is some circumstantial evidence that the magnetic fields of solitary neutron stars do not decay even over timescales very much larger than their lifetimes as radio pulsars.

(ii) Very low field pulsars ($B < 10^{10.5}$ G) occur almost exclusively in binaries. This suggests a causal connection between the field decay mechanism and the evolution of a neutron star in a binary.

(iii) Whereas the magnetic fields of neutron stars in binaries may decay, this decay does not seem to continue indefinitely. There appears to be an asymptotic or residual value of the field beyond which no further decay occurs.

(iv) For neutron stars with very low mass companions this residual field is $\sim 10^8$ G. For binary neutron stars with massive companions this field is $\sim 10^{10}$ G.

7.3 Mechanism of Field Decay

Over the years there have been several suggestions regarding the underlying mechanism for field decay. These range from MHD instabilities in the core to convective instability in the core. In all these scenarios the fluid core of the neutron star is assumed to be "normal" rather than in a superfluid state. Even if one does not insist that the arguments for superfluidity in the core is very compelling, these mechanisms do not distinguish between solitary neutron stars and those that have had a binary history. In view of our discussion of the phenomenology of the problem we shall not discuss these any further (for a more detailed discussion see the review by Bhattacharya and Srinivasan 1995).

In the superconducting picture, the first concrete suggestion for a mechanism to expel the fluxoids from the interior was due to Muslimov and Tsygan (1985). They pointed out that in a self-gravitating body such as a neutron star, because of the pressure gradient the quantized fluxoids will migrate towards the crust due to buoyancy. Their outward drift will, however, be retarded by a viscous force arising due to the magnetic scattering of the degenerate electrons with the vortices. Thus, over a period of time, the magnetic flux frozen in the superconducting core will be deposited in the subcrustal region. The field can either decay there, or be transported further upwards due to magnetic buoyancy and decay in the outer crust. While this is an ingenious mechanism, it suffers the same defect that we outlined before, viz. that it does not explain why field decay occurs only in binary systems.

Field Decay Due to Mass Accretion

As mentioned earlier, Bisnovatyi-Kogan and Komberg (1974) had anticipated that pulsars in binaries may have low fields. The particular suggestion they made is that the accreted matter may "screen" or "bury" the field. This scenario has been revisited by several others in recent times. From evolutionary models Taam and van den Heuvel (1986) have argued that there is a correlation between the amount of matter accreted from the companion and the magnitude of the field decay. Romani (1990) has recently argued that accretion may actually be able to destroy the crustal magnetic field. For example, heating of the crust due to accretion may hasten ohmic decay in the crust due to a reduction in the electrical conductivity. The compression of the current carrying layers may also result in the reduction of the dipole moment. And there are other variations on the theme. But we wish to point out two basic difficulties with this approach. First, the neutron star in the X-ray binary 4U1626-67 may be a counter-example; this may have accreted a considerable amount of mass and yet has a magnetic field $\sim 10^{12}$ G. The more basic difficulty is that the various mechanisms just mentioned may destroy or modify the crustal field, but accretion is unlikely to have any significant influence on the magnetic field that resides in the core of the neutron star.

Field Decay Due to Spin-Down

We shall now outline a novel suggestion for the expulsion of the flux from the superconducting interior of a neutron star and its subsequent decay in the crust. In this mechanism, the flux expulsion is related to the slowing down of the neutron star (Srinivasan et al. 1990).

To recall, the magnetic flux in the core is confined to quantized fluxoids. The core field can only decrease if the fluxoids migrate outwards. But as we have already remarked, the timescale for the spontaneous expulsion of the flux frozen into a superconductor is much longer than the age of the Universe. So one is left speculating about a mechanism that will transport the fluxoids to the crust. What we need to do is essentially "fish out" the fluxoids. The question is 'what does one use for the fishing rods?'

Before proceeding further, let us again recall the nature of the interior fluids (Fig. 60). As we have discussed, the neutron superfluid is perforated by a number of quantized vortices − the Onsager-Feynman vortices − parallel to the rotation axis of the star. The proton superconductor is threaded by the quantized fluxoids parallel to the magnetic axis. Thus, there are two families of vortices in the core. Although both kinds of vortices have been separately invoked in the past to understand some aspects of the rotational dynamics or the decay of the magnetic field, the *interaction* between these two families of vortices and the consequence of such an interaction were not explored. The new physics that we would like to introduce is that the two families of vortices − the vortices and the fluxoids − may be strongly interpinned. It is well known that the vortices in a superfluid and the fluxoids in a Type II superconductor pin to point defects, dislocations etc. The reason for this may be understood as follows. The cores of these vortices are normal, i.e., superfluidity is destroyed inside a quantized vortex, and similarly superconductivity inside the fluxoid. Therefore it costs a certain amount of energy to create a vortex of unit length. This depends on the local energy gap. Pinning is a consequence of the fact that it might be energetically more favourable to create a vortex in a certain local environment than in another environment. This is an extremely well studied subject because pinning of fluxoids in a superconductor has implications for potential technological applications of superconductors. The point that had been overlooked before is that

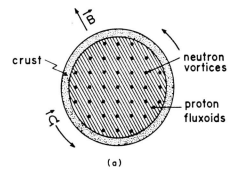

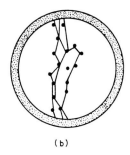

Fig. 60. (a) A view of the equitorial plane of a rotating, magnetized neutron star showing an idealized geometry of the arrangement of the Onsager-Feynman vortices in the neutron superfluid (parallel to the rotation axis) and the Abrikosov quantized fluxoids (parallel to the magnetic axis) in the proton superconductor. For the purpose of illustration the magnetic axis has been assumed to be perpendicular to the rotation axis. For the magnetic field trapped in the core to decay, the fluxoids must migrate towards the crust where the field can, in principle, decay due to ohmic effects. (b) A more realistic state of affairs in the presence of strong interpinning between the fluxoids and the vortices.

for the very same reason the *fluxoids and vortices will pin to each other*. Pinning a line defect to another line defect is a lot easier than pinning to a point defect. There are several independent physical reasons for expecting such interpinning between these two families of vortices, and estimates suggests that the pinning energy may be as large as $\sim$ 0.1 to 1 MeV. We shall not elaborate on this any further, but refer to arguments given in Srinivasan et al. (1990) and Sauls (1989) Instead we shall now argue that if such interpinning is important, then one can construct a simple and elegant model for the field decay based upon flux expulsion from the interior as the neutron star slows down. It is now well established that solid crust and the core superfluid are strongly coupled. Because of this, as the crust slows down (in response to the energy and angular momentum radiated away) the superfluid will slow down, too. But in order to do this it must destroy the required number of vortices. This will happen by a radially outward migration of the vortices and their annihilation at the interface between the inner crust and the superfluid core. If the interpinning between vortices and the fluxoids is sufficiently strong, as various estimates suggest, then as the vortices move out radially the fluxoids will also be *dragged* towards the crust (as shown in Fig. 61), and the field will eventually decay there due to ohmic dissipation. As we shall presently see this very simple idea of using the Onsagar-Feynman vortices as fishing rods to pull out the fluxoids from the interior of the crust is able to explain at least at a qualitative level all the observational features that one set out to explain.

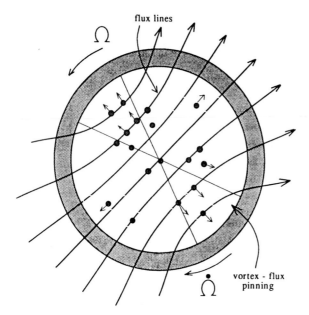

Fig. 61. Spin-down induced flux expulsion from the superconducting interior. As the neutron star slows down the vortices in the core superfluid will move radially outwards. In the process they will transport the interpinned fluxoids towards the crust (from Sauls 1989).

Solitary Pulsars

In principle, the mechanism outlined above will operate even in solitary neutron stars. But it would be rather ineffective. Since the energy of the electromagnetic radiation and the relativistic wind emitted by a pulsar comes at the expense of its rotational energy, the equation governing the slowing down of the neutron star is the following:

$$-\frac{d}{dt}\left(\frac{1}{2}I\Omega^2\right) = \frac{1}{6c^3}B^2 R^6 \Omega^4.$$

If the evolution of the magnetic field is directly related to the secular slowing down as assumed in the present discussion, then one can substitute $B(t) \propto \Omega(t)$ in the above formula. Integrating this one finds that at late times

$$B(t) \propto (t/\tau)^{-1/4}.$$

Here τ is the characteristic spindown timescale of the pulsar at late times (typically $\sim 10^6 - 10^7$ yr). This is a very slow rate of decay indeed. Thus, the magnetic fields of solitary neutron stars will not decay significantly even over a timescale comparable to the age of the Galaxy. The underlying reason is that an isolated neutron star will not spin down significantly for a substantial fraction of the flux to be dragged out from its interior.

Neutron Stars in Binaries

This difficulty will not be encountered if the neutron star is in an interacting binary system. While discussing the rotational history of the first-born neutron star in a binary system we argued that during the stellar wind phase of the companion the neutron star will be slowed down to very long periods (Fig. 46). The main evidence for this comes from the twenty or so binary X-ray pulsars whose rotation periods range from a few seconds to 835 s. During this spin down phase a substantial fraction of the field trapped in the interior can be expelled from the core. Whether or not the expelled field will decay in the crust while the star is spinning down will depend on whether the spindown timescale is short or long compared to the ohmic dissipation timescale in the crust. Both circumstances can be obtained, but in either case given sufficient time the field will decay in the crust.

The Residual Field

When the companion fills its Roche Lobe and heavy mass transfer sets in, the neutron star will be spun up. Consequently, the degree to which the flux is expelled from the interior will be determined by the *maximum period* to which the neutron star was spun down prior to spin up. In other words, the magnetic field in the core would have attained its asymptotic value at the end of the spin down phase.

Regarding why there is no further decay of the magnetic field, there are two straightforward explanations. First, during the rapid spin up of the crust the superfluid core will respond by creating new vortices which will now move radially *inward* thus pushing the residual cold field even deeper into the superconducting core. Second, after the spindown phase is over, the slowing down rate of the recycled pulsar will be determined by its electromagnetic luminosity appropriate to the reduced value of the field. Hence, further field decay due to slowing down will be negligible (we saw that it is negligible even for high field pulsars).

The last puzzle that needs to be explained is why the residual fields of millisecond pulsars are much smaller than that of pulsars with massive companions?. This can be easily understood if neutron stars with low mass companions are spun down to much longer periods than their counterparts with massive companions (as schematically shown in Fig. 62). Intuitively this seems very plausible because the lifetime of the main sequence phase of a low mass companion is very large compared to that of a massive companion. This is borne out by detailed computations of Jahan Miri and Bhattacharya (1994). Even more significantly, they find that regardless of the orbital parameters of a low mass binary it is very difficult to spin down a neutron star to periods longer than ~ 1000 s. This limiting period naturally explains why all the millisecond pulsars have roughly the same value of the asymptotic field.

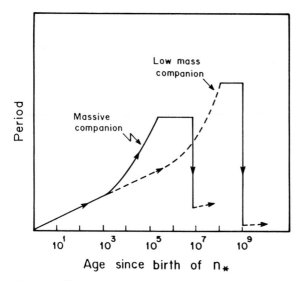

Fig. 62. The rotational history of neutron stars with massive and low mass companions. It is likely that a neutron star with a companion of mass $\sim 1 M_\odot$ will be spun down to much longer periods before mass accretion begins. Since the residual fields of recycled pulsars are determined by the *maximum* period to which they are spun down, this offers a simple way to understand why millisecond pulsars have fields much smaller than binary pulsars with massive companions.

8 Glitches

In this section we shall discuss some aspects of the rotational dynamics of neutron stars.

As we have already discussed, the periods of pulsars lengthen with time, and the period derivative is proportional to the square of the magnetic field and inversely proportional to the period. Soon after pulsars were identified with neutron stars it occurred to the early observers to look for such pulsars at the centres of known supernova remnants. This led to the discovery of the Crab pulsar, and one in the Vela SNR. In late February 1969 the Vela pulsar exhibited a remarkable 'glitch' or a sudden spinning up (Radhakrishnan and Manchester 1969). The jump in its rotation period was enormous, and the jump in the slow-down rate was even larger:

$$\frac{\Delta\Omega}{\Omega} \sim 10^{-6} \qquad \frac{\Delta\dot{\Omega}}{\dot{\Omega}} \sim 10^{-2}$$

As the pulsar was monitored after this event it became clear that the recovery timescale was quite large and of the order of months (a schematic description of a glitch is shown in Fig. 63). Very soon after that, Baym and his collaborators

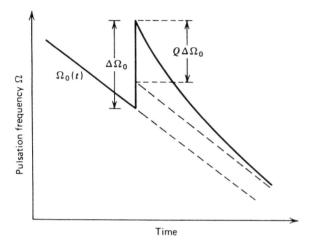

Fig. 63. A schematic representation of a *glitch* (from Manchester and Taylor 1977).

advanced the hypothesis that the glitch might be due to a *starquake* resulting in a very slight reduction in the moment of inertia of the crust with a consequent spinning up of the crust (Baym et al. 1969). Quite remarkably, a change in the radius of the star by 0.5 cm can account for the enormous spin up! The basic idea was rather simple. As the star slows down its oblateness must decrease. Because of the rigidity of the crust the oblateness does not follow the slowing down. Consequently, enormous elastic stresses are built up in the crust. When the stresses built up exceed a critical value, the crust cracks and readjusts to the required value of the oblateness, which results in a small reduction in the moment of inertia. These authors were also very quick to point out that the long recovery timescale after the glitch can best be understood in terms of the presence of a superfluid interior. It turns out that every kind of coupling one can think of between the core and the crust leads one to expect a relaxation timescale of the order of minutes or less if the interior was not a superfluid. Very soon after the glitch in the Vela pulsar, the Crab pulsar glitched, although it was less dramatic. This glitch could also be understood in terms of the simple two-component model of Baym et al. This model consists of a normal component (the crust) with a moment of inertia I_c, weakly coupled to the superfluid interior with a moment of inertia I_s, with $I_c + I_s =$ total moment of inertia I. The crust (*plus everything that is strongly coupled to it*) rotates at the *observed* angular velocity $\Omega_c(t)$, and the superfluid has an angular velocity Ω_s which is larger than Ω_c. The coupling time between the normal and superfluid component is described by a parameter τ. As already mentioned, in this model the sudden spinning up of the crust is due to a starquake resulting in a reduction of the moment of inertia of the crust. It is clear that the long post-glitch relaxation of the crust is then due to the *gradual recoupling* of the superfluid component, i.e., the interior torque due to the superfluid becoming slowly effective. Thus this model had the right ingredients to explain glitches. Unfortunately this simple and elegant explanation for the glitch had to be abandoned when the Vela pulsar glitched again two years later in August 1971! The reason for abandoning this picture was rather simple. The amount of energy released in the Vela glitch is approximately $\sim 10^{-6}$ of its stored rotational energy. It is simply not possible to store and release this much energy in the solid crust over two years. The entire elastic energy of the crust is only $\sim 10^{-6}$ of the rotational energy. Or to put it differently, starquakes induced by oblateness changing cannot recur in two years. Therefore one had to look elsewhere for the origin of the glitch.

In a seminal paper written in 1975 Anderson and Itoh suggested a novel mechanism in which the neutron superfluid – the superfluid in the crust, to be precise – was itself responsible for the glitch! To explain this, as well as the modern scenario for the post-glitch response, we have to digress and develop the necessary background about the dynamics of the superfluid.

We saw earlier that a superfluid cannot be in rigid body rotation. As pointed out by Onsager and Feynman, it *mimics* rigid body rotation by forming an array of vortices each with a quantum of circulation or vorticity $h/2m$. The rotation rate of the superfluid, Ω_s, is related to the number n of vortices per unit area by $n\kappa = 2\Omega$, where κ is the quantum of vorticity. In order for the superfluid to spin down, vortices must move radially outwards from the rotation axis, and reduce the density of vortices. The equation of motion governing the slowing down of the superfluid is

$$\dot{\Omega}_s = -n\,\kappa\,\frac{V_r}{r}$$

where V_r is the radial velocity of the vortices at a distance r from the rotation axis. Microscopically such a motion of the vortices is driven by the spindown of the crust.

Pinning of Vortices

What we have outlined so far applies equally to the superfluid in the core as well as the crustal superfluid. But there is an essential difference in the way the two superfluids respond to the slowing down of the crust. You will recall that the crustal superfluid coexists with a lattice – a lattice of exotic very neutron-rich nuclei. It is believed that the neutrons *in the nuclei* will also be in a superfluid state. Therefore the vortices in the crustal superfluid are in a very inhomogeneous medium. In such a situation they may prefer to *pin* to the lattice nuclei or may thread the space between the nuclei. As already remarked, the vortices have normal cores. Thus it costs energy to produce a vortex. Typically this is Δ^2/E_F per particle in the core, where Δ is the superfluid energy gap and E_F is the Fermi energy of the neutrons. Both these quantities depend on the local density of neutrons, which is different within the neutron-rich nuclei and the interstitial region. A careful consideration shows that it may be energetically favourable for the vortices to thread the nuclei in the inner crust – or be "pinned to them". Such a pinning leads to *pinning forces*. Let f_p be the pinning force per unit length of the vortex. As remarked earlier, such pinning of vortices in superfluids or fluxoids in superconductors to defects or dislocations is a very common phenomenon.

Pinning of vortices has an interesting consequence. As the crust slows down, vortex pinning will lead to a lag between the superfluid rotation rate Ω_s, and that of the crust Ω_c. Pinning prevents the radial outward motion of the crust necessary for the superfluid to slow down. Since the pinned vortices are rotating at Ω_c, there is a relative velocity between the vortex line and the superfluid. This results in a force on the vortex line known as the Magnus force:

$$\mathbf{f}_M = n\boldsymbol{\kappa} \times (\mathbf{V}_l - \mathbf{V}_s)$$

where κ is the vorticity vector along the line and V_l, and V_s are the velocities of the line and the superfluid respectively. As may be seen, a lag between the vortex and the superfluid results in a force on the vortex along the radial direction. As the lag increases – due to the immobility of the vortex lines – the Magnus force increases till eventually it can overcome the pinning force on the vortex. At this stage the vortex will unpin and move radially, resulting in a slowing down of the superfluid.

Actually the situation is a little more interesting and subtle than this. Let us depict the pinning potential as a random potential. At finite temperatures the vortex lines can "climb over" the potential barriers and "creep". If the crust is not slowing down, and the superfluid and the crust are rotating at the same rate i.e., there is no lag, then there is no *bias* for a radial outward motion of the vortices. In such a situation a creep induced average radial velocity $<V_r> = 0$.

But the situation is different when there is a lag. Because of the Magnus force the random potential is biased towards a radial outward thermally activated motion of the vortices. Such a *vortex creep* theory gives the following characteristic expression for V_r is:

$$<V_r> = 2V_0 e^{-\frac{E_p}{kT}} \sinh\left(\frac{E_p}{kT} \cdot \frac{\omega}{\omega_{cr}}\right)$$

where V_0 is a microscopic velocity ($\sim 10^7$ cm s^{-1}) and ω_{cr} is the critical lag defined by equating the pinning force to the Magnus force

$$\omega_{cr} = \frac{E_p}{n\kappa r b \xi}.$$

Let us briefly summarize the above discussion. (1) The superfluid slows down by a radial outward motion of the vortices. (2) This motion is driven by the lag between the angular velocities of the superfluid and the crust. (3) Such a lag arises due to the pinning of the vortices to the nuclei in the inner crust. (4) The motion of the vortices may be described as thermally activated creep.

Once we have such physical picture of the slowing down of the crustal superfluid in response to the slowing down of the crust, one can write down the equation of motion for the crustal rotation rate:

$$I_c \dot{\Omega}_c = N_{ext} - I_s \dot{\Omega}_s$$

Here N_{ext} is the external torque due to radiation reaction, and I_p is the moment of inertia of the pinned crustal superfluid. The meaning of this equation is clear. The crust spins down because the energy it radiates comes at the expense of the stored rotational energy. This is partly compensated by the internal torque due to the superfluid spinning down – the radial outward motion of the vortices results in a small amount of angular momentum being transferred from the superfluid to the crust. Before proceeding further a couple of comments are called for concerning the *core superfluid* which accounts for much of the moment of inertia of the star.

1. The vortices in the core superfluid do not pin to the crustal nuclei at the interface between the fluid core and the crust.

2. During the past decade it has been well established that the core superfluid is strongly coupled to the crust through the electrons. This arises due to the fact that the vortices in the core neutron superfluid are strongly magnetized, and electron scattering off magnetized vortices is very efficient. Theoretical estimates suggest a coupling timescale ~ 400 seconds. Therefore "crust" in all the previous discussion really means "crust plus core". Thus I_c represents the moment of inertia of essentially the entire star since the contribution of the crustal superfluid to the moment of inertia is expected to be $\sim 10^{-2}$ (see Sauls 1989 for a discussion as well as references to original work).

The Glitch

After much digression let us return to the origin of the glitch or the sudden spinning up of the star. As already mentioned, a starquake followed by a slow recoupling of the loosely coupled crustal superfluid would provide a simple minded and yet adequate description of the observations. But since starquake won't work (at least for Vela) one was left speculating on other possibilities.

The suggestion by Anderson and Itoh was simply the following: the glitch arises due to a sudden and catastrophic unpinning of the vortices in a certain part of the star, their radial motion outwards, and either repinning or destruction in the boundary between the crust and the core. The reason why such a catastrophic unpinning would occur is not clear. But for lack of anything better let us accept this. Let I_p be the net moment of inertia of the superfluid which has undergone such a sudden decrease in the angular velocity $\delta\Omega_s$. From the conservation of angular momentum it follows that

$$I_c \Delta\Omega_c(\text{glitch}) = I_s \delta\Omega_s$$

where $\Delta\Omega_c$ (glitch) is the change in the angular velocity of the "crust" due to the glitch. Such a sudden spinning down of the superfluid, and the consequent spinning up of the rest of the star perturbs the vortex creep rate from the steady state value. For one thing, the lag between the superfluid and the rest of the star decreases at the glitch – the superfluid has suddenly spun down and the crust has spun up. It may be recalled that the radial velocity of a vortex depends critically on the lag. This will drastically affect the slowing down rate of the superfluid, which in turn will affect the spin down rate of the rest of the star following the glitch. This may be seen as follows. Let us recall the equation of motion governing the star:

$$I_c \dot{\Omega}_c = N_{ext} - I_s \dot{\Omega}_s$$

Clearly, in the steady state $\dot{\Omega}_s = \dot{\Omega}_c$, i.e., although $\Omega_c > \Omega_c$ the slow down *rates* must be the same. Otherwise the lag will increase; clearly this is not the steady state solution. In the steady state,

$$(I_c + I_s) \dot{\Omega} = N_{ext}$$

Therefore, given an external torque (which is determined by the magnetic field and the angular velocity of the star) the spin down rate is determined by the *total moment of inertia* of the star $I = I_c + I_s$. After the glitch, since vortex creep is less effective $\dot{\Omega}_s \ll \dot{\Omega}_c$, the spin down rate of the "crust" is determined by the moment of inertia of the "crust" alone, viz., I_c. Consequently the spin down rate of the "crust" is larger than the steady state value. It relaxes back to the steady state value as the lag increases back to the steady state value, and the crustal superfluid fully recouples again.

Let us now summarize the "standard model" for the glitch event. In the steady state the crustal superfluid spins somewhat faster than the rest of the star – there is a lag. But the slow down rate of the superfluid is the same as that of the rest of the star, for otherwise the lag will increase. The superfluid slows down through a radially outward "creep" of the vortices. The glitch or sudden spinning up of the star occurs when a large fraction of the vortices collectively unpin and move outwards. The long relaxation back to the steady state is understood in terms of the gradual recoupling of the crustal superfluid to the rest of the star.

This is the most popular explanation for the glitch and post-glitch relaxation. (For a more detailed account of the standard model we refer to Pines and Alpar 1992, and the references cited therein.) Although this scenario is probable, even possible, there are many difficulties. Quite apart from the details, the most glaring is the lack of physical understanding of why there is a catastrophic unpinning of the vortices, and how repinning occurs.

9 Plate Tectonics

Finally we turn to a qualitative account of some very recent and fascinating developments. These may be grouped together and categorized as 'Plate Tectonics'. This is a very active area of research in geophysics, and it deals with the motion of plates on the Earth. Continental drift, earthquakes and many other important phenomena are in some way related to the fact that the crust of the Earth consists of several "plates" and they move and slip with respect to each other. Such a phenomenon may be occuring in the crust of the neutron star also, and this may have several dramatic astrophysical consequences. The ideas described here are largely due to Ruderman (1991a, b, c).

The first question that comes to one's mind is "why is the crust of a neutron star not a monolithe, why should it break into plates, and why should they move?". There are two underlying physical mechanisms that may be responsible for this, and both of them are intimately related to the superfluidity in the interior of the star. Imagine a piece of concrete with reinforcing rods going through it (Fig. 64). Suppose one pulls at the rods sideways with considerable force then what does one expect to happen? There are two possibilities. If the binding energy of the steel rods to the concrete is smaller than the elastic yield strength of the concrete then the rods will simply cut through the concrete. On the other hand, if the energy of binding of the rods to the concrete is much greater than the elastic yield strength then the concrete will yield as one pulls at the rods. It may yield in a continuous manner or in a brittle manner. This is the basic idea. *The concrete slab in our case is the crust of the neutron star, and the rods that are frozen into them are the vortices in the crustal superfluid on the one hand or the quantized fluxoids in the superconducting interior on the other.* To be specific, let us consider the latter case. As the neutron star slows down (or spins up) the vortices will move radially outwards (or inwards). If the fluxoids are interpinned to the vortices as we argued earlier, then they too will be dragged along. This is the origin of elastic stresses building up in the crust of the star. Let us now look at this in a little more detail. Let us first recall the expected elastic properties of the crust. To a first approximation the melting temperature of the crust will be $\sim 10^9$ K.

$$T_m = \frac{(Ze)^2}{180 b_Z k_B} \sim 4 \times 10^9 \text{ K}$$

Here b_Z is the internuclear spacing $\sim 5 \times 10^{-12}$ cm, and $Z \sim 40$. The shear modulus μ is

$$\mu = 0.3 \frac{(Ze)^2}{b_Z^4} \sim 2 \times 10^{29} \text{ dyne cm}^{-2}$$

If the crust may be treated as a classical solid then the dimensionless quantity Θ defined as the change in length per unit length under tension or compression will be related to the stress σ in the form

$$\sigma = \mu \Theta f(\Theta, T/T_m, \chi)$$

Fig. 64. A schematic representation of the stress on the inner crust due to flux expulsion.

where χ is a dimensionless measure of the crystal dislocation distribution. Very careful laboratory experiments show the following types of behaviour (see Fig. 65):

1. $\sigma = \mu\Theta$ until the elastic strain limit is reached.
2. when $\Theta > \Theta_{max}$, σ remains near $\mu\Theta_{max}$. The crystal yields and continues to deform without further increase in stress.
3. When $T \leq 0.1 T_m$ the lattice yields discontinuously. It "breaks" erratically on a scale $\Delta\Theta < \Theta_{max}$. In this case the yield is not by plastic flow but the response is brittle.

This is what laboratory experiments on small crystal grains tell us. It is a dangerous business to extrapolate this to a crust as thick as a kilometre. Under extreme strain will such a thick crust yield by numerous microscopic cracks in which case on a macroscopic scale the response will resemble plastic flow or yield through substantial cracks? Frankly, one does not know. But it is more interesting to speculate that the crust will break and major "faults" will develop. This is the point of view that Ruderman has taken during the last couple of years. Assuming that such a cracking of the crust will result due to stresses built up during the slowing down of the neutron star — due to the strong pinning of the vortices and fluxoids to the crust — one can ask 'do we have a reasonable estimate of the maximum strain angle that the neutron star crust will support before yielding?'. Estimates suggest that this maximum strain angle may be $\sim 10^{-5} - 10^{-3}$, although this is highly uncertain.

9.1 Stresses on the Crustal Lattice

As already indicated, two kinds of shear stresses are especially important in forcing the solid crust to break or move.

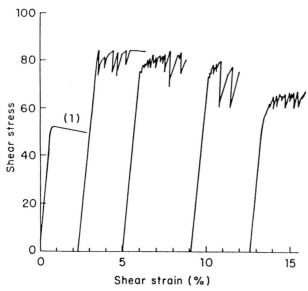

Fig. 65. Shear stress *versus* strain for Li crystals (stabilized with 57% Mg) under tension. Curve (1) is for $T = 10K$, while the others are different samples at $T = 77K$ (Siedersleben and Taylor 1989).

Magnetic Stresses

The quantized magnetic flux tubes in the superconducting interior terminate at the base of the crust. There they open out and permeate as normal field. We also saw that these quantized flux tubes may be severely entangled with the vortices in the neutron superfluid. As the neutron star slows down (or spins up) the superfluid vortices have to move out radially (or in). Because of the entanglement the fluxoids will also move outwards. Indeed, this was the mechanism we discussed for flux expulsion from the superconducting interior, and formed the basis for a model for magnetic field decay. Since the crust has very high electrical conductivity the magnetic flux would be frozen into it and this will result in stesses building up in the solid as one tries to move the fluxoids. In the language of the example we gave before, the fluxoids are the steel rods, and the crust is the piece of concrete. Each quantized flux tube which terminates at the base of the crust contains an average magnetic field $B_c \sim 10^{15}$ G. Therefore, if the crust were to remain rigid and immobile, shear stresses at the bottom of the crust could grow to

$$S(B) \sim \frac{BB_c}{8\pi} \sim \left(\frac{B}{3 \times 10^{12} \text{ G}}\right) 10^{26} \text{ dy cm}^{-2}$$

where B is the average field through the crust.

The maximum possible stress that the crust could bear before yielding is

$$S_{max} \sim \frac{l}{R} \mu \Theta_{max} \sim 10^{26} \left(\frac{\Theta_m}{10^{-2}}\right) \text{ dy cm}^{-2}.$$

Here $l \sim 10^4$ cm is the thickness of the crust, and R is the radius of the star. Therefore if B is greater than $\sim 10^{12}$ G, as in most pulsars, there is a real possibility of the crust yielding by cracking as a response to the slowing down of the stars.

Vortex Pinning Shear Stresses

While trying to understand glitches we discussed at some length the crustal superfluid and pinning of the vortices in the superfluid to the crustal nuclei. As the crust slows down due to external torque the superfluid also attempts to slow down by a radial "creep" of the vortices. In our previous discussion we assumed that if the Magnus force acting on the vortices exceeds the pinning force then the vortices will unpin. But there is another interesting possibility. Let us assume that the pinning force is very large. Then, as the crust slows down the force density on the lattice due to the pinned vortices will increase. A fraction of this could be absorbed in deforming the lattice etc. But the rest of it must be balanced by the lattice shear strength. Ruderman has argued that *provided the vortices remain pinned* to the crustal lattice, the lattice yield strength will be exceeded when the lag between the angular velocity of the crust and the crustal superfluidity exceeds

$$\omega \sim 0.1 \left(\frac{P}{ms}\right) \left(\frac{\Theta_{max}}{10^{-2}}\right) \text{ s}^{-1}$$

Beyond this point the lattice will have to yield either through a plastic flow or by crumbling and cracking. In our previous discussion of the glitch we did not entertain the possibility that the lattice will be stressed to the breaking point; we assumed that the vortices will unpin long before this.

It turns out that the pinning force on the vortices depends to a large extent on whether there is a relative alignment between the vortex core and the axis of the crystal lattice. If the crystal axis and the vortex are aligned then pinning is much stronger, and the critical differential rotation rate before unpinning takes place is much larger than in an unaligned case. At this stage one might ask 'what determines whether the crystal axis and the vortex array are aligned or not?'. If the lattice forms before superfluidity sets in then there need not be any correlation between the lattice axis and the vortex direction. On the other hand, if the lattice forms after superfluidity sets in, then the energetics themselves may align the two axes.

Let us summarize our discussion so far. When a neutron star is spun down or spun up, enormous stresses will build up in the inner crust. There are two contributing factors: (i) the evolution of the core field due to the entanglement with the core superfluid vortices, (ii) stress due to the pinning of the vortices in the crustal lattice.

9.2 Plate Tectonics

When the deep crustal matter is subjected to a stress σ then it will either compress or stretch resulting in a dimensionless strain Θ. This will continue till the elastic yield limit is reached. At this point $\sigma_{max} = \mu\Theta_m$. We will now assume that provided the temperature of the lattice is not too close to its melting temperature, further increase in the strain on the lattice will result in the crust cracking. As a result of the cracking the stress will suddenly be relieved. Therefore it is conceivable that as a result of growing stresses and strain the lattice will continuously crack into plates (see Fig. 66). In response to the slowing down of the neutron star these plates will "flow" like a viscous fluid and move towards the equatorial zone over very long timescales. At the equator, the downward moving plates from the northern hemisphere will meet the upward moving plates from the southern hemisphere. Thus, the equatorial region will be a subduction zone where the plates are pushed into the core and will dissolve into the superfluid sea of neutrons and protons.

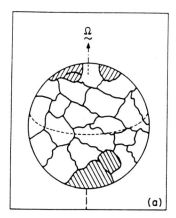 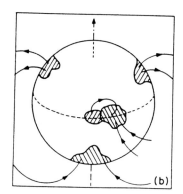

Fig. 66. Evolution of surface *platelets* and magnetic field after crust cracking sets in. (a) An initial configuration of cracks; the magnetic field is continuously distributed through the surface. (b) Separation of original platelets (hatched) after further spin-down and equatorial zone subduction (Ruderman 1991c).

9.3 Some Astrophysical Consequences

The ingenious ideas of Ruderman that we have just presented are admittedly very speculative and must be examined more critically. But the reason for outlining them is that they may have many exciting astrophysical consequences. For example, glitches in pulsars, X-ray and gamma-ray bursts from neutron stars, magnetic field configurations in recycled puslars etc.

Glitches

Without going into the details we shall merely hint at how this mechanism may provide an elegant way out of the major difficulty we had in explaining glitches. You will recall the major stumbling block was that the pinned vortices in the crustal superfluid had to collectively unpin, move outwards and re-pin again. In the present scenario this difficulty is neatly solved. When the lattice breaks as a result of this spin-down induced crustal stresses, the plates will move radially outwards with respect to the rotation axis. This will result in a sudden slowing down of the superfluid causing a glitch. We have side-stepped our major difficulty by physically transporting the vortices pinned to a plate without any catastrophical unpinning.

Gamma-Ray Bursts

Formation of major faults may also give rise to X-ray and γ-ray bursts. Let us first get a feeling for the elastic energy released when the crust cracks. Let R_p be the characteristic lateral size of the platelet and l its thickness. Then the elastic energy released is

$$\varepsilon_g \sim (lR_p^2)\,\mu\Theta_{max}\Delta\Theta$$
$$\sim 10^{40} R_p^2 \text{ ergs}$$

Such a crack is expected to propagate with a shear velocity $V_s = \left(\frac{\mu}{\rho}\right)^{1/2} \sim 10^8$ cm s^{-1}. Therefore the *cracking timescale* (l/V_s or R_p/V_s) will be less than a millisecond. After the crack the energy released will initially be in the form of *shear vibrations* of the platelet with frequencies (reciprocal of the cracking time) in the kilo Hertz region. As Blaes et al. (1989) have shown, such high frequency elastic vibrations will reach the surface as Alfvén Waves, greatly amplified in amplitude.

A magnetized platelet oscillating with a frequency ω will give rise to a potential drop between the field lines:

$$\Delta V = \left(\frac{8P_A}{c}\right)^{1/2} \cos\omega t \sim 10^{17} \cos\omega t \text{ Volts}$$

Here P_A is the Alfvén wave power ($\sim 10^{38}$ erg s^{-1}). Such a large oscillating voltage is expected to result in electron-positron pair production, and their acceleration to ultrarelativistic speeds. Although the details are not clear this is entirely plausible. Given this, and the strong surface magnetic field, X-ray and gamma-ray radiation through various processes such as synchotron radiation will occur.

There are several attractive features of this model:

(i) The energy released in each gamma ray burst will be of the order of the elastic energy released when the crust cracks ($\sim 10^{38}$ erg). Assuming the source to be at a distance of 1 kiloparsec, this will imply a fluence

$$\frac{\varepsilon_g}{4\pi d^2} \; 10^{-6} \text{ erg cm}^{-2}$$

consistent with γ-ray bursts.

(ii) The rise-time of the burst should be the timescale of cracking. This is $\sim$ millisecond, again consistent with observations.

Therefore, cracking of the crust provides a nice way of making gamma ray bursters.

Evolution of the Magnetic Field Structures

We shall end by making a few remarks about how the ideas outlined above can profoundly influence the configuration of the magnetic fields of neutron stars. This is a matter of considerable importance because there is growing evidence that millisecond and other binary pulsars are predominantly "orthogonal rotators", i.e., the magnetic axis is nearly perpendicular to the rotation axis. It is tempting to speculate whether such a field configuration resulted as a consequence of their rotational history. Although it is premature to make detailed arguments, it may be worth giving a couple of examples. Consider a neutron star which is rapidly spinning down. At some stage, as a result of stresses, there will be large scale cracking of the crust into "platelets". New matter from the core will flow into the cracks between the plates. These new plates will not have any of the original field which is frozen into the original platelets. The newly formed plates may be more susceptible to cracking in response to continued spin-down. The kind of rearrangement of the stellar field due to plate motions is schematically sketched in the Fig. 66. At some period, say P_{crit}, the surface cracks. Thereafter, the plates begin to move towards the equator, new plates are added etc. Each of the original platelet will have frozen into it the field $B(P_c)$ it had when cracking first began. Therefore, the magnetic fields of old neutron stars deduced from observations could be in conflict with a simple dipole geometry. For example, the far-field deduced from spin-down torques or accretion torques could yield small values representing the *average field*, while "local fields" such as deduced from cyclotron lines may give a much larger value. This is may be the case in Her X-1.

The magnetic field evolution of recycled pulsars which are spun up after initial spin down is also interesting subject but we won't go into it because there is a whole zoo of possibilities. Two possibilities are shown in Fig. 67; figure (a) shows the possible secular evolution of the magnetic field structure of a *spun-down* pulsar, while (b) refers to the case of a *spun-up* pulsar.

Epilogue

The subject of *astrophysics* began with the study of the physics of gaseous stars. To Eddington must go the credit for laying the foundations, as well as developing the subject of the structure and evolution of the stars. The second half of this

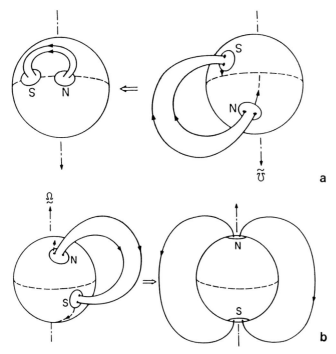

Fig. 67. (a) Evolution of the surface magnetic field of a spinning-down neutron star when equatorial zone magnetic field line reconnection is ignored. (b) Evolution of the surface magnetic field of a short period spinning-up neutron star when flux lines initially connect the two hemispheres (from Ruderman 1991a).

century has been the era of neutron stars, supernovae and black holes. As we saw in the first chapter, much of what we understand well was already anticipated in the 1930s. As we came closer to the present time we saw that very little is understood in detail. For example, nearly thirty years after their discovery one does not understand very much about how pulsars work, why they suddenly spin up, or why or how their magnetic field decays. We ventured to discuss several speculative but nevertheless exciting recent ideas. It is appropriate that we end on the following modest note:

Perhaps to move
His laughter at their quaint opinions wide
Hereafter, when they come to model heaven
And calculate the stars: how they will wield
The mighty frame: how build, unbuild, contrive
To save appearances.

– Paradise Lost

Acknowledgements

Naturally, I deem it an honour to be asked to lecture in one of the "Saas-Fee" courses, and I would like to express my appreciation to Georges Meynet and Daniel Schaerer for inviting me and making the stay in Les Diableret such a memorable experience. They were so thoughtful about everything; they even ordered fresh snow for the participants!

Over the years I have benefited enormously through discussions with a number of colleagues. The interaction with many of them has been very inspiring and enabled me to make the transition from condensed matter physics into astrophysics. While it is not possible to name all of them I certainly would like to record my indebtedness to V.Radhakrishnan, Ed van den Heuvel, Mal Ruderman, Lo Woltjer, Franco Pacini and Martin Rees. I also had the privilege of learning a lot from many of my present and past students, as well as my colleagues at the Raman Research Institute, in particular Dipankar Bhattacharya and Avinash Deshpande. I wish to thank all of them profusely.

Finally, with all humility I wish to dedicate these lectures to the memory of my teacher Subrahmanyan Chandrasekhar.

References

Alpar, M.A., Cheng, A.F., Ruderman, M.A. and Shaham, J.: 1982, *Nature*, **300**, 728.
Anderson, B. and Lyne, A.G.: 1983, *Nature*, **303**, 597.
Anderson, P.W. and Itoh, N.: 1975, *Nature*, **256**, 25.
Baade, W. and Zwicky, F.: 1934, *Phys. Rev.*, **45**, 138.
Backer, D.C., Kulkarni, S.R., Heiles, C., Davis, M.M. and Goss, W.M.: 1982, *Nature*, **300**, 615.
Backer, D.C.: 1973, *Astrophys. J.*, **182**, 245.
Bardeen, J., Cooper, L.N. and Schrieffer, J.R.: 1957, *Phys. Rev.*, **108**, 1175.
Baym, G.: 1991, *Neutron Stars: Theory and Observations*, eds. J. Ventura and D. Pines, Kluwer, Dordrecht-Holland.
Baym, G., Bethe, H.A. and Pethick, C.J.: 1971b, *Nuclear Physics*, **A175**, 225.
Baym, G. and Pethick, C.J.: 1979, *Ann. Rev. Astron. Astrophys.*, **17**, 415.
Baym, G., Pethick, C. and Pines, D.: 1969, *Nature*, **224**, 673.
Baym, G., Pethick, C.J. and Sutherland, P.: 1971a, *Astrophys. J.*, **170**, 299.
Bhattacharya, D. and Srinivasan, G.: 1986, *Current Science*, **55**, 327.
Bhattacharya, D. and Srinivasan, G.: 1991, *J. Astrophys. Astron.*, **12**, 17.
Bhattacharya, D. and Srinivasan, G.: 1995, in *X-ray Binaries*, eds. W.H.G. Lewin, J. van Paradijs and E.P.J. van den Heuvel, Cambridge University Press.
Bhattacharya, D., Wiejers, R.A.M., Hartman, J.W. and Verbunt, F.: 1992, *Astron. Astrophys.*, **254**, 198.
Bisnovatyi-Kogan, G.S. and Komberg, B.V.: 1974, *Soviet Astron.*, **18**, 217.
Bisnovatyi-Kogan, G.S. and Komberg, B.V.: 1976, *Soviet Astron. Lett.*, **2**, 130.
Blaauw, A.: 1961, *Bull. Astron. Inst. Neth.*, **15**, 265.
Blaauw, A.: 1985, in *Birth and Evolution of Massive Stars and Stellar Groups*, eds. W. Boland and H. van Woerden (Dordrecht: D.Reidel).
Blaes, O., Blandford, R.D., Goldreich, P. and Madav, P.: 1989, *Astrophys. J.*, **343**, 839.
Blandford, R.D., Applegate, J.H. and Hernquist, L.: 1983, *MNRAS*, **204**, 1025.

Burrows, A. and Woosley, S.E.: 1986, *Astrophys. J.*, **308**, 680.
Burton, W.B.: 1971, *Astron. Astrophys.*, **10**, 76.
Chandrasekhar, S.: 1931a, *Phil. Mag.*, **11**, 592.
Chandrasekhar, S.: 1931b, *Astrophys. J.*, **74**, 81.
Chandrasekhar, S.: 1932, *Zeitschrift für Astrophysik*, **5**, 321.
Chandrasekhar, S.: 1934, *The Observatory*, **57**, 373.
Chandrasekhar, S.: 1937, *Nature*, **139**, 757.
Chandrasekhar, S.: 1939a, *An Introduction to the Study of Stellar Structure*, University of Chicago Press.
Chandrasekhar, S.: 1939b, in *Colloque International d'Astrophysics* XIII – *Novae and White Dwarfs*, Herman and Cie.,Editeurs, Paris (1941).
Chandrasekhar, S.: 1964, *Phys. Rev. Lett.*, **12**, 114.
Chao, N.C., Clark, J.W. and Yang, C.H.: 1972, *Nuclear Physics*, **A179**, 320.
Cheng, K.S., Ho, C. and Ruderman, M.A.: 1986, *Astrophys. J.*, **300**, 500.
Chiu, H.Y. and Salpeter, E.E.: 1964, *Phys. Rev. Lett.*, **12**, 413.
Clark, D.H. and Caswell, J.L.: 1976, *MNRAS*, **174**, 267.
Clark, D.H. and Stephenson, F.R.: 1977, *MNRAS*, **179**, 87p.
Davidson, K. and Ostriker, J.P.: 1973, *Astrophys. J.*, **179**, 583.
Deshpande, A.A., Ramachandran, R. and Srinivasan, G.: 1995, *J. Astrophys. Astron.*, **16**, 69.
Downs, G.S.: 1981, *Astrophys. J.*, **249**, 687.
Duin, R.M. and Strom, R.G.: 1975, *Astron. Astrophys.*, **39**, 33.
Eddington, A.S.: 1926, *The Internal Constitution of the Stars*, Cambridge, England.
Eddington, A.S.: 1935, *The Observatory*, **58**, 37.
Fabian, A.C., Pringle, J.E., Verbunt, F. and Wade, R.: 1983, *Nature*, **301**, 222.
Feynman, R.P.: 1955, *Prog. Low Temp. Phys.*, **1**, 17.
Fichtel, C.E., Simpson, G.A. and Thompson, D.J.: 1978, *Astrophys. J.*, **222**, 833.
Fowler, R.H.: 1926, MNRAS, **87**, 114.
Friedman, J.L., Imamura, J.N., Durisen, R.H. and Parker, L.: 1988, *Nature*, **336**, 560.
Ginzburg, V.L. and Kirzhnits, D.A.: 1964, *Zh. Eksperim. Theor. Fiz.*, **47**, 2006.
Goldriech, P. and Julian, W.H.: 1969, *Astrophys. J.*, **157**, 869.
Gunn, J.E. and Ostriker, J.P.: 1969, *Phys. Rev. Letters*, **22**, 728.
Harding, A.K., Tademaru, E. and Esposito, L.W.: 1978, *Astrophys. J.*, **225**, 226.
Harrison, E.R. and Tademaru, E.P.: 1975, *Astrophys. J.*, **201**, 447.
Harrison, P.A., Lyne, A.G. and Anderson, B.: 1993, *MNRAS*, **261**, 113.
Harrison, B.K. and Wheeler, J.A.: 1958, in *La Structure et l'evolution de l'univers*, eds. B.K. Harrison, K.S. Thorne, M. Wakano and J.A. Wheeler, Onzième Conseil de Physique Solvay, Brussels.
Helfand, D.J., Taylor, J.H., Backus, P.R. and Cordes, J.M.: 1980, *Astrophys. J.*, **237**, 206.
Hewish, A., Bell, S.J., Pilkington, J.D.H., Scott, P.F. and Collins, R.A.: 1968, *Nature*, **217**, 709.
Hoffberg, M., Glassgold, A.E., Richardson, R.W. and Ruderman, M.A.: 1970, *Phys. Rev. Lett.*, **24**, 175.
Hulse, R.A. and Taylor, J.H.: 1975, *Astrophys. J.*, **195**, L51.
Illarionov, A.F. and Sunyaev, R.A.: 1975, *Astron. Astrophys.*, **39**, 185.
Jahan Miri, M. and Bhattacharya, D.: 1994, *MNRAS*, **269**, 455.
Joss, P.C. and Rappaport, S.A.: 1983, *Nature*, **304**, 419.
Kerr, F.J. and Lynden-Bell, D.: 1986, *MNRAS*, **221**, 1023.

Kippenhahn, R. and Weigert, A.: 1990, *Stellar Structure and Evolution*, Springer-Verlag.
Kulkarni, S.R.: 1986, *Astrophys. J.*, **306**, L85.
Kulkarni, S.R. and Narayan, R.: 1988, *Astrophys. J.*, **335**, 755.
Lyne, A.G.: 1995, *J. Astrophys. Astron.*, **16**, 97.
Lyne, A.G. and Graham-Smith, F.: 1990, *Pulsar Astronomy*, Cambridge University Press.
Lyne, A.G. and Lorrimer, D.R., 1994, *Nature*, **369**, 127
Manchester, R.N.: 1993, in *Pulsars as Physics Laboratories*, eds. R.D. Blandford, A. Hewish, A.G. Lyne and L. Mestel, Oxford University Press.
Manchester, R.N. and Taylor, J.H.: 1977, *Pulsars*, Freeman, San Francisco.
McKee, C.F. and Ostriker, J.P.: 1977, *Astrophys. J.*, **218**, 148.
Melrose, D.B.: 1993, in *Pulsars as Physics Laboratories*, eds. R.D. Blandford, A. Hewish, A.G. Lyne and L. Mestel, Oxford University Press.
Migdal, A.B.: 1959, *Zh. Eksperim. Theor. Fiz.*, **37**, 249.
Morris, D., Radhakrishnan, V. and Shukre C.S.: 1976, *Nature*, **254**, 676.
Murdin, P. and Murdin, L.: 1985, *Supernovae*, Cambridge University Press.
Muslimov, A.G. and Tsygan, A.I.: 1985, *Soviet Astron. Lett.*, **11**, 80.
Narayan, R.: 1987, *Astrophys. J.*, **319**, 162.
Onsager, L.: 1949, *Nuovo Cimento Suppl.*, **6**, 249.
Oppenheimer, J.R. and Snyder, H.: 1939, *Phys. Rev.*, **56**, 455.
Oppenheimer, J.R. and Volkoff, G.M.: 1939, *Phys. Rev.*, **55**, 374.
Ostriker, J.P. and Gunn, J.E.: 1969, *Astrophys. J.*, **157**, 1395.
Pacini, F.: 1971, *Astrophys. J.*, **163**, L17.
Pacini, F.: 1983, *Astron. Astrophys.*, **126**, L11.
Pacini, F. and Rees, M.J.: 1970, *Nature*, **226**, 622.
Pacini, F. and Salvati, M.: 1973, *Astrophys. J.*, **186**, 249.
Pethick, C.J. and Ravenhall, D.G.: 1991, *Neutron Stars: Theory and Observations*, eds. J. Ventura and D. Pines, Kluwer, Dordrecht-Holland.
Phinney, E.S., Evans, C.R., Blandford, R.D. and Kulkarni, S.R.: 1988, *Nature*, **333**, 832.
Piddington, J.H.: 1957, *Australian J. Phys.*, **10**, 530.
Pines, D. and Alpar, M.A.: 1992, in *The Structure and Evolution of Neutron Stars*, eds. D. Pines, R. Tamagaki and S. Tsuruta, Addison-Wesley.
Pringle, J.E. and Rees, M.J.: 1972, *Astron. Astrophys.*, **21**, 1.
Radhakrishnan, V. and Cooke, D.J.: 1969, *Astrophys. Lett.*, **3**, 225.
Radhakrishnan, V. and Manchester, R.N.: 1969, *Nature*, **222**, 228.
Radhakrishnan, V. and Srinivasan, G.: 1981, in *The Proceedings of the 2nd Asia-Pacific Regional Meeting of the IAU*, eds. B. Hidayat and M.W. Feast, Tira Pustaka, Jakarta (1984).
Radhakrishnan, V. and Srinivasan, G.: 1982, *Current Science*, **51**, 1096.
Ramachandran, R. and Deshpande, A.A.: 1994, *J. Astrophys. Astron.*, **15**, 69.
Rappaport, S.A. and Joss, P.C.: 1983, in *Accretion-driven Stellar X-ray Sources*, eds. W.H.G. Lewin and E.P.J. van den Heuvel, Cambridge University Press.
Rees, M.J. and Gunn, J.E.: 1974, *MNRAS*, **167**, 1.
Rhoades, C.E. and Ruffini, R.: 1974, *Phys. Rev. Lett.*, **32**, 324.
Romani, R.W.: 1990, *Nature*, **347**, 741.
Ruderman, M.A.: 1987, in *High Energy Phenomena Around Collapsed Stars*, ed. F. Pacini, D. Reidel, Dordrecht-Holland.
Ruderman, M.A.: 1991a, *Astrophys. J.*, **366**, 261.

Ruderman, M.A.: 1991b, *Astrophys. J.*, **382**, 576.
Ruderman, M.A.: 1991c, *Astrophys. J.*, **382**, 587.
Ruderman, M.A. and Sutherland, P.G.: 1975, *Astrophys. J.*, **196**, 51.
Ruderman, M.A., Shaham, J. and Tavani, M.: 1989, *Astrophys. J.*, **336**, 507.
Salpeter, E.E.: 1961, *Astrophys. J.*, **134**, 669.
Sauls, J.: 1989, in *Timing Neutron Stars*, eds. H. Ogelman and E.P.J. van den Heuvel (Dordrecht: Kluwer).
Shapiro, S.L. and Teukolsky, S.A.: 1983, *Black Holes, White Dwarfs and Neutron Stars*, John Wiley & Sons.
Shklovskii, I.S.: 1970, *Astr. Zu.*, **46**, 715.
Siedersleben, M. and Taylor, E.: 1989, *Phil. Mag.*, **A60**, 631.
Smarr, L.L. and Blandford, R.D.: 1976, *Astrophys. J.*, **207**, 574.
Spitzer, L. (Jr.): 1978, *Physical Processes in the Interstellar Medium*, Wiley-Interscience, New York.
Srinivasan, G.: 1989, *Astron. Astrophys. Rev.*, **1**, 209.
Srinivasan, G., Bhattacharya, D. and Dwarakanath, K.S.: 1984, *J. Astrophys. Aston.*, **5**, 403.
Srinivasan, G., Bhattacharya, D., Muslimov, A.G. and Tsygan, A.I.: 1990, *Current Science*, **59**, 31.
Srinivasan, G. and van den Heuvel, E.P.J.: 1982, *Astron. Astrophys.*, **108**, 143.
Stokes, G.H., Segelstein, D.J., Taylor, J.H. and Dewey, R.J.: 1986, *Astrophys. J.*, **311**, 694.
Sturrock, P.A.: 1971, *Astrophys. J.*, **164**, 529.
Taam, R.E. and van den Heuvel, E.P.J.: 1986, *Astrophys. J.*, **305**, 235.
Taylor, J.H.: 1995, *J. Astrophys. Astron.*, **16**, 307.
Taylor, J.H. and Cordes, J.M.: 1993, *Astrophys. J.*, **411**, 674.
Taylor, J.H. and Stinebring, D.R.: 1986, *Ann. Rev. Astron. Astrophys.*, **24**, 285.
Tinsley, B.M.: 1977, in *Supernovae*, ed. D.N. Schramm, D.Reidel Publishing Co. (Dordrecht-Holland).
Trimble, V.: 1968, *Astrophys. J.*, **73**, 535.
Urpin, V.A. and Yakovlev, D.G.: 1980, *Soviet Astron.*, **24**, 425.
Usov, V.V.: 1983, *Nature*, **305**, 409.
van den Heuvel, E.P.J.: 1977, *Ann. N.Y. Acad. Sci.*, **302**, 14.
van den Heuvel, E.P.J.: 1983, in *Accretion-driven Stellar X-ray Sources*, eds. W.H.G. Lewin and E.P.J. van den Heuvel, Cambridge University Press.
van den Heuvel, E.P.J.: 1985, in *Supernovae, Their Progenitors and Remnants*, eds. G. Srinivasan and V. Radhakrishnan, Indian Academy of Sciences, Bangalore, India.
van den Heuvel, E.P.J., van Paradijs, J.A. and Taam, R.E.: 1986, *Nature*, **322**, 153.
Wheeler, J.C., Miller, G.M. and Scalo, J.M.: 1980, *Astron. Astrophys.*, **82**, 152.
Wilson, A.A.: 1972, *MNRAS*, **157**, 229.
Wiringa, R.B., Fiks, V. and Fabrochini, A.: 1988, *Phys. Rev.*, **C38**, 1010.
Woltjer, L.: 1964, *Astrophys. J.*, **140**, 1309.
Woodward, J.F.: 1978, *Astrophys. J.*, **225**, 574.
Yarmchuk, E.J., Gordon, M.J.V. and Packard, R.E.: 1979, *Phys. Rev. Lett.*, **43**, 214.
Zwicky, F.: 1957, *Morphological Astronomy*, Springer, Berlin.

Black Holes

Igor Novikov

Theoretical Astrophysics Center, Blegdamsvej 17, Copenhagen DK-2100, Denmark
Nordita, Blegdamsvej 17, DK-2100 Copenhagen, Denmark
University Observatory, Øster Voldgade 3, DK-1350, Copenhagen K, Denmark
Astro Space Center of P.N.Lebedev Physical Institute, Profsoyuznaya 84/32, Moscow, 117810, Russia

1 Astrophysics of Black Holes

1.1 Introduction

Modern astrophysics considers three types of black holes in the Universe: a) Black holes of stellar masses, which were born when massive stars died, b) Supermassive black holes with masses up to $10^9 M_\odot$ and more ($M_\odot = 2 \times 10^{33}$g is mass of the Sun) at centers of galaxies, c) Primordial black holes which might appear from the large-scale inhomogeneities at the very beginning of expansion of the Universe. Their masses can be arbitrary, but primordial black holes with $M \leq 10^{15}$g will have radiated away their mass by the Hawking quantum process in a time $t \leq 10^{10}$ years (the age of the Universe). Only primordial black holes with mass $M > 10^{15}$g could exist in the contemporary Universe.

The history of the idea of the black holes and their astrophysics is given in Israel (1987). General problems of the astrophysics of the black holes are discussed in: Zeldovich and Novikov (1971), Novikov and Thorne (1972), Shapiro and Teukolsky (1983), Blandford (1987), Lamb (1991), Novikov and Frolov (1989), Frolov and Novikov (1995). For the preparation of these notes I used partly the materials which will be included in our book Frolov and Novikov (1995), materials of my reviews Novikov (1995abc), and our book Novikov and Frolov (1989).

1.2 The Origin of Stellar Black Holes

"When all the thermo-nuclear sources of energy are exhausted a sufficiently heavy star will collapse" – this is the first phrase of the abstract of a wonderful paper by Oppenheimer and Snyder (1939). Every statement of this paper accords with ideas that remain popular today (including terminology). The authors conclude in the abstract: "... an external observer sees the star shrinking to its gravitational radius."

This was the modern prediction of the formation of the black holes when massive stars die. How heavy should a star be to turn into a black hole? The answer is not simple. A star light enough ends up as a white dwarf or a neutron star. Both these types of celestial bodies have the upper limits of their masses. For white dwarfs it is the Chandrasekhar limit which is about 1.2–1.4 $M_\odot$ [see Shapiro and Teukolsky 1983, Kippenhahn and Weigert 1990]. For neutron stars it is the Oppenheimer-Volkoff limit [Oppenheimer and Volkoff 1939]. The exact value of this limit depends on the equation of state at matter density larger than the density of nuclear matter $\rho_0 = 2.8 \times 10^{14}$g cm^{-3}. The modern theory gives for this maximum mass $M_{max,OV} \approx (2-3)M_\odot$ [see Baym and Pethick 1979, Lamb 1991]. Some authors discussed the possibility of existence of so called "quark stars" and "hadronic stars" [see Alcook, Farhi and Olinto 1986, Bahcall, Lynn, and Selipsky 1990, Madsen 1993]. At present, there is no evidence for such stars.

Rotation can increase $M_{max,OV}$ only slightly [up to 25%; Friedman and Ipser 1987, Haswell et al. 1993]. Thus one can believe that the upper mass limit for neutron stars should not be greater than $M_0 \approx 3M_\odot$ [Lamb 1991, Cowley 1992, McClintock 1992]. If a star in the very end of its evolution has a mass greater than M_0 it must turn into a black hole. It does not mean that black hole progenitors are all normal stars (on "main sequence" of the Herzsprung-Russell diagram, see for example Bisnovaty-Kogan (1989), Kippenhahn and Weigert (1990)) with masses $M > M_0$. The point is that the final stages of evolution of massive stars are poorly understood. Steady mass loss, catastrophic mass ejections and even disruption at supernovae explosions are probably possible [see for example Kippenhahn and Weigert 1990]. These processes can essentially reduce the initial stellar mass at the end of its evolution. Thus the initial mass of black hole progenitors could be essentially greater than M_0.

There are different estimates for the minimal masses of progenitor stars M_* that form black holes. For example $M_* \approx 10M_\odot$ [Shapiro and Teukolsky 1983] and $M_* \approx 30M_\odot$ [Lipunov 1987] or even $\geq 40M_\odot$ [van den Heuvel and Habets 1984, Schild and Maeder 1985]. Note that the evolution of stars in close binary systems differs from the evolution of sole stars because of mass transfer from one star to another [see Novikov 1974, Masevich, Tutukov 1988], and conclusions about masses of black hole progenitors in this case could be essentially different. In our further discussion in this section we shall focus our attention on the fate of single stars.

One can estimate how many black holes were created by stellar collapse in our Galaxy during its existence. The present spectrum of stellar masses for stars on the "main sequence" in the solar neighborhood is roughly known from the observational data together with theory of stellar structure.

The lifetime of massive stars is less than T_0 – the age of the Galaxy ($T_0 \approx 10^{10}$ years). We assume the constant birthrate and the same constant deathrate for massive stars during the lifetime of the Galaxy. Now if we suppose that all stars with $M > M_*$ on the "main sequence" (progenitors) must turn into black holes, we can calculate the birthrate of black holes in the solar neighborhood. If this rate is the same everywhere we can estimate the total number of black holes in the Galaxy and the total mass of all stellar black holes in it.

Many works were devoted to such estimates. But because of great uncertainties the progress here is very slow, starting from pioneering works by Zwicky

(1958), Schwarzschild (1958), Hoyle and Fowler (1963), Novikov and Ozernoy (1964), Hoyle, Fowler, Burbidge and Burbidge (1964). A review of more recent estimates is given by Shapiro and Teukolsky 1983. Because of the large uncertainty in the lower mass limit M_* for the stars collapsing to black holes we give the following expression for the rate of black hole formation in the Galaxy [see Novikov 1974]

$$\frac{dN}{dt} \approx 0.1 \left(\frac{M_*}{3M_\odot}\right)^{-1.4} \text{year}^{-1}.$$

At present, we can probably repeat the conclusion of Novikov and Thorne (1973): "For stars with masses greater than ~ 12 to $30 M_\odot$ the [supernovae] explosion may produce a black hole. If this tentative conclusion is correct, then no more than ~ 1 per cent of the [visible] mass of the Galaxy should be in the form of black holes today; and new black holes should be created at a rate not greater than ~ 0.01 per year." Thus, the total number of stellar black holes in the Galaxy can be of the order of $N \approx 10^8$ or less.

2 A Nonrotating Black Hole

2.1 Introduction

The formulae describing the gravitational field of the spherical black hole were the first exact solution to the equations of the General Relativity.

The spherically symmetric gravitational field (spacetime with spherical three-dimensional space) is described in every textbook on general relativity [see, e.g., Landau and Lifshitz 1973, Misner, Thorne, and Wheeler 1973]. Therefore, here we will only reproduce the necessary results. Let us write the expression for a squared interval far from strong gravitational fields (i.e., where special relativity is valid), using the spherical spatial coordinate system (r, θ, ϕ):

$$ds^2 = -c^2 dt^2 + dr^2 + r^2(d\theta^2 + \sin^2\theta \, d\phi^2), \qquad (2.1)$$

where c is the speed of light, and dl is the distance in three-dimensional space.

We will consider a curved spacetime but preserve the condition of spatial spherical symmetry. Spacetime is not necessarily empty, it may contain matter and physical fields (which are, of course, also spherically symmetric if their gravitation is considered). Calculations show [see, e.g., Misner, Thorne, and Wheeler 1973] that in this case the interval can always be written (after a suitable coordinate system has been chosen) in the form

$$ds^2 = g_{00}(x^0, x^1) \, dx^{0^2} + g_{11}(x^0, x^1) \, dx^{1^2} + (x^1)^2 (d\theta^2 + \sin^2\theta \, d\phi^2). \qquad (2.2)$$

This form is written in some special coordinates. The same components of the metric tensor are nonzero in (2.2) as in the expression for flat space, (2.1). The components g_{00} and g_{11} are functions only of x^0 and x^1 and are independent of θ and ϕ.

The coordinates in which the expression for g_{22} is written in the form $(x^1)^2$ are called *curvature coordinates*. Usually the x^1 coordinate is denoted by r [by analogy to (2.1)] and $x^0/c \equiv t$. We will see that this choice of symbols is not always logically justified inside a black hole (see Section 2.5).

If a spherical gravitational field is considered not in vacuum, then in the general case matter moves radially in the three-dimensional coordinate system defined by the coordinates x^1, θ, ϕ, that is, energy flows exist. Sometimes it is more convenient to choose a different frame of reference, for example, a co-moving one, but which is also spherically symmetric. All such reference frames possess the following property. The points of which one such reference frame is composed move radially with respect to some other frame. The relation between different reference frames which preserve the spherical symmetry is given by the transformations

$$\tilde{x}^0 = \tilde{x}^0(x^0, x^1), \qquad (2.3)$$

$$\tilde{x}^1 = \tilde{x}^1(x^0, x^1), \qquad (2.4)$$

plus any rotation of a coordinate system

$$\tilde{\theta} = \theta, \quad \tilde{\varphi} = \varphi. \qquad (2.5)$$

The coordinates in the new reference frame are marked with a tilde. Expression (2.4) describes the radial motion of the points of the new reference frame (its coordinates are $\tilde{x}^1 = $ const) with respect to the older one. Once (2.4) has been chosen, thus defining the new frame of reference, it is always possible to choose (2.3), which defines the time coordinate in the new system, in such a way that the component $\tilde{g}_{01}$ would not arise and the general expression for ds^2 would have the form

$$ds^2 = \tilde{g}_{00}(\tilde{x}^0, \tilde{x}^1) d\tilde{x}^{0^2} + \tilde{g}_{11}(\tilde{x}^0, \tilde{x}^1) d\tilde{x}^{1^2} + \tilde{g}_{22}(\tilde{x}^0, \tilde{x}^1)(d\tilde{\theta}^2 + \sin^2 \tilde{\theta} \, d\tilde{\phi}^2). \qquad (2.6)$$

Note that the expression for g_{22} can be written in the form

$$\sqrt{\tilde{g}_{22}} = x^1(\tilde{x}^0, \tilde{x}^1), \qquad (2.7)$$

where $x^1 = x^1(\tilde{x}^0, \tilde{x}^1)$ is the solution of (2.3) and (2.4) for x^1. It describes the radial motion of the points of the older reference frame (with the coordinates $x^1 = $ const) with respect to the new one.

2.2 Schwarzschild Gravitational Field

Let us consider the gravitational field of a black hole. It means that we should consider a spherical field in vacuum. The solution to Einstein's equations for this case were found by Schwarzschild (1916); it has the following form:

$$ds^2 = -\left(1 - \frac{2GM}{c^2 r}\right) c^2 dt^2 + \left(1 - \frac{2GM}{c^2 r}\right)^{-1} dr^2 + r^2(d\theta^2 + \sin^2 \theta \, d\phi^2), \qquad (2.8)$$

where G is Newton's gravitational constant and M is the mass of the field source.

A most important property of this solution is that it is independent of the temporal coordinate t and depends only on r, and that it is determined by a single

Black Holes

parameter M, that is, the total mass of the gravitational source which produces the field. Even if the field source involves radial motions (which preserve spherical symmetry), the field beyond the region occupied by matter remains constant [this statement is known as *Birkhoff's theorem* 1923]. Far from the center of gravity (as $r \to \infty$) spacetime converts to the flat Minkowski spacetime with metric (2.1). The coordinates t, r, θ, ϕ in which (2.1) is written are called the *Schwarzschild coordinates*, and the frame of reference they form is called the *Schwarzschild reference frame*. For ordinary measurements of length in a small neighborhood of each spatial point, we can use a local Cartesian coordinate system (x, y, z):

$$\delta x = \sqrt{g_{11}}\, dr = \left(1 - \frac{2GM}{c^2 r}\right)^{-1/2} dr, \tag{2.9}$$

$$\delta y = \sqrt{g_{22}}\, d\theta = r\, d\theta, \tag{2.10}$$

$$\delta z = \sqrt{g_{33}}\, d\phi = r \sin\theta\, d\phi. \tag{2.11}$$

The factor $(1 - 2GM/c^2 r)^{-1/2}$ in (2.9) reflects the curvature of the three-dimensional space.

Let us consider now the physical time. The physical time τ at a given point r is given by the expression

$$d\tau = \frac{\sqrt{-g_{00}}}{c}\, dx^0 = \sqrt{-g_{00}}\, dt = \left(1 - \frac{2GM}{c^2 r}\right)^{1/2} dt. \tag{2.12}$$

Far from the gravitational center (as $r \to \infty$), we have $d\tau = dt$, that is, t is the physical time of the observer removed to infinity.

At smaller r, the time τ runs progressively slower in comparison with the time t at infinity. As $r \to 2GM/c^2$, we find $d\tau \to 0$.

Let us now calculate the acceleration of free fall of a body which is at rest (or moves at a low velocity $v \ll c$) in the Schwarzschild reference frame. Using the formula for the calculation of the acceleration (see Appendix in Novikov and Frolov 1989) we find

$$F = \sqrt{F_i F^i} = \frac{GM}{r^2(1 - 2GM/c^2 r)^{1/2}}. \tag{2.13}$$

The acceleration points along the radius. As $r \to 2GM/c^2$, the acceleration tends to infinity. The singularity both in time flux arising as $r \to 2GM/c^2$ [see (2.12)] and that in the expression for acceleration F [see (2.13)] demonstrate that at this value of r, the Schwarzschild reference frame has physical singularity.[1] The

---

[1] Expression (2.13) gives the acceleration, and $\mathcal{F}_0 = Fm$ gives the force acting on a body of mass m, measured by an observer located near this body at a point r_0. If the body is suspended on a weightless absolutely rigid string, then the force applied to the free end of the string at the point r_1 is

$$\mathcal{F}_1 = \mathcal{F}_0 \sqrt{\frac{g_{00}(r_0)}{g_{00}(r_1)}}.$$

As r_0 tends to $2GM/c^2$, $\mathcal{F}_0 \to \infty$, while $\mathcal{F}_1$ remains finite.

quantity $r = r_g = 2GM/c^2$ is called the *Schwarzschild radius* (or gravitational radius), and the sphere of radius r_g is said to be the *Schwarzschild sphere*. We will later give a detailed analysis of the physical meaning of the singularity at $r = r_g$. The following aspects are emphasized here.

The Schwarzschild reference frame is static and nondeformable [$g_{\alpha\beta}$ is independent of t, and $g_{0i} = 0$, $D_{ik} = 0$]. It can be thought of as a coordinate lattice 'welded' of weightless rigid rods which fill the space around the black hole. We can study the motion of particles relative to this lattice, the evolution of physical fields at different points of this lattice, and so on. The Schwarzschild lattice thus, to some extent, resembles the lattice of fixed coordinates in the invariable Newtonian space of nonrelativistic physics. Of course, the geometry of the three-dimensional Schwarzschild space around a gravitational center is non-Euclidean, in contrast to the Euclidean Newtonian space of nonrelativistic physics. In other respects the properties of these frames are very similar.[2] This factor is a great help in our intuition.

When we speak of the motion of particles in the Schwarzschild field, we mean the motion and evolution of fields in this analogue of absolute Newtonian space.[3] As a result of the presence of the critical radius, $r_g = 2GM/c^2$, in the spherical field in vacuum, where the free-fall acceleration becomes infinite, such a rigid nondeformable lattice cannot be extended to $r \leq r_g$, since this region contains no nondeformable space (no analogue of the Newtonian space). The fact that F tends to infinity at r_g is an indication that at $r \leq r_g$ all systems must be nonrigid in the sense that $g_{\alpha\beta}$ must be a function of time and all systems must be deformed (all bodies must fall centerward). We will see that this is indeed the case.

Note that these specificities at $r = r_g$ do not indicate that a singularity of the type of infinite curvature, or something similar, exists in the geometry of the four-dimensional spacetime. We shall see later that here the spacetime is quite regular, and the singularity at r_g points to a physical singularity only in the Schwarzschild reference frame, that is, it signifies the impossibility of extending this reference frame as a rigid and nondeformable one (not falling to the center) to $r \leq r_g$.

Note in conclusion that r_g is extremely small, even for heavenly bodies. Thus, $r_g = 0.9$ cm for the Earth's mass and $r_g = 3$ km for the Sun's mass. If $r \gg r_g$, the Schwarzschild field is the ordinary Newtonian gravitational field with the free fall acceleration $F = GM/r^2$ and negligible curvature of the three-dimensional space. Outside the typical heavenly bodies (and all ordinary bodies as well), the gravitational field is the Newtonian field, because their sizes are typically much greater than r_g. The only known exceptions are neutron stars and black holes. The Schwarzschild solution within these bodies is not valid, and, obviously, the gravitational field is also Newtonian, with enormously high accuracy.

---

[2] These questions are discussed in detail in Section 3.2.

[3] Recall that in the general case of nonspherical time-dependent gravitational fields, it is not possible to introduce a rigid three-dimensional space; this fact stands in the way of clear intuitive concepts and inhibits calculations.

2.3 Motion of Photons Along the Radial Direction

Let us consider the radial motion of the photon which always propagates at the fundamental velocity c. The radial motion of an ultrarelativistic particle has the same properties. For this particle, $ds = 0$. For the radial motion, $d\theta = d\phi = 0$. Substituting $ds = d\theta = d\phi = 0$ into (2.8), we find the equation of motion

$$\frac{dr}{dt} = \pm c \left(1 - \frac{r_g}{r}\right). \tag{2.14}$$

This expression describes the rate at which the coordinate r changes with the time t of a distant observer (and not with respect to the physical time τ at a given point), that is, this is the coordinate (not the physical) velocity. The physical velocity is the rate of change of physical distance, dx [see (2.9)], in the physical time τ [see (2.12)]

$$\frac{dx}{d\tau} = \pm \frac{\sqrt{g_{11}}\, dr}{\sqrt{-g_{00}}\, dt} = \pm c. \tag{2.15}$$

The physical velocity of the photon (in any reference frame) is always equal to c.

From the standpoint of a distant observer (and according to his clock), the change dx in the physical radial distance with t is

$$\frac{dx}{dt} = \pm c \left(1 - \frac{r_g}{r}\right)^{1/2}. \tag{2.16}$$

Therefore, a distant observer finds that a light ray close to r_g moves slower, and as $r \to r_g$, $dx/dt \to 0$. Obviously, this behavior reflects the slowing down of time close to r_g [see (2.12)].

What time does a photon take, seen by the clock of a distant observer, in order to reach the point r_g, if the motion starts radially from $r = r_1$? We integrate equation (2.14) and obtain

$$t = \frac{r_1 - r}{c} + \frac{r_g}{c} \ln\left(\frac{r_1 - r_g}{r - r_g}\right) + t_0, \tag{2.17}$$

where r_1 is the position occupied by the photon at the moment t_0. Expression (2.17) shows that $t \to \infty$, as $r \to r_g$. Whatever coordinate r_1 from which the photon starts its fall, seen by the clock of the distant observer, the time t taken by the photon to reach r_g is infinite.

How does the photon energy change in the course of the radial motion? Energy is proportional to frequency. Let us look at the evolution of frequency. Suppose that light flashes at a point $r = r_1$ at an interval Δt. The field being static, the flashes will reach the observer at $r = r_2$ after the same interval Δt. The ratio of the proper time intervals at these two points is

$$\frac{\Delta \tau_1}{\Delta \tau_2} = \frac{\sqrt{-g_{00}(r_1)}\,\Delta t}{\sqrt{-g_{00}(r_2)}\,\Delta t}; \tag{2.18}$$

hence, the ratio of frequencies is

$$\frac{\omega_1}{\omega_2} = \frac{\Delta \tau_2}{\Delta \tau_1} = \sqrt{\frac{g_{00}(r_2)}{g_{00}(r_1)}} = \sqrt{\frac{1 - r_g/r_2}{1 - r_g/r_1}}. \tag{2.19}$$

The frequency of a quantum decreases as it leaves the gravitational field and increases as it moves centerward. This effect is called the red and blue gravitational shift, respectively.

2.4 Radial Motion of Nonrelativistic Particles

Let us now look at the radial motion of nonrelativistic particles in vacuum. We begin with free motion in which no nongravitational forces act on a particle (free fall, motion along a geodesic). The integration of the equation for a geodesic in the case $d\theta = d\phi = 0$ [see Bogorodsky 1962] yields the expression

$$\frac{dr}{dt} = \pm \frac{(1 - r_g/r)[(E/mc^2)^2 - 1 + r_g/r]^{1/2}}{E/mc^2}\,c, \tag{2.20}$$

where E is the constant of motion describing the total energy of a particle, including its mass m. If the particle is at rest at infinity where the gravitational field vanishes, then $E = mc^2$. In the general case, the value E/mc^2 may be greater or smaller than unity, but E for a particle moving outside the sphere of radius r_g is invariably positive.

At a large distance $r \gg r_g$, we find that for nonrelativistic particles $|(E - mc^2)/mc^2| \ll 1$, and expression (2.20) is rewritten in the form

$$\frac{m(dr/dt)^2}{2} = (E - mc^2) + \frac{GmM}{r}. \tag{2.21}$$

The quantity $\mathcal{E} = E - mc^2$ is the energy of a particle in Newtonian theory (where the rest, or proper, mass is not included in the energy), and thus expression (2.21) converts to the energy conservation law in Newtonian theory.

As we emphasised a few times dr/dt in (2.20) is the coordinate (not the physical) velocity. The physical velocity measured by an observer which is at rest in the Schwarzschild reference frame situated in the neighborhood of the freely moving body is

$$\frac{dx}{d\tau} = \sqrt{\frac{g_{11}}{|g_{00}|}}\,\frac{dr}{dt} = \pm \frac{[(E/mc^2)^2 - 1 + r_g/r]^{1/2}}{E/mc^2}\,c. \tag{2.22}$$

If the falling body approaches r_g, the physical velocity constantly increases: $dx/d\tau \to c$, as $r \to r_g$. By the clock of the distant observer, the velocity dx/dt

Black Holes

tends to zero as $r \to r_g$, as in the case of the photon. This fact reflects the slowing down of time as $r \to r_g$.

What is the time required for a body falling from a point $r = r_1$ to reach the gravitational radius r_g (by the clock of the distant observer)?

The time of motion from r_1 to r_g is given by the integral of (2.20). This integral diverges as $r \to r_g$. This result is not surprising because $\Delta t \to \infty$ as $r \to r_g$ even for light, and nothing is allowed to move faster than light. Furthermore, the divergence of Δt for a falling body is of the same type as for light because the velocity of the body, v, always tends to c as $r \to r_g$. Obviously, whatever force acts on a particle, the time Δt of reaching r_g is always infinite, because in this case again $v < c$. We conclude that both the free fall and motion towards r_g at any acceleration always take infinite time measured by the clock of the distant observer.

Let us return to a freely moving particle. What is the time ΔT for reaching r_g measured by the clock of the falling particle itself? It is found from the formula

$$\Delta T = \frac{1}{c} \int_{r_g}^{r_1} |ds|, \qquad (2.23)$$

where ds is taken along the world line of the particle. Here, using the expression for ds from (2.8), for $d\theta = d\phi = 0$, we find

$$\Delta T = \frac{1}{c} \int_{r_g}^{r_1} \sqrt{\left| \frac{g_{00}}{(dr/c\, dt)^2} + g_{11} \right|}\, dr. \qquad (2.24)$$

In order to calculate ΔT, we substitute into (2.24) the expression for dr/dt from (2.20). It is easy to show that the integral converges and the interval ΔT is finite. In the particular case of $E = mc^2$, when the particle falls at the parabolic (escape) velocity (i.e., $dr/dt = 0$ at $r \to \infty$), we find for the time of fall from r_1 to r

$$\Delta T = \frac{2}{3} \frac{r_g}{c} \left[\left(\frac{r_1}{r_g} \right)^{3/2} - \left(\frac{r}{r_g} \right)^{3/2} \right]. \qquad (2.25)$$

We thus conclude that while the duration Δt of falling for the distant observer is infinite, the time ΔT measured by the clock of the particle itself is finite. This result, which at first glance seems quite unexpected, can be given the following physical interpretation. The clock on the particle falling toward r_g is slowed down relative to the clock at infinity; first, because time is slowed down in the gravitational field [see (2.12)], and second, because of the Lorenz contraction of time when the velocity of the clock $v \to c$ as $r \to r_g$. As a result, the interval in t is infinite, but it becomes finite when clocked in T. This is a good example of the relativity of the infinity of time.

2.5 The Puzzle of the Gravitational Radius

Can one investigate a spacetime within the Schwarzschild sphere? The fact that the proper time of fall to the Schwarzschild sphere is finite, suggests a method of constructing a reference frame which can be extended to $r < r_g$. The reference frame must be fixed to the falling particles. No infinite accelerations F and no corresponding infinite forces will arise at the gravitational radius of this system, because the particles of the system fall freely, are weightless, and F is identically zero everywhere. The simplest such frame of reference consists of freely falling particles that have zero velocity at spatial infinity [Lemaitre's reference frame (1933); see also Rylov (1961)]. The motion of these particles is described by equation (2.25).

In order to introduce this reference frame, we choose for the time coordinate the time T measured by a clock fixed to the falling particles. At a certain instant $T = $ const that we take for $T = 0$, the ensemble of freely falling particles are at different r_1. We can choose these values of r_1, which mark the particles and remain unchanged for each of them, as the new radial coordinate in this reference frame.

The squared interval in the frame of freely falling particles is written in the form

$$ds^2 = -c^2 dT^2 + \frac{r_1 \, dr_1^2}{r_g \, B} + B^2 \, r_g^2 \left(d\theta^2 + \sin^2\theta \, d\phi^2\right), \tag{2.26}$$

where

$$B = \left[\left(\frac{r_1}{r_g}\right)^{3/2} - \frac{3}{2}\frac{cT}{r_g}\right]^{2/3}.$$

It is convenient to use instead of r_1, the following radial coordinate in the frame considered

$$R = \frac{2}{3} r_g \left(\frac{r_1}{r_g}\right)^{3/2}. \tag{2.27}$$

The squared interval (2.26) is now transformed to the form

$$ds^2 = -c^2 dT^2 + \frac{dR^2}{B^*} + (B^*)^2 \, r_g^2 \left(d\theta^2 + \sin^2\theta \, d\phi^2\right), \tag{2.28}$$

where

$$B^* = \left[\frac{3}{2}\frac{(R - cT)}{r_g}\right]^{2/3}.$$

The reference frame with interval (2.28) (the *Lemaitre reference frame*) indeed has no singularity on the Schwarzschild sphere. In order to show that this is the case, we write the explicit relation between the Schwarzschild and Lemaitre coordinates

$$r = B^* r_g, \tag{2.29}$$

$$t = \frac{r_g}{c} \left\{ -\frac{2}{3} \left(\frac{r}{r_g}\right)^{3/2} - 2 \left(\frac{r}{r_g}\right)^{1/2} + \ln\left|\frac{(r/r_g)^{1/2} + 1}{(r/r_g)^{1/2} - 1}\right| + \frac{R}{r_g} \right\}. \tag{2.30}$$

Setting $r = r_g$ in (2.29), we obtain the equation for position of the Schwarzschild sphere in the Lemaitre reference frame:

$$r_g = \frac{3}{2}(R - cT). \tag{2.31}$$

The expressions for all $g_{\alpha\beta}$ in (2.28) on the Schwarzschild sphere are quite regular, without any singularity. The calculation of all nonzero invariants of the curvature of four-dimensional spacetime also reveals no singularities on the Schwarzschild sphere. The Lemaitre reference frame extends to $r < r_g$. The spacetime in the Lemaitre R and T coordinates is shown in Figure 1 (by virtue of symmetry, the angular coordinates θ and ϕ are irrelevant).

The reference frame can be extended up to $r = 0$, that is [see (2.29)], up to $R = cT$. Here we find the true singularity of spacetime, namely infinite curvature. We find this, for example, in the fact that the curvature invariant $R_{\alpha\beta\gamma\delta}R^{\alpha\beta\gamma\delta}$ tends to infinity as $R - cT \to 0$. The infinity in this invariant signifies the infinity of gravitational tide forces.

Each freely falling particle with $R = $ const in the Lemaitre reference frame moves in time T to smaller r. The particle reaches r_g over a time T, keeps falling, and reaches the true singularity $r = 0$. Spacetime cannot be extended beyond the singularity at which gravitational tide forces grow infinitely and particles would be destroyed. In the neighborhood of $r = 0$, quantum effects of the gravitational field become essential. But we will not discuss them here.

The world lines of radial light rays can be found from (2.28) by imposing the conditions $ds = 0$, $d\theta = d\phi = 0$:

$$c\frac{dT}{dR} = \pm \left[\frac{r_g}{\frac{3}{2}(R - cT)}\right]^{1/3}. \tag{2.32}$$

The position of light cones immediately indicates why the Schwarzschild sphere plays a special role in the spherical gravitational field in general and in the Schwarzschild reference frame in particular. Indeed, if $r > r_g$, then the $r = $ const world lines (here and below we assume $\theta = $ const, $\phi = $ const) lie within the light cone, that is, they are time-like; the $r = r_g$ line coincides with the photon's world line, that is, it is light-like; and finally, if $r < r_g$, the $r = $ const world lines are space-like.

This is why the Schwarzschild reference frame formed by particles with $r = $ const cannot be extended to $r < r_g$.

This situation is found to be typical for general relativity and constitutes the difference between it and the ordinary field theory in flat space. Special coordinates must be chosen for solving Einstein's equations. Correspondingly, additional conditions are to be introduced in order to fix the form of metrics. In general, it is impossible to guarantee that the chosen coordinates cover the entire spacetime, because the spacetime of general relativity may have a complicated global structure (e.g., have a nontrivial topology). That was the situation encountered above in the attempts of describing the entire spherically symmetric spacetime in curvature coordinates (2.2). The general method of establishing whether the obtained solution indeed describes the entire spacetime or only its part is to analyze the behavior of test particles and light rays. If some of the particles reach the 'boundary' of the chosen coordinate system in a finite proper time (or for the finite value of the affine parameter for photons), and there are no physical singularities at the 'final' points of particles trajectories, then this coordinate system is incomplete. By changing the coordinate constrains and switching to metric (2.26), we were able to cover a greater part of the spacetime and, among other things, to describe the possible events below gravitational radius. A discussion of whether the Lemaitre coordinate system is indeed complete and whether metric (2.26) describes the entire spacetime will be delayed until Section 2.9, but now we return to considering the properties of the Schwarzschild sphere and the region of spacetime within it. [For a general discussion of the relevant aspects, see Hawking and Ellis 1973].

The most striking feature of the Schwarzschild sphere lies in the following. An Outward-going light ray from a point with $r > r_g$ travels to greater r and escapes to spatial infinity. For points with $r < r_g$, both rays, travel towards smaller r; they do not escape to the spatial infinity but are 'stopped' at the singularity $r = 0$. The world line of any particle necessarily lies within the light cone. For this reason, if $r < r_g$, all particles have to move toward $r = 0$: this is the direction into the future. The motion toward greater r is impossible in the region $r < r_g$ [see Finkelstien 1958]. It should be emphasized that this is true not only for freely falling particles (i.e., particles moving along geodesics) but also for particles moving at arbitrary acceleration. Neither radiation nor particles can escape from within the Schwarzschild sphere to the distant observer.

2.6 R and T Regions

In (2.8), we defined r in such a way that $g_{22} = r^2$, that is, as the radial coordinate in curvature coordinate system [see (2.8)]. Formally, within the Schwarzschild sphere r is defined in the same manner, although here the $r =$ const line is space-like and cannot serve as the radial spatial coordinate. If $r < r_g$, the quantity g_{22} is always a function of time, and a monotone function in any reference frame determined by relations (2.3)–(2.5). At $r < r_g$ all reference frames are nonstatic and both radial rays travel only to smaller r (and hence, to smaller g_{22}). Spacetime regions possessing this property are referred to as T-regions [Novikov 1962a,b, 1964a]. The spacetime region outside the Schwarzschild sphere is said to form the R-region.

Let us give a more exact definition of R- and T-regions. Consider a spherically symmetric spacetime. It may contain matter ($T_{\alpha\beta} \neq 0$) or it may be empty. By the definition of the spherically symmetric gravitational field, its metric can always be written in the form (2.2). If the $x^1 =$ const, $\theta =$ const, $\phi =$ const world line in the neighborhood of a given point is time-like, this point belongs to the R-region. If this line is space-like, the point in question belongs to the T-region.

Let us return to the case of spherically symmetric gravitational field in vacuum.

Apart from the already described Lemaitre reference frame, other reference frames are employed for analyzing a region both inside and outside the Schwarzschild sphere. Here and in the sections that follow, we describe some of these frames.

First of all, we again turn to the coordinate system (2.8). As we have shown in Section 2.2, this system has a singularity on the Schwarzschild sphere. But if r is strictly less than r_g, the metric coefficients are again regular. Is there a straight-forward physical interpretation of this system if $r < r_g$? Indeed, there is [Novikov 1961]. As was demonstrated above, now the coordinate r ($r < r_g$) cannot be a radial spatial coordinate. It can play, however, the role of the temporal coordinate, as follows from (2.8) where the coefficient with dr^2 reverses its sign on crossing the Schwarzschild sphere and is negative where $r < r_g$. On the other hand, now the coordinate t can be used as the spatial radial coordinate, the coefficients with dt^2 being positive for $r < r_g$. The coordinates r and t thus changed their roles when r became less than r_g. We change the variables, $r = -c\widetilde{T}$, $t = \widetilde{R}/c$, and rewrite (2.8) in the form

$$ds^2 = -\left[\frac{r_g}{(-c\widetilde{T})} - 1\right]^{-1} c^2\, d\widetilde{T}^2 + \left[\frac{r_g}{(-c\widetilde{T})} - 1\right] d\widetilde{R}^2 + c^2\widetilde{T}^2(d\theta^2 + \sin^2\theta\, d\phi^2),$$
(2.33)

$$0 < -c\widetilde{T} < r_g, \qquad -\infty < \widetilde{R} < \infty.$$
(2.34)

The frame of reference (2.33)–(2.34) can be realized by free test particles moving along geodesic inside the sphere $r = r_g$. A three-dimensional section $\widetilde{T} = $ const has an infinite spatial extension along the coordinate $\widetilde{R}$, while along the coordinates θ and ϕ it is closed, constituting on the whole a topological product of the sphere by a straight line. The three-dimensional volume of this section is infinite. The system is nonstationary, it contracts along θ and ϕ (the radius of the sphere decreases from r_g to 0) and expands along $\widetilde{R}$. Its proper lifetime is finite:

$$\tau = \int_{r_g/c}^{0} \sqrt{-g_{00}}\, d\widetilde{T} = \frac{\pi}{2} r_g.$$
(2.35)

The world lines of the particles with $\widetilde{R} = $ const that form the frame (2.28) in are within the Schwarzschild sphere and the system is by no means an extension of the Schwarzschild system to $r < r_g$. Time and the spatial radial direction undergo a peculiar change of roles in these systems.

Now we shall describe a reference frame which historically was the first constructed system of coordinates without singularities on r_g [Eddington 1924, Finkelstein 1958]. This reference frame is fixed to photons which move freely along the radius. The equation of motion of photons is given by (2.17). The photons moving towards the center are characterized by r decreasing with t. Expression (2.17) for such photons can be rewritten in the form

$$t = -\frac{r}{c} - \frac{r_g}{c}\ln\left|\frac{r}{r_g} - 1\right| + \frac{V}{c},$$
(2.36)

where V is a constant characterizing the radial coordinate of the photon at a fixed instant t.

The logarithm in (2.36) is taken of the modulus of the difference $r/r_g - 1$, so that (2.36) remains valid for $r < r_g$.[4] If we take a set of photons at a fixed t and assign each photon a number V which remains unchanged in the motion of the photon, this V can be chosen as another new coordinate. This is similar to our choice of r_1 for a new radial coordinate in the case of nonrelativistic particle [see (2.25)]. In fact, an essential difference must not be overlooked. Namely, no observer can move together with a photon, so that in this sense the new reference frame does not fall, strictly speaking, under the definition of a reference frame. Nevertheless, such 'system' of test photons proves to be convenient. One needs to remember, though, that V is a light coordinate (neither spatial nor temporal). For a second coordinate, we can choose the familiar coordinate r. Differentiating now (2.36) and substituting the obtained expression for dt into (2.8), we find

$$ds^2 = -\left(1 - \frac{r_g}{r}\right) dV^2 + 2\, dV\, dr + r^2(d\theta^2 + \sin^2\theta\, d\phi^2). \quad (2.37)$$

Expression (2.37) is regular on $r = r_g$. Indeed, the coefficient with dV^2 vanishes on r_g, but the presence of the term $2\, dV\, dr$ ensures that this coordinate system remains nondegenerate.

2.7 Two Types of T-Regions

Now we will discuss the principal difference between the spacetime inside and outside the Schwarzschild sphere. The properties discussed above of reference frames within the Schwarzschild sphere in the T-region are quite peculiar. Indeed, we notice that all these coordinate systems must contract along the θ and ϕ directions, and the coefficient g_{22} must decrease in time (this is equivalent to r decreasing with time). This fact can be rephrased as the inevitable motion of all light rays and all particles in the T-region toward the singularity. We know that Einstein's equations are invariant under time reversal. All the formulas given above remain the solution of Einstein's equations if the following change of variables is made: $t \to -t$, $T \to -T$, $\widetilde{T} \to -\widetilde{T}$, $V \to -U$, where U enumerates the outgoing rays ($U = 2t - V$). However, this change is equivalent to time reversal. Hence, reference frames are possible (e.g., the Lemaitre, Eddington, etc. reference frames) which expand from below the Schwarzschild sphere and are formed by particles escaping from the singularity in the T-region, intersecting later the Schwarzschild sphere, and escaping to infinity.

Is this conclusion of the escape of particles from below the Schwarzschild sphere compatible with the statement, emphasized on several occasions above, that no particle can escape from this sphere? The situation is as follows. No particle can escape from a T-region (or from within the Schwarzschild sphere) if this (or some other) particle *entered* it before. In other words, if the Schwarzschild

---

[4] One never should forget the modification in the meaning of the coordinates r and t when $r < r_g$ (see above).

sphere *can be entered*, it *cannot be escaped*. The T-region which appears in solution with reversed time is a *quite different* T-region, with very different properties. While only contraction was possible in the former T-region, only expansion is possible in the latter one, so that nothing can fall into it.

Note that the external space (beyond the $r = r_g$ sphere) is essentially the same in both cases. Its metric is reduced to (2.8) by a transformation of coordinates, but it can be extended into the Schwarzschild sphere *in two ways*; either as a contracting T-region, or as an expanding T-region (but these modes are incompatible!). It depends on boundary or initial conditions which type of T-region is realized in a specific situation. This aspect is treated in detail in the next section. The contracting T-region is usually denoted by T_-, and the expanding one, by T_+.

2.8 Gravitational Collapse and White Holes

So far we discussed the static black hole. In this section we analyze the process of formation of a black hole as a result of the contraction of a spherical mass to a size less than r_g. In order to eliminate the effects which are not directly relevant to the formation of a black hole and can only make the solution more difficult to obtain, we consider the contraction of a spherical cloud of matter at zero pressure $p = 0$ (dust cloud). There is no need then to include in the analysis the hydrodynamic phenomena due to the pressure gradient. All dust particles move along the geodesic, being subjected only to the gravitational field. The solution of Einstein's equations for this case was obtained by Tolman (1934). In the solution below, the reference frame is comoving with the matter, that is, dust particles have constant R, θ, ϕ:

$$ds^2 = -c^2\,dT^2 + g_{11}(T,R)\,dR^2 + r^2(T,R)(d\theta^2 + \sin^2\theta\,d\phi^2)\,, \qquad (2.38)$$

$$\dot{r}^2 = f(R) + \frac{F(R)}{r}\,, \qquad (2.39)$$

$$g_{11}(T,R) = \frac{(r')^2}{1 + f(R)}\,, \qquad (2.40)$$

$$\frac{8\pi G\rho}{c^2} = \frac{F'(R)}{r'r^2}\,. \qquad (2.41)$$

Here a dot denotes differentiation with respect to cT, and a prime, differentiation with respect to R; $f(R)$ and $F(R)$ are two arbitrary functions of R (subject to the condition $1 + f(R) > 0$). Equation (2.39) defines the function r after $f(R)$ and $F(R)$ have been fixed, and ρ is the density of matter. Tolman's solution can describe, for example, the compression of a spherical dust cloud of finite size. For this description, we choose an initial moment $T = \text{const}$. Then (2.41) gives density distribution. If the coordinate R_1 determines the boundary of the sphere, then beyond the sphere (at $R > R_1$) we find $\rho = 0$ and $F = \text{const}$. The evolution of r with T for the particles of the sphere is described by equation (2.39). The equation shows that each particle with fixed R and $\dot{r} < 0$ reaches the point $r = 0$, where spacetime has a true singularity, in a finite time T.

Outside the sphere, the metric of spacetime is determined in a unique manner by the value of F at the boundary R_1. In vacuum this metric is the Schwarzschild metric (see Section 2.2).

The particles on the surface of the sphere are freely falling in this outer metric, so that their motion can be described as the motion along radial geodesic in the Schwarzschild metric [see (2.20)]. We can thus consider the contraction of a spherical cloud on whose surface the particles fall at the parabolic (escape) velocity. The motion of such particles is given by especially simple formulas [see (2.25)]. The equation of the world line of these particles in the Lemaitre reference frame is $R = $ const. In Eddington-Finkelstein reference frame the equation of the same line is given parametrically by expression (2.36), (2.29), (2.30), provided we set $R = R_1 = $ const in the last two formulas.

Figure 1 represents the spacetime in the case of the contracting sphere in the Eddington-Finkelstein coordinates.

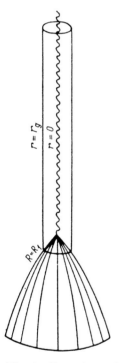

Fig. 1. Spacetime of a contracting spherical cloud creating a black hole: Eddington-Finkelstein coordinates

Figure 1, which also shows one of the rotational degrees of freedom, is especially illustrative. The surface of the contracting spherical cloud reaches the Schwarzschild sphere $r = r_g$ in a finite proper time and then contracts to a point at $r = 0$. This process is known as the relativistic *gravitational collapse*. As a result of the collapse, a spacetime region is formed within the Schwarzschild sphere from which no signals can escape to the spatial infinity. This region is defined as the *black hole*. The relativistic gravitational collapse of a spherical nonrotating body thus generates a spherical black hole.

Note now that the assumption made above on the absence of pressure produces no qualitative changes in the picture of the birth of a spherical black hole. We find the same behavior in the general case of the contraction of a sphere at a nonzero pressure ($p \neq 0$). When the surface of the contracting spherical cloud approaches the Schwarzschild sphere, no pressure can prevent the formation of the black hole [for details see Zel'dovich and Novikov 1971]. These aspects do not directly concern us here, so the specifics are omitted.

The gravitational collapse produces a contracting T-region within the Schwarzschild sphere. This result follows from the condition of continuity of the coefficient g_{22} of the metric (the coefficient with the angular spatial term) in the transition from the contracting surface of the sphere to the vacuum at a fixed instant of time.[5] As time goes on, this coefficient decreases on the surface of the contracting sphere for $r < r_g$. Therefore, it will decrease, owing to the continuity, also outside the spherical cloud (for $r < r_g$), that is, the region within the $r = r_g$ sphere is indeed the contracting T-region.

What are the conditions necessary for the formation of an expanding T_+-region? Reversing the time arrow in Figures 1, we obtain Figure 2.

They represent the expansion of the spherical cloud from within the Schwarzschild sphere. Now the continuity of g_{22} at the boundary of the sphere implies that the vacuum beyond the spherical cloud but within the Schwarzschild sphere $r = r_g$ contains the expanding T_+-region. Recall that the $r = 0$ line is space-like so that a reference frame exists in which all events on this line are simultaneous. Therefore, one cannot say (as one would be tempted to conclude at first glance) that *first* the singularity $r = 0$ existed in vacuum and *then* the matter of the spherical cloud began to expand from the singularity. These events are not connected by a time-like interval. It is more correct to say that the nature of the space-like singularity at $r = 0$ is such that it produces the expansion in vacuum (expanding T_+-region) at the lower part of Figures 2 and the expanding matter of the spherical cloud at the upper part of Figures 2. Note that no particle coming from the spatial infinity (or from any region of $r > r_g$) can penetrate the expanding T-region. Such regions of spacetime are called *white holes* [Novikov 1964b, Ne'eman 1965]. These objects cannot appear in the Universe as a result of collapse of some body, but could be formed, in principle, in the expanding Universe at the moment the expansion set in. This range of problem is discussed in detail in Novikov and Frolov (1989).

To conclude the section, we again emphasize that it is mathematically impossible to extend the solution beyond the true spacetime singularity at $r = 0$. Therefore, general relativity cannot answer the question on what will happen after the contraction to $r = 0$ in a T_--region, or what was there before the start of the expansion from $r = 0$ in a T_+-region (or even say whether these questions are correctly formulated). It is physically clear that in the neighborhood of

---

[5] Discontinuities in g_{22} result in discontinuities in spacetime.

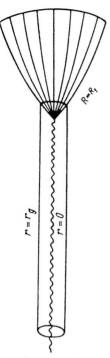

Fig. 2. Expansion of the spherical cloud from within the Schwarzschild sphere in expanding Lemaitre coordinates

$r = 0$ quantum processes become essential for spacetime itself (this effect is not described by general relativity). In this paper we do not consider the processes near the singularity.

2.9 Eternal Black Hole?

At first glance it seems that an eternal black hole might exist in empty space, that is, a black hole which does not appear as a result of contraction of mass but exists perpetually. The spacetime always contains a Schwarzschild sphere but has no contracting material cloud.

Surprisingly, it is found that the existence of such 'pure' eternal black hole is forbidden in principle. The reason is as follows. The picture (or rather, map) of the spacetime in the Lemaitre coordinates does not cover the entire spacetime. In order to demonstrate this, consider a particle which moves freely along the radius away from the Schwarzschild sphere. Its world line in the contracting Lemaitre coordinate system is given by the expression

$$const = -\frac{R}{r_g} + \frac{2cT}{r_g} - 4\left(\frac{3}{2}\frac{R-cT}{r_g}\right)^{1/3} + 2\ln\left\{\frac{\left[\frac{3}{2}(R-cT)\right]^{1/3} + r_g^{1/3}}{\left[\frac{3}{2}(R-cT)\right]^{1/3} - r_g^{1/3}}\right\}. \tag{2.42}$$

Continued into the past, this line approaches asymptotically the line $r = r_g$, without intersecting it. In the time T of the Lemaitre frame the particle exists from $T = -\infty$ on. But we know that in proper time the path from r_g to any finite r takes a finite interval of time. Therefore, the Lemaitre coordinates does not cover the entire past of the particle in question from $\tau = -\infty$ in its proper clock. Indeed, the history of a lone particle (i.e., a particle that does not interact with other particles, say, a particle that is not created in an interaction) does not terminate on the Schwarzschild sphere. The world line of such a particle must either continue indefinitely in its proper time or must terminate on the true singularity of spacetime, where new physical laws take over. Consequently, the map is incomplete and does not cover the entire space.

Is it possible to construct an everywhere empty spacetime with an eternal black hole which is complete in the sense that it covers the histories of all particles moving in this space? The answer was found to be affirmative, although it includes not only an eternal black hole but also an eternal white hole. In order to approach this construction in a natural way, consider a white hole with an expanding dust ball. Assume that the energy of motion of particles in the dust ball is such that the surface of the cloud does not escape to infinity but reaches a maximal radius and then again contracts to the size r_g and subsequently collapses to $r = 0$.

According to formula (2.20), the specific energy E/mc^2 of a particle on the surface of the dust cloud must be less than unity to ensure that $dr/dt = 0$ for a certain r. In Tolman's solution, (2.38)–(2.41), this expansion to a finite radius corresponds to the choice $f(R) < 0$. This spacetime first contains a white hole and then develops a black hole.

Let us begin to reduce the specific energy E/mc^2 of the particles on the surface of the dust ball, assuming the total mass M of the ball and, hence, the value of r_g to be fixed. In other words, by reducing E/mc^2 we reduce the share of the kinetic energy of outward motion in the total energy Mc^2 of the dust ball. As a result, the cloud will expand to gradually smaller radii. Finally, the cloud expands to $r = r_g$ when $E/mc^2 = 0$.

What if the constant E is reduced further and made negative? At first glance, this is physically meaningless; formally, it leads to enhanced maximum expansion radius r which is found by equating dr/dt in (2.20) to zero. Actually, there is nothing meaningless in this operation. In order to clarify the situation, let us look at formulas (2.38)–(2.41). Assume that the dust cloud whose evolution we analyze is homogeneous. Then the spacetime metric within the cloud corresponds to the metric of the homogeneous isotropic Universe. In solution (2.38)–(2.41), this metric corresponds to choosing the functions

$$f(R) = -\sin^2 R, \tag{2.43}$$

$$F(R) = a \sin^3 R, \tag{2.44}$$

where a is the scale factor determined by the density ρ_{max} within the spherical cloud at the moment of its maximum expansion:

$$a = \sqrt{\frac{3c^2}{8\pi G \rho_{max}}}. \tag{2.45}$$

The matter of the cloud stretches up to the boundary value of the coordinate $R = R_1$. The value R_1 may be in the range $0 < R_1 \leq \pi$. In the vacuum outside the cloud (at $R > R_1$), the particles that constitute the reference frame move freely along radial geodesics. The metric is determined by the following functions [Novikov 1963, 1964a]:

$$f(R) = -\frac{1}{(R + \cotan R_1 - R_1)^2 + 1}, \tag{2.46}$$

$$F(R) = r_g. \tag{2.47}$$

In this situation,

$$r_g = a \sin^3 R_1. \tag{2.48}$$

The sum of the masses of the particles making up the cloud equals the product of density by the sphere volume:

$$M_* = \frac{3}{4} \frac{ac^2}{G} \left(R_1 - \frac{1}{2} \sin 2R_1 \right). \tag{2.49}$$

The quantity M (the gravitational mass) characterizes the total energy of the particles in the cloud, including the gravitational energy. If the boundary coordinate R_1 lies in the range $\pi/2 < R_1 < \pi$, the inner region of the sphere is the so-called *semiclosed world* [see Zel'dovich and Novikov 1975 and the bibliography of earlier work therein]. In these conditions an increase in R_1 (addition of new layers of matter) increases M_* but diminishes M (because of a large gravitational defect of mass).

Our objective is to analyze the evolution of the spherical cloud when we supply progressively smaller and smaller specific energy to its particles. This means that we will take progressively smaller ratios M/M_*. In order to find the result of this change, we can take different values of the ratio M/M_* while fixing either M or M_*. This choice is of methodological significance only. When we are interested in the metric outside the dust cloud, we fix M which determines the outer metric.

Let us follow the evolution in the time of cloud boundary $r(R_1, T)$ for each fixed R_1 and find the metric outside the cloud. The ratio M/M_* is determined by (2.48) and (2.49):

$$\frac{M}{M_*} = \frac{2}{3} \sin^3 R_1 (R_1 - \sin R_1 \cos R_1)^{-1}. \tag{2.50}$$

The evolution of the boundary of the cloud is described by the ratio of the maximum radius of expansion, r_{max}, to the gravitational radius r_g:

$$r_{max}/r_g = \sin^{-2} R_1. \tag{2.51}$$

When $R_1 \ll \pi/2$, the ratio M/M_* is only slightly less than unity; a qualitative picture of the evolution is shown in Figure 11 ($r_{\max}/r_g \gg 1$). If $R_1 = \pi/2$, then

$$M/M_* = \frac{4}{3\pi} \quad \text{and} \quad r_{max}/r_g = 1; \qquad (2.52)$$

the corresponding situation is shown in Figure 12. If $R_1 > \pi/2$, the dust ball is a semiclosed world and the ratio M/M_* decreases as R_1 approaches π. This corresponds to $E < 0$ in (2.20).

A qualitatively new feature has emerged. The ratio r_{max}/r_g is again greater than unity. But now the boundary of the cloud does not break out from under the sphere of radius r_g in the space of the distant observer R'. A new region R'' has appeared, perfectly identical to R' outside the spherical cloud.

As $R_1 \to \pi$, the boundary R_1 gradually shifts leftward along R coordinate, leaving a progressively greater fraction of R'' free. The ratio r_{max}/r_g tends to infinity, and the ratio M/M_* to zero.

In the limit $R_1 = \pi$, the region occupied by matter vanishes, leaving the entire spacetime empty. It contains a white hole T_+, a black hole T_-, and two identical outer spaces R' and R'' which transform into Euclidean spaces at spatial infinity. This spacetime is complete in the sense that any geodesic now either continues indefinitely or terminates at the true singularity.

The reference frame covering the entire spacetime is described by a solution of type (2.46)–(2.47), where it is now convenient to place the origin of R at the minimum of the function $f(R)$. Then

$$f(R) = -\frac{1}{R^2 + 1}, \qquad F = r_g, \qquad -\infty < R < \infty, \qquad (2.53)$$

The complete, everywhere empty spacetime shown in Figure 14 was first constructed by Synge (1950) and then by Fronsl (1959), Kruskal (1960), Szekeres (1960). The physical arguments given above and the solution (2.53) were obtained by Novikov (1962b, 1963).

We have thus obtained an everywhere empty space with a white and a black hole (essentially together). These holes can be described as 'eternal' because for distant observers which are at rest in R' and R'' these holes are eternal.

The physical meaning of the second 'outer space' R'' has become clear above, where we described the evolution of a spherical cloud with progressively lower specific energy M/M_*. In the book by Novikov and Frolov (1989) we discuss whether eternal black and white holes (similar to those in Figure 14) completely devoid of matter can really exist; this is related to the problem of stability of white holes.

To conclude this section, we give the coordinate system suggested by Kruskal (1960) and Szekeres (1960). As system (2.53), this one covers the entire spacetime of eternal white and black holes. In these coordinates, the interval is written in the form

$$ds^2 = \frac{4r_g^3}{r} e^{-r/r_g} \left(-d\tilde{v}^2 + d\tilde{u}^2\right) + r^2(d\theta^2 + \sin^2\theta\, d\phi^2), \qquad (2.54)$$

where r is a function of $\tilde{v}$ and $\tilde{u}$:

$$\left(\frac{r}{r_g} - 1\right) e^{r/r_g} = \tilde{u}^2 - \tilde{v}^2. \tag{2.55}$$

The following formulas give the relation of the coordinates $\tilde{v}$ and $\tilde{u}$ with r and t in the regions R' and T_-:

for $r > r_g$
$$\begin{cases} \tilde{u} = (r/r_g - 1)^{1/2} \, e^{r/2r_g} \cosh(ct/2r_g) \\ \tilde{v} = (r/r_g - 1)^{1/2} \, e^{r/2r_g} \sinh(ct/2r_g) \end{cases} \tag{2.56}$$

for $r < r_g$
$$\begin{cases} \tilde{u} = (1 - r/r_g)^{1/2} \, e^{r/2r_g} \sinh(ct/2r_g) \\ \tilde{v} = (1 - r/r_g)^{1/2} \, e^{r/2r_g} \cosh(ct/2r_g) \end{cases} \tag{2.57}$$

Similar relations in the regions R'' and T_+ are obtained by the change of variables $\tilde{u} \to -\tilde{u}$, $\tilde{v} \to -\tilde{v}$. The Kruskal coordinate system's convenience lies in that the radial null geodesics are always plotted by straight lines inclined at an angle of 45^0 to coordinate axes.

2.10 Black Hole Celestial Mechanics

The spacetime within the Schwarzschild sphere cannot be observed by the external observer. Let us return to processes in the space which is exterior with respect to the Schwarzschild sphere of the black hole. In this section, we consider the motion of test particles along geodesics in the gravitational field of a black hole. These phenomena were analyzed in detail a long time ago and included in text books and monographs [see e.g., Zel'dovich and Novikov 1971, Misner, Thorne, and Wheeler 1973]. Here we will briefly discuss those features of motion which are specific for black holes, not just for the strong gravitational field (say, the field around the neutron star).

We consider the motion of particles in the space exterior with respect to the Schwarzschild reference frame, clocked by an observer at infinity (section 2.2). The gravitational field being spherically symmetric, the trajectory of a particle is planar; we can assume it to lie in the plane $\theta = \pi/2$. The equations of motion have the form

$$\left(\frac{dr}{c\,dt}\right)^2 = \frac{(1 - r_g/r)^2 [\tilde{E}^2 - (1 - r_g/r)(1 + \tilde{L}^2 r_g^2/r^2)]}{\tilde{E}^2}, \tag{2.58}$$

$$\frac{d\phi}{c\,dt} = \frac{(1 - r_g/r)\tilde{L} r_g}{\tilde{E} r^2}, \tag{2.59}$$

Here $\tilde{E}$ is the specific energy of a particle (per unit proper energy mc^2, m being the mass of the particle) and $\tilde{L}$ is the specific angular momentum (measured in units of mcr_g). These two quantities are conserved in the course of motion. The physical velocity v of the particle, measured by a nearby observer by his clock, τ, is directly related to the energy $\widetilde{E}$, as we find from (2.58)–(2.59):

$$\widetilde{E}^2 = (1 - r_g/r)(1 - v^2/c^2)^{-1}. \qquad (2.60)$$

The qualitative features of the motion are revealed in the following way. Setting dr/dt equal to zero, we find the points of maximum approach of the particle to the black hole and the maximum distance from it. The right-hand side (2.58) vanishes when the equality

$$\widetilde{E}^2 = (1 - r_g/r)(1 + \widetilde{L}^2\, r_g^2/r^2) \qquad (2.61)$$

is satisfied. This expression is sometimes called the *effective potential*. A typical curve (2.61) for a fixed $\widetilde{L}$ is plotted in Figure 3.

The specific energy of a moving particle remains constant; in Figure 3 this motion is shown by a horizontal line. Since the numerator of (2.58) must be positive, the horizontal segment representing the motion of the particle lies above the (2.61) curve. The intersection of the horizontal line with the effective potential gives the points of the maximum approach to, and the maximum separation from, the black hole. The trajectory of the motion of a particle is not a conic section, and, in general, is not closed. Figure 3 demonstrates horizontal for typical motions. The horizontal $\widetilde{E}_1 < 1$ corresponds to the motion in a bounded region in space between r_1 and r_2; this is an analogue of the elliptic motion in Newtonian theory (an example of such a trajectory is shown in Figure 4a).[6]

The segment $\widetilde{E}_2 > 1$ corresponds to a particle arriving from infinity and again moving away to infinity (an analogue of the hyperbolic motion). An example of this trajectory is plotted in Figure 4b.

Finally, the segment $\widetilde{E}_3$ does not intersect the potential curve but passes above its maximum $\widetilde{E}_{max}$; it corresponds to a particle falling into the black hole (gravitational capture). This type of motion is impossible in Newtonian theory and is typical for the black hole. The trajectory of this motion is shown in Figure 4c. The gravitational capture (see the next section) becomes possible owing to the maximum in the effective potential. No such maximum appears on the corresponding curve of Newtonian theory.

In addition, another type of motion is possible in the neighborhood of a black hole, namely that corresponding to the horizontal segment $\widetilde{E}_4$ in Figure 4. This line may lie below or above unity (in the latter case, for $\widetilde{E}_{max} > 1$), stretching from r_g to the intersection with the curve $\widetilde{E}(r)$. This segment represents the motion of a particle which, for example, first recedes from the black hole and reaches r_{max} [at the point of intersection of $\widetilde{E}_4$ and $\widetilde{E}(r)$] but then again falls toward the black hole and is absorbed by it (Figure 4d).

A body can escape to infinity if its energy $\widetilde{E} \geq 1$. From equation (2.60) we find that the expression for the escape velocity (v_{esc} corresponds to $\widetilde{E} = 1$) is

---
[6] If the orbit as a whole lies far from the black hole, it is an ellipse which slowly rotates in the plane of motion.

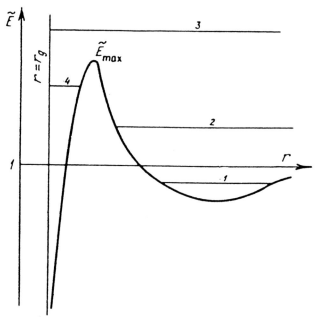

Fig. 3. Effective black hole potential. $1 - \widetilde{E} = \widetilde{E}_1$, $2 - \widetilde{E} = \widetilde{E}_2$, $3 - \widetilde{E} = \widetilde{E}_3$, $4 - \widetilde{E} = \widetilde{E}_4$

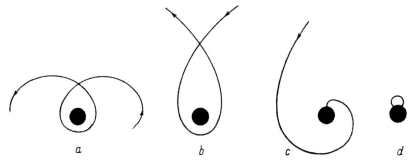

Fig. 4. Trajectories of particles with energies (a) $\widetilde{E}_1$, (b) $\widetilde{E}_2$, (c) $\widetilde{E}_3$, and (d) $\widetilde{E}_4$

$$v_{esc} = c\sqrt{r_g/r} = \sqrt{2GM/r}, \qquad (2.62)$$

which coincides with the expression given by Newtonian theory.

Let us consider the following important point. Note that in Newtonian theory, the escape velocity in the field of a point-like mass guarantees the escape to infinity regardless of the direction of motion. The case of the black hole is different. Here trajectories are possible that terminate in the black hole (of type $\widetilde{E}_4$ or $\widetilde{E}_3$ in Figure 4, the latter occurring if the particle moves towards the black hole). We have already called this effect the *gravitational capture*.

2.11 Circular Motion Around a Black Hole

Circular motion around a black hole is an important particular case of motion of a particle, in which $dr/dt \equiv 0$. This motion is shown in Figure 4 by a point at the extremum of the effective potential curve. A point at the minimum corresponds to a stable motion, and that at the maximum, to an unstable motion. The latter mode has no analogue in Newtonian theory and is specific of black holes. Of course, no real motion of a point with $\widetilde{E}$ equal to $\widetilde{E}_{max}$ for a given $\widetilde{L}$, that is, motion along an unstable curvilinear orbit, is possible, just as any motion along unstable trajectories would be impossible. If, however, the motion of a particle is represented by a horizontal line $\widetilde{E} = $ const very close to $\widetilde{E}_{max}$, then the particle makes many turns around the black hole at an r corresponding to $\widetilde{E}_{max}$ before the orbit moves far away from this value of r. An example of this motion is shown by the orbit of Figure 4b. The shape and position of the potential $\widetilde{E}(r)$ are different for different $\widetilde{L}$: the corresponding curves for some values of $\widetilde{L}$ are shown in Figure 5.

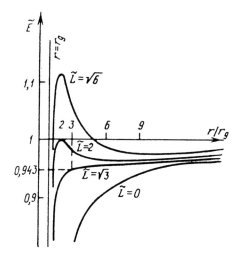

Fig. 5. Effective potentials for different angular momentums $\widetilde{L}$

The maximum and minimum appear on $\widetilde{E}(r)$ curves when $\widetilde{L} > \sqrt{3}$. If $\widetilde{L} < \sqrt{3}$ the $\widetilde{E}(r)$ curve is monotonous. Hence, the motion on circular orbits is possible only if $\widetilde{L} > \sqrt{3}$. The minima of the curves then lie at $r > 3r_g$. Stable circular orbits thus exist only for $r > 3r_g$ [Hagihara 1931]. At smaller distance, there are only unstable circular orbits corresponding to the maximum of $\widetilde{E}_{max}$ curves. If $\widetilde{L} \to \infty$, the coordinates of the maximum on $\widetilde{E}_{max}$ curves decrease to $r = 1.5\,r_g$. Even unstable circular motion becomes impossible at r less than $1.5\,r_g$.

The critical circular orbit that separates stable motions from unstable ones corresponds to $r = 3r_g$. Particles move along it at a velocity $v = c/2$, the energy of a particle being $\widetilde{E} = \sqrt{8/9} \approx 0.943$. This is the motion with the maximum possible binding energy $E \approx 0.057\, mc^2$. The velocity of motion on (unstable) orbits with $r < 3r_g$ increases, as r decreases, from $c/2$ to c on the last circular orbit with $r = 1.5\, r_g$. When $r = 2r_g$, the particle's energy is $\widetilde{E} = 1$, that is, the circular velocity is equal to the escape velocity. If r is still smaller, the escape velocity is smaller than the circular velocity. There is no paradox in it, since the circular motion here is unstable and even the tiniest perturbation (supplying momentum away from the black hole) transfers the particle to an orbit removing it to infinity, that is, an orbit corresponding to hyperbolic motion.

Now we consider the motion of an ultrarelativistic particle or a photon. In (2.58), (2.59) it corresponds to the limit $v \to c$, so that $\widetilde{E} \to \infty$ and $\widetilde{L} \to \infty$. One must keep in mind that the ratio $\widetilde{E}/\widetilde{L}$ is always equal to r_g/b, where b is the impact parameter of a particle at infinity. In view of this remark, we obtain, instead of (2.58), (2.59),

$$\left(\frac{dr}{c\,dt}\right)^2 = \left(1 - \frac{r_g}{r}\right)^2 \left[1 - \frac{b^2}{r^2}\left(1 - \frac{r_g}{r}\right)\right], \tag{2.63}$$

$$\frac{d\phi}{c\,dt} = \left(1 - \frac{r_g}{r}\right)\frac{b}{r^2}. \tag{2.64}$$

Formulas (2.63), (2.64) describe the bending of the trajectories of an ultrarelativistic particle and a light beam moving close the black hole. Setting the expression in brackets in (2.63) equal to zero, we find the position of the extremum points on the trajectory as a function of radius r. The corresponding $b(r)$ curve is shown in Figure 6. The sign of b depends on the sense of motion; we assume

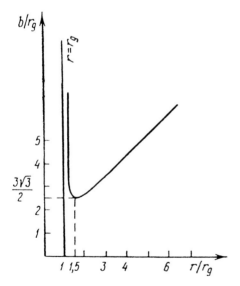

Fig. 6. The position of extrema in r on the trajectory of an ultrarelativistic particle as a function of impact parameter b

that b is positive. In this figure the motion of an ultrarelativistic particle with a given b is shown by a horizontal line $b = \text{const}$. A particle approaches the black hole, passes by it at the minimal distance corresponding to the point of intersection of $b = \text{const}$ on the right-hand branch of the $b(r)$ curve, and again recedes to infinity. If the intersection occurs close to the minimum $b_{min} = 3\sqrt{3} \times r_g/2$, the particle may go through a large number of turns before it flies away to infinity. The exact minimum of the curve $b(r)$ corresponds to (unstable) motion on a circle of radius $r = 1.5 \, r_g$ at the velocity $v = c$. Note that the left-hand branch of $b(r)$ in Figure 6 corresponds to the maximum distance between the ultrarelativistic particle and the black hole; the particle first recedes to $r < 1.5 \, r_g$ but then again falls on the black hole. Obviously, the parameter b does not have the literal meaning of the impact parameter at infinity since the particle never recedes to infinity. For a given coordinate r, this parameter can be found as a function of the tangent of the angle ψ between the trajectory of the particle and the direction to the center of the black hole:

$$b = \frac{r |\tan \psi|}{\sqrt{(1 - r_g/r)(1 + \tan^2 \psi)}}. \tag{2.65}$$

If an ultrarelativistic particle approaches the black hole on the way from infinity, and the parameter b is less than the critical value $b_{min} = 3\sqrt{3} \, r_g/2$, this particle falls into the black hole.

2.12 Gravitational Capture of Particles by a Black Hole

In the Newtonian physics gravitational capture of a test particle by a gravity center is impossible. In this section we deal with such a motion of test particle in which its trajectory terminates in the black hole. Two types of such motion are possible. First the trajectory of particle starts at infinity and ends in the black hole; second, the trajectory starts and ends in the black hole. Of course, a particle cannot be ejected from the black hole. Hence, the motion on second-type trajectory becomes possible either if the particle was placed on this trajectory via a nongeodesic curve or the particle was created close to the black hole.[7]

The gravitational capture of a particle coming from infinity is of special interest. Let us have a better look at this case.

It is clear from the analysis of motion given in the preceding section that a particle coming from infinity can be captured if its energy is greater, for a given $\tilde{L}$, than the maximum $(\tilde{E}_{max})$ on the curve $\tilde{E}(r)$. Let us consider the gravitational capture in two limiting cases, one for a particle whose velocity at infinity is much lower than the speed of light ($v_\infty/c \ll 1$) and another for a particle which is ultrarelativistic at infinity.

In the former case, $\tilde{E} = 1$. The curve $\tilde{E}(r)$, which has $\tilde{E}_{max} = 1$, corresponds to $\tilde{L}_{cr} = 2$ (see Figure 5). The maximum of this curve lies at $r = 2r_g$. Hence, this radius is minimal for the periastra of the orbits of the particles with $v_\infty = 0$

---

[7] Of course, a particle may escape from a white hole and fall into a black one; this case was discussed in Section 2.9.

which approach the black hole and again recede to infinity. If $\tilde{L} \leq 2$, gravitational capture takes place. Therefore, the impact parameter corresponding to the capture is $b_{cr} = \tilde{L}_{cr}/\tilde{E} = 2r_g(v_\infty/c)$. The capture cross-section for a nonrelativistic particle is

$$\sigma_{nonrel} = \pi b_{cr}^2 = 4\pi (v_\infty/c)^2 \, r_g^2 . \tag{2.66}$$

For an ultrarelativistic particle, $b_{cr} = 3\sqrt{3}\, r_g/2$, and the capture cross-section is

$$\sigma_{rel} = \frac{27}{4} \pi \, r_g^2 . \tag{2.67}$$

Owing to a possible gravitational capture, not every particle whose velocity exceeds the escape limit flies away to infinity. In addition, it is necessary that the angle ψ between the direction to the black hole center and the trajectory be greater than a certain critical value ψ_{cr}. For the velocity equal to the escape threshold this critical angle is given by the expression

$$\tan \psi_{cr,esc} = \pm \frac{2\sqrt{(1 - r_g/r)r_g/r}}{\sqrt{1 - 4r_g/r(1 - r_g/r)}} . \tag{2.68}$$

The plus sign is chosen for $r > 2r_g$ ($\psi_{cr} < 90^\circ$), and the minus sign is chosen for $r < 2r_g$ ($\psi_{cr} > 90^\circ$).

For an ultrarelativistic particle, the critical angle is given by the formula

$$\tan \psi_{cr,rel} = \pm \frac{\sqrt{1 - r_g/r}}{\sqrt{r_g/r - 1 + \frac{4}{27} (r/r_g)^2}} . \tag{2.69}$$

The plus sign is taken for $r > 1.5\, r_g$ and the minus for $r < 1.5\, r_g$.

2.13 Corrections for Gravitational Radiation

In the Newtonian theory there is not gravitational radiation. In relativistic theory, celestial mechanics differs from the Newtonian one in an additional factor not yet discussed above: emission of gravitational waves by accelerated bodies. As a result, the energy $\tilde{E}$ and angular momentum $\tilde{L}$ are not strictly the integrals of motion.

The emission of gravitational waves decelerates the moving particle (it loses energy and angular momentum). The force of deceleration is related to the interaction of the test particle of mass m with its own gravitational field and is proportional to m^2, while the interaction with the external field is proportional to mM. Therefore, if m/M is small, the force of 'radiative friction' is a small correction to the main force and the motion of the test particle is almost indistinguishable from the motion along a geodesic.

However these small corrections may accumulate over long periods of time and thus cause appreciable deviations of the motion from the initial trajectory.

One can evaluate the change in the cross-section of the capture of a test particle approaching the black hole from infinity, taking into account the emission of gravitational waves and the process of gradual capture of the body circling the center [Zel'dovich and Novikov 1964, 1971].

The gravitational radiation is calculated by analyzing the small perturbations of the Schwarzschild metric [Zerilli 1970a, Davis et al. 1971]. The analysis shows that the evaluation of the changes in motion within the framework of the weak field theory and for nonrelativistic velocities gives a good approximation in all interesting cases.[8]

Let us take the case of the cross-section for a particle approaching from infinity. As a result of the gravitational radiation, which mostly occurs at the periastron of the orbit, the particle cannot escape to infinity after it passes by the black hole (as would happen without radiation) but switches to a bound elongated orbit which brings the particle back to the black hole; emission will occur again at the periastron, etc., until the particle falls into the hole. Taking this into consideration, one comes to the following approximate formula for the capture cross-section of a particle of mass m and velocity v_∞ at infinity:

$$\sigma_{grav,rad} \approx \pi \left(\frac{c}{v_\infty}\right)^2 (2x)^{2/7} r_g^2, \qquad x \gg 2^6, \qquad (2.70)$$

where $x = (c^2/v^2)m/M$. If $x \leq 2^6$, the cross-section coincides with (2.66) for nonrelativistic particles.

After the first passage past the black hole at a distance r_1 at the periastron, the particle recedes to a maximum distance (apoastron) given by the approximate formula

$$l_{max} \approx \frac{r_g}{2\left[\frac{m}{M}\left(\frac{r_g}{r_1}\right)^{3/5} - \frac{v_\infty^2}{2c^2}\right]}. \qquad (2.71)$$

If r_1 is small, l_{max} rapidly decreases after subsequent passages. Ultimately, the particle falls on the black hole.

The effect of gravitational radiation is probably even more important for the analysis of the circular motion of particles. If a particle moves at $r \gg r_g$, the gradual decrease of the orbital radius obeys the following law [Landau and Lifshitz 1975]:

$$\frac{dr}{dt} = \frac{8}{5} c \left(\frac{m}{M}\right) \left(\frac{r_g}{r}\right)^3. \qquad (2.72)$$

This process lasts until the limiting stable circular orbit at $r = 3r_g$ is reached. At this orbit the binding energy $E \approx 0.057\, mc^2$ (see page 262). This energy is emitted during the entire preceding motion. The energy emitted by the particle during one revolution on the critical circle $r = 3r_g$ is $\Delta E \approx 0.1\, mc^2 (m/M)$. Then the particle slips into a spiral fall into the black hole; this takes about $(M/m)^{1/3}$ additional revolutions. The amount of energy radiated away at this stage is much less than that lost before $r = 3r_g$ was reached.

---

[8] The process of radiation in the theory of the strong field of the black hole is treated in Novikov and Frolov (1989).

3 A Rotating Black Hole

3.1 Introduction

What kind of a black hole will be formed in the case of gravitational collapse of a rotating star? In the preceding chapters we have demonstrated that the gravitational collapse of a spherical nonrotating mass produces a spherically symmetric black hole when the radius of the body becomes less than the gravitational radius. In Section 3.4 we show that after a black hole has been formed in the collapse of a body slightly deviating from spherical symmetry, all deviations from spherical symmetry rapidly vanish, except those due to small angular momentum J. Angular momentum remains practically unaltered through the collapse. If the collapsing body has an electric charge, then its total charge and electric field due to this charge are also unaltered, while all other components of the electromagnetic field in the external space also rapidly vanish.

Suppose the collapsing body deviates considerably from spherical symmetry and its angular momentum and electromagnetic field are large. Will a black hole form? If it does, what will its properties be?

It will be argued in subsequent chapters that the contraction of an arbitrary rotating mass possessing an electromagnetic field to a sufficiently small size, produces a black hole, and that all the properties of this black hole and its external gravitational field are completely determined by three parameters: mass M, angular momentum J, and electric charge Q.[9] The other properties of the collapsing body, such as its composition, asymmetry in the distribution of mass and electric charge, the magnetic field and its characteristics, and so forth, do not influence the properties of the resultant black hole.

This conclusion follows qualitatively from an analysis of the behavior of small perturbations in the course of the formation of the spherical black hole (Section 3.4). Radiative multipoles of all fields rapidly vanish, and only nonradiative modes survive, being determined by the three parameters M, J, Q. Physics knows no other 'classical' physical fields with other (nonradiative) modes. Gravitational radiation carries off a part of the energy and angular momentum of the collapsing mass in the course of the gravitational collapse when the deviations from symmetry are large. As a result, M and J of the black hole become slightly smaller than those the body had before the collapse (this will be discussed later). This reduction could not be found in the analysis of small perturbations because the back-reaction of the perturbations on the metric was assumed to be negligible. In astrophysics, the total electric charge of a body can typically be treated as small and accordingly neglected. Therefore, we consider the case of $Q = 0$.

What is the gravitational field of a black hole with nonzero angular momentum J? This field is described by the stationary axisymmetric solution of Einstein's equations which was found by Kerr (1963). We begin by describing the physical properties of the external space of a rotating black hole.

---

[9] It will be clear hereafter that the black hole forms only if the inequality $M^2 \leq J^2/M^2 + Q^2$ is satisfied (the system of units is such that $G = c = 1$).

3.2 Gravitational Field of a Rotating Black Hole

For the analysis of the external gravitational field of a nonrotating black hole and the motions of the bodies in this field we used the Schwarzschild reference frame. This reference frame is static, independent of time, and uniquely defined for each black hole. It can be thought of as a lattice 'welded' of weightless rigid rods. The motion of particles was defined with respect to this lattice. For the time variable, we used the time t of an observer placed at infinity. True, the rate of the flow of the physical (proper) time τ at each point of our lattice did not coincide with that of t (time is slowed down in the neighborhood of a black hole), but this 'parametrization' in $t = $ const meant simultaneity in our entire frame of reference.

The Schwarzschild reference frame in a certain sense resembles the absolute Newtonian space in which objects move, and t resembles the absolute Newtonian time of the equations of motion.

Of course, important differences exist. Our 'absolute' space is curved (curved very strongly close to the black hole) and the 'time' t is not the physical time.

This reference frame is used not only to facilitate mathematical manipulations in solving, say, the equations of motion, but also to increase the graphical clarity. We make use of the habitual concepts of the Newtonian physics (the 'absolute' rigid space as the scene on which events take place, and the unified time) and thereby help our intuition. Although the Schwarzschild reference frame has a singularity at r_g we choose this reference frame for the spacetime outside the black hole and not, say, the Lemaitre frame which has no singularity at r_g but is everywhere deformable.

Obviously, a rigid reference frame can be chosen only because the spacetime outside the black hole is static. In the general case of a variable gravitational field, this choice is impossible since the spatial grid would be deformed with time.

In the case of a rotating black hole (Kerr metric), the spacetime outside it is stationary and one can choose a time-independent reference frame which asymptotically transforms in the Lorentz frame at infinity. This frame is unique. The coordinates proposed by Boyer-Lindquist (1967) represent such a frame of reference. Let us express the Kerr metric in these coordinates:

$$ds^2 = -\frac{\rho^2 \Delta}{A} c^2\, dt^2 + \frac{A \sin^2 \theta}{\rho^2}\left(d\phi - \frac{2aGMr}{c^2 A}\,dt\right)^2 + \frac{\rho^2}{\Delta}\, dr^2 + \rho^2\, d\theta^2, \quad (3.73)$$

where

$$\rho^2 \equiv r^2 + \frac{a^2 \cos^2 \theta}{c^2}, \qquad \Delta \equiv r^2 - \frac{2GMr}{c^2} + \frac{a^2}{c^2},$$

$$A = \left(r^2 + \frac{a^2}{c^2}\right)^2 - \frac{a^2 \Delta \sin^2 \theta}{c^2}; \qquad (3.74)$$

a is the specific angular momentum ($a = J/M$) and M is the black hole mass. In what follows we use the system of units in which $c = G = 1$. Presumably, the physically meaningful solutions are those with $M^2 > a^2$ (see note 9 to page 266).

The properties of the three-dimensional space $t = $ const in the (3.73) metric, which is external with respect to the black hole, do not change with time. This means that there exists a Killing vector field directed along the lines of time t; shifting the spatial section along this field, we pass from one section to another identical to it. We can thus 'trace' in the space a grid which remains invariant in the transition from one section to another along the Killing vector field.[10] The variable t, that is, the time of the distant observers, can serve as the universal 'time' enumerating the spatial sections, as was the case for the Schwarzschild spacetime.

Important differences must be mentioned, of course.

(1) In the case of the Schwarzschild field, the transition from one three-dimensional section to another, preserving the coordinate grid, is carried out by shifting along the time lines perpendicular to the spatial section. The situation in the Kerr field is different, the Killing vector field being tilted with respect to the section $t = $ const; the tilting angle is different for different r and θ.

(2) The Killing vector that realizes the transition from one section to another becomes space-like at points close to the boundary of the black hole [inside ergosphere, see page 271]. This means that a three-dimensional rigid grid *cannot be made, in such regions, of material bodies* (cannot be 'welded' of rods). In the neighborhood of a black hole, this grid would move at superluminal velocity with respect to any observer (on a time-like world line).

Despite these specifics, we can still operate with our space sections $t = $ const as with 'absolute' rigid space (resembling the Newtonian case) and with t, as with 'time' which is universal in the entire 'space' (of course, subject to all the qualifications given above).

In general relativity, the splitting of spacetime in an arbitrary gravitational field into a family of three-dimensional spatial sections (in general, their geometries vary from section to section) and the universal 'time' that enumerates these sections, is referred to as '3+1' split of spacetime,[11] or the kinematic method [Vladimirov 1982]. This method is especially useful when all spatial sections are identical and the motion of particles, electromagnetic processes, etc., that unfold on this invariant 'scene' can be described in terms of a universal 'time' t.

---

[10] We have chosen the Killing vector field in such a way that far from the black hole (at infinity) the Killing vector is directed along the time lines of the Lorentz coordinate system into which the system (3.73) converts. This clarification is necessary because in addition to stationarity, metric (3.73) has the property of axial symmetry (is independent of angle ϕ). Hence, there is another Killing vector field due to the invariance of space under rotations around the symmetry axis. A linear combination of two Killing vectors is always a Killing vector (i.e., in our case we can combine a shift of section in time with a rotation around a spatial axis). We single out a Killing vector that corresponds to a lack of any rotation of the space grid around the symmetry axis at large distances from the black hole ($r \to \infty$).

[11] Another approach to the '3+1' split is possible, where one chooses first not three-dimensional sections but congruences of time-like lines. This approach is known as the chronometric method [Zel'manov 1956].

Black Holes

We have already mentioned that in this case our intuition is supported by our 'visual' images of space and time supplied by everyday experience.

Studying the processes in the vicinity of stationary black holes, we employ the kinematic method. As spatial sections, we choose the $t = \text{const}$ section in (3.73) metric; t is the time coordinate.

3.3 Specific Reference Frames

Let us consider the physical properties of the 'absolute' space introduced on the previous section. The geometric properties are described by a three-dimensional metric obtained from (3.73) by setting $dt = 0$. In this three-dimensional 'absolute' space we can, at a fixed moment of the universal 'time' $t = \text{const}$, analyze the distribution of three-dimensional vector fields, or calculate, say, the three-dimensional divergence of the vector field $\mathbf{A}$, and so on. The change in $\mathbf{A}$ with 'time' t at a fixed point of the 'absolute' space is given by the derivative $\partial \mathbf{A}/\partial \mathbf{t}$.

Consider now the reference frame of the observers which are at rest in the 'absolute' space $t = \text{const}$, that is, observers who 'sit still' on our rigid nondeformable lattice. This frame of reference is called the chronometric [Vladimirov 1982], Lagrange [Thorne and Macdonald 1982, Macdonald and Thorne 1982, Thorne, Price, and Macdonald 1986] or Killing reference frame. Let us look at the forces acting in this frame owing to the presence of a rotating black hole.

The three-dimensional components of the acceleration vector $\widetilde{F}_i$ in the coordinates r, θ, ϕ are given by the expressions [Vladimirov 1982]

$$\widetilde{F}_r = \frac{M(\rho^2 - 2r^2)}{\rho^2(\rho^2 - 2Mr)}, \qquad \widetilde{F}_\theta = \frac{Mra^2 \sin 2\theta}{\rho^2(\rho^2 - 2Mr)}, \qquad \widetilde{F}_\phi = 0. \qquad (3.75)$$

We mark all quantities in this chronometric reference frame by a tilde to avoid confusing them with the quantities used hereafter.

The physical components of acceleration are[12]

$$\widetilde{F}_{\hat{r}} = \frac{M(\rho^2 - 2r^2)\sqrt{\Delta}}{\rho^3(\rho^2 - 2Mr)}, \qquad \widetilde{F}_{\hat{\theta}} = \frac{Mra^2 \sin 2\theta}{\rho^3(\rho^2 - 2Mr)}, \qquad \widetilde{F}_{\hat{\phi}} = 0. \qquad (3.76)$$

The reference frame of our observers is rigid, so that the deformation rate tensor vanishes:

$$\widetilde{D}_{ik} = 0. \qquad (3.77)$$

The angular velocity tensor is

$$\widetilde{A}_{r\phi} = -\frac{Ma(\rho^2 - 2r^2)\sin^2 \theta}{\rho(\rho^2 - 2Mr)^{3/2}}, \qquad \widetilde{A}_{\theta\phi} = -\frac{Mra \sin 2\theta}{\rho(\rho^2 - 2Mr)^{3/2}}, \qquad \widetilde{A}_{r\theta} = 0. \qquad (3.78)$$

---

[12] $\widetilde{F}_{\hat{i}}$ are the components of the acceleration vector, measured directly by an observer at rest in the given reference frame. In formula (3.76), they are given in a local Cartesian coordinate system whose axes are aligned along the directions of r, θ, and ϕ.

The nonvanishing tensor $\tilde{A}_{ik}$ signifies that gyroscopes that are at rest in our reference frame are precessing with respect to it and, hence, with respect to distant objects, because at a large distance our rigid reference frame becomes Lorentzian. The tensor $\tilde{A}_{ik}$ is proportional to the specific angular momentum of the black hole and reflects the presence of the 'vortex' gravitational field due to its rotation.

The following important difference between the external fields of a rotating and a nonrotating black holes must be emphasized.

If a black hole is not rotating, the condition $t = $ const signifies physical simultaneity in the entire external space for the observers that are at rest in it (with respect to a rigid reference frame). In the case of a rotating black hole, a nonvanishing component g_{0i} in the rigid reference frame forbids [see Landau and Lifshitz 1975] the introduction of the concept of simultaneity. Usually, the events with equal t are said to be simultaneous in the time of a distant observer. But this does not mean at all the physical simultaneity of these events which is determined by the synchronization of clocks via sending and receiving light signals.

Note that the components $\widetilde{F}_r$, $\widetilde{F}_\theta$, the components $\tilde{A}_{ik}$, and the angular velocity of precession Ω_{pr} of a gyroscope, calculated using these components, tend to infinity, while the component g_{00} in (3.73) (which determines the rate of flow of time) vanishes at

$$\rho^2 - 2Mr = r^2 + a^2 \cos^2\theta - 2Mr = 0 \tag{3.79}$$

or at $r = r_1$, where r_1 is given by the relation

$$r_1 = M + \sqrt{M^2 - a^2 \cos^2\theta}. \tag{3.80}$$

These properties signified that a *physical singularity* exists at this point in the reference frame, and this frame cannot be prolonged closer to the black hole, that is, observers cannot be at rest relative to our grid.[13] Formally, the reason for this is the same as in the Schwarzschild field at $r = r_g$. Namely, the world line of the observer, $r = $ const, $\theta = $ const, $\phi = $ const, ceases to be time-like, as indicated by the reversal of sign of g_{00} at $r < r_1$. In fact, an essential difference in comparison with the Schwarzschild field must be emphasized.

In order to obtain a world line inside the light cone in a nonrotating black hole for $r < r_1$, it was sufficient to perform the transformation

$$r = r(\tilde{r}, t), \qquad \frac{\partial r}{\partial t} = A_1 \neq 0. \tag{3.81}$$

With a suitable choice of $A_1 = A_1(r)$, the $\tilde{r} = $ const, $\phi = $ const, $\theta = $ const line becomes time-like. This means that at $r < r_g$, a body necessarily moves centerward *along the radius*, and that r_g is the boundary of an isolated black hole.

---

[13] Of course, the coordinate grid can be extended closer to the black hole but it cannot be constructed of material bodies.

Black Holes

In the case of a rotating black hole [we assume $\Delta > 0$; see (3.73)] a transformation of the type of (3.81) cannot generate a time-like world line. But the transformation of the type

$$\phi = \phi(\tilde{\phi}, r, \theta, t), \qquad \frac{\partial \phi}{\partial t} = \Omega_1 \neq 0 \qquad (3.82)$$

makes this possible (Ω_1 is a function of r and θ). This fact signifies that if $r < r_1$ and $\Delta > 0$, all bodies necessarily participate in the rotation around the black hole (the sense of rotation is determined by the sign of a; see below) with respect to a rigid coordinate grid that stretches to infinity. As for the motion along the radius r, bodies can move in the range $r < r_1$, $\Delta > 0$ both increasing and decreasing the value of r.

Therefore, the *static (stationary) limit* r_1 has quite different nature for a rotating black hole than in the Schwarzschild field. Inside it, bodies are unavoidably dragged into rotation, although r_1 is not the event horizon because a body can escape from this region. In metric (3.73), the event horizon lies at $\Delta = 0$, that is, at $r = r_+$, where

$$r_+ = M + \sqrt{M^2 - a^2}. \qquad (3.83)$$

The region $r_+ < r < r_1$ is called the *ergosphere*.

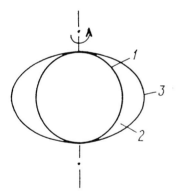

Fig. 7. A rotating black hole: 1—horizon, 2—ergosphere, 3—static limit

A rigid, static frame of reference (i.e., one that is at rest relative to a distant observer) made of material bodies does not stretch to r_+. The static limit lies beyond the horizon and coincides with it at the pole (Figure 7).

An important feature of a static reference frame is the precession of gyroscopes in it, as we have mentioned above. Our reference frame rotates at each of its points with respect to the local Lorentz frame. This is, of course, a reflection of the fact that the rotation of the black hole changes the state of motion of local Lorentz frames, dragging them into the rotation around the black hole.

This effect has been known qualitatively for quite a long time for the case of the weak gravitational field of a rotating body [Thirring and Lense 1918].

So far we discussed the properties of the gravitational field of a rotating black hole using a reference frame which does not change with time. Now we introduce into the external space of a rotating black hole a reference frame which does not rotate, in the sense given above, with respect to the local Lorentz frame. This frame of reference is said to be the reference frame of locally nonrotating observers [also known as the reference frame of zero angular momentum observers (ZAMO)]. Obviously, such a frame cannot be rigid. To introduce it, we trace a congruence of world lines which are everywhere orthogonal to the spatial sections $t = \text{const}$ that we chose. By definition, these time-like lines are not twisted and form the sought reference frame. Observers that are at rest in this frame are said to be locally nonrotating observers [sometimes this reference frame is referred to as Eulerian; see Thorne and Macdonald 1982]. These observers move with respect to the Boyer-Lindquist coordinate system, that is, they move in the 'absolute' space.[14] This motion takes place at constant r and θ, at a constant (in time) angular velocity in ϕ. If the angular velocity ω is determined with respect to the universal time t (time of the distant observer), then

$$\omega \equiv \frac{d\phi}{dt} = -\frac{g_{\phi t}}{g_{\phi\phi}} = \frac{2Mar}{(r^2+a^2)^2 - \Delta a^2 \sin^2\theta}, \qquad (3.84)$$

where $g_{\phi t}$ and $g_{\phi\phi}$ are taken from (3.73).

If angular velocity is measured by the clock of the locally nonrotating observer, then

$$\Omega_\tau = \frac{\omega}{\sqrt{-g_{tt} - 2\omega\, g_{t\phi} - \omega^2\, g_{\phi\phi}}}. \qquad (3.85)$$

The linear physical velocity of locally nonrotating observers with respect to a rigid reference frame is

$$v_\phi = \frac{2Mra\sin\theta}{\rho^2\sqrt{\Delta}}. \qquad (3.86)$$

As could be expected, this velocity becomes equal to the speed of light at the static limit $r = r_1$ and exceeds it in the ergosphere.

We again emphasize that the proper time of locally nonrotating observers, τ, is not equal to the universal 'time' t. The relation between them is equal by

---

[14] Several remarks on terminology. In fact, it is not unified among different authors and even in different papers of one author. Thus, Macdonald and Thorne (1982) refer to the space that we describe in Section 3.2 as 'absolutely rigid' and specify that locally nonrotating observers move in this space [see Section 2 of their paper prior to formula (2.6)]. In another paper, Thorne (1985) refers to the space comoving with locally nonrotating observers as absolute, and says that the observers are at rest in the 'absolute' space [see Thorne 1985, p. 11; see also: Thorne, Price, and Macdonald 1986]. We invariably hold to the former viewpoint.

We have mentioned that several terms are used to indicate the chronometric reference frame (see page 269). This 'discord' has historic roots.

the 'lapse' function α:

$$\left(\frac{d\tau}{dt}\right)_{ZAMO} \equiv \alpha = \left(\frac{\rho^2 \Delta}{A}\right)^{1/2}. \tag{3.87}$$

The expression for the vector **F** of acceleration of free fall in the reference frame of locally nonrotating observers are:

$$F_r = \frac{M}{\rho^2 \Delta \Delta_1} [(r^2 + a^2)^2 (a^2 \cos^2\theta - r^2) + 4Mr^3 a^2 \sin^2\theta],$$

$$F_\theta = a^2 \sin 2\theta \frac{Mr(r^2 + a^2)}{\rho^2 \Delta}, \qquad F_\phi = 0, \tag{3.88}$$

where $\Delta_1 = \rho^2(r^2 + a^2) + 2Mra^2 \sin^2\theta$. This vector is related to α as follows:

$$\mathbf{F} = -\nabla \ln \alpha. \tag{3.89}$$

The tensor of deformation rates of the reference frame is written in the form

$$D_{rr} = D_{r\theta} = D_{\theta\theta} = D_{\phi\phi} = 0,$$

$$D_{r\phi} = -Ma[2r^2(r^2 + a^2) + \rho^2(r^2 - a^2)] \sin^2\theta (\rho^3 \sqrt{\Delta \Delta_1})^{-1}, \tag{3.90}$$

$$D_{\theta\phi} = 2Mra^3 \sin^3\theta \cos\theta \sqrt{\Delta} (\rho^3 \sqrt{\Delta_1})^{-1},$$

and the tensor $A_{ik} = 0$.

The very important point is that the introduced frame of reference has no singularities at the static limit and extends into the ergosphere up to the boundary of the black hole, $r = r_+$. At $r \leq r_+$, falling along r necessarily occurs in addition to the rotation around the black hole. At $r = r_+$, the reference frame of locally nonrotating observers has a physical singularity $F_r \to \infty$ as $r \to r_+$ [see formula (3.88)].

As we approach the event horizon, the angular velocity of rotation of locally nonrotating observers tends to a limit

$$\omega_+ = c^3 a / 2GM r_+. \tag{3.91}$$

This limit is constant at the horizon, being independent of θ. It is called the angular velocity of rotation of the black hole (or horizon), Ω^H.

At the spatial infinity, the reference frame of locally nonrotating observers transforms into the same Lorentz frame as the Boyer-Lindquist coordinate system (the chronometric reference frame) does.

To conclude the section, consider the 'rotation' of localy nonrotating observers and the precession of gyroscopes in a reference frame fixed to these observers.

On the one hand, the reference frame of these observers was chosen to be nonrotating, that is, chosen such that $A_{ik} = 0$. This means that there is no rotation of the reference frame with respect to the locally Lorentz frame and, hence, no precession of gyroscopes in the reference frame of locally nonrotating observers. On the other hand, it is said in, for example, the monograph of Misner, Thorne, and Wheeler (1973) that gyroscopes precess with respect to locally nonrotating observers at an angular velocity

$$\Omega_{\mathrm{pr}} = \frac{1}{2} \sqrt{\frac{g_{\phi\phi}}{g_{tt} - \omega^2 g_{\phi\phi}}} \left[\frac{\omega_{,\theta}}{\rho} \mathbf{e}_{\hat{r}} - \frac{\Delta^{1/2} \omega_{,r}}{\rho} \mathbf{e}_{\hat{\theta}} \right], \qquad (3.92)$$

where $\mathbf{e}_{\hat{r}}$ and $\mathbf{e}_{\hat{\theta}}$ are unit vectors along r and θ, respectively, and the quantities $g_{\alpha\beta}$ are taken from (3.73). Is it possible to make these statements compatible?

The paradox is solved in the following way. Recall that the motion of a small element of an arbitrary frame of reference with respect to a locally comoving Lorentz frame consists of a rotation around the instantaneous rotation axis and of a deformation along the principal axes of the deformation rate tensor. In the case of no rotation ($A_{ik} = 0$), we have only deformation. A gyroscope whose center of mass is at rest in the reference frame, does not precess with respect to the principal axes of the deformation rate tensor. If lines comoving with the frame of reference (lines 'glued' to it) are traced along these directions, a gyroscope cannot change its orientation with respect to these lines. But this does not mean that the gyroscope does not change its orientation with respect to any line traced in a given element of volume in the comoving reference frame. Indeed, the anisotropic deformation tilts the lines traced, say, at an angle of 45° to the principal axes of the deformation tensor, so that they turn closer the direction of greatest extension.

The gyroscope precesses with respect to these lines even though $A_{ik} = 0$. It is this situation that we find in the case of locally nonrotating observers in the Kerr metric.

Consider locally nonrotating observers in the equatorial plane. Everywhere $A_{ik} = 0$, and formulas (3.90) imply that only the component $D_{r\phi}$ is nonzero. This means that the instantaneous orientations of the principal axes of the deformation tensor lie at an angle of 45° to the vectors $\mathbf{e}_{\hat{r}}$ and $\mathbf{e}_{\hat{\phi}}$. Note that the coordinate lines are 'glued' to the reference frame. A gyroscope does not rotate with respect to the principal axes but, in view of the remark made above, does rotate relative to the ϕ coordinate line and, hence, relative to $\mathbf{e}_{\hat{\phi}}$ (and therefore, relative to the vector $\mathbf{e}_{\hat{r}}$, perpendicular to $\mathbf{e}_{\hat{\phi}}$, which is not 'glued' to the reference frame; see below).

If a locally nonrotating observer always orients its vectors frame along the directions $\mathbf{e}_{\hat{r}}$, $\mathbf{e}_{\hat{\phi}}$ and $\mathbf{e}_{\hat{\theta}}$ the gyroscope thus precesses with respect to this frame as given by formula (3.92), although in the observers frame of reference we have $A_{ik} = 0$. The $\mathbf{e}_{\hat{r}}$, $\mathbf{e}_{\hat{\phi}}$, $\mathbf{e}_{\hat{\theta}}$ frame is a natural one; the precession of a gyroscope must be determined with respect to this frame because it is dictated by the symmetry of the space around the observer. But we could introduce a different frame as well, for example, a frame which is also fixed to locally nonrotating observers but does not rotate with respect to the instantaneously comoving Lorentz frame. Obviously, gyroscopes do not precess in such a frame.

Finally, we notice that if we choose, at some instant of time, one system of coordinate lines 'glued' to locally nonrotating observers and oriented strictly

Black Holes   275

along r, and another system oriented along ϕ, the coordinate lines directed along ϕ slide with time in the 'absolute' space along themselves while the lines perpendicular to them 'wind' on the black hole and become helical, because they are dragged by a faster motion of the observers located closer to the black hole; hence, these lines rotate with respect to the ϕ lines.

3.4 General Properties of the Spacetime of a Rotating Black Hole; Spacetime Inside the Horizon

For the discussion of the general properties of the spacetime of a nonrotaring black hole we introduced in the Chapter 2 the reference frame which did not have singularities at the event horizon.

Let us introduce in the Kerr Space Time a coordinate system which does not have coordinate singularities at the event horizon r_+ in the same manner as it was done in the Schwarzschild spacetime.[15] In that case we could use the world lines of photons moving centerward along the radii as coordinate lines [see (2.37)]. The world lines of photons moving toward a rotating black hole can also be chosen, but now the trajectories of the photons wind up on the black hole in its neighborhood, because they are dragged into rotation by the 'vortex' gravitational field. Therefore, if the black hole rotates, we have to supplement the substitution of coordinates [like (2.36)] with a 'twist' in the coordinate ϕ.

The simplest expression for the metric is obtained if we use the world lines of the photons that move at infinity at constant θ and whose projection of the angular momentum on the rotation axis of the black hole is $L_z = aE \sin^2 \theta$ (see the next section), where E is the particle energy at infinity. It can be shown that a transition to such a reference frame of 'freely falling' photons is achieved by a change of coordinates:

$$d\widetilde{V} = dt + (r^2 + a^2)\frac{dr}{\Delta}, \qquad d\widetilde{\phi} = d\phi + a\frac{dr}{\Delta}. \qquad (3.93)$$

The resulting system of coordinates is known as the Kerr (1963) coordinate system:

$$ds^2 = -[1 - \rho^{-2}(2Mr)]\, d\widetilde{V}^2 + 2\, dr\, \widetilde{V} + \rho^2 d\theta^2 +$$

$$+ \rho^{-2}[(r^2 + a^2)^2 - \Delta a^2 \sin^2 \theta]\sin^2 \theta\, d\widetilde{\phi}^2 -$$

$$- 2a \sin^2 \theta\, d\widetilde{\phi}\, dr - 4a\rho^{-2}Mr \sin^2 \theta\, d\widetilde{\phi}\, d\widetilde{V}. \qquad (3.94)$$

The general properties of the geometry of a rotating black hole are best seen on a spacetime diagram in Kerr coordinates (Figure 8). Here the time coordinate $\tilde{t}$ is substituted for the coordinates $\widetilde{V}$:

$$\tilde{t} = \widetilde{V} - r. \qquad (3.95)$$

---

[15] Of course, we ignore the trivial coordinate singularity at the pole of the spherical coordinate system: Everyone is used to it and its meaning is obvious.

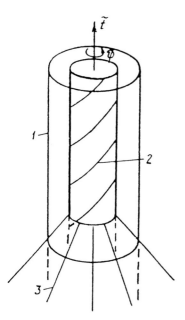

Fig. 8. Spacetime of a rotating black hole: 1—null world line along the static limit, 2—'outgoing' photons forming the horizon, 3—photons falling into the black hole

We have already employed such diagrams in the Eddington coordinates in Chapter 2. The case we are considering now is essentially different in that the Kerr metric has the axial but not the spherical spatial symmetry.

Since one of the rotational degrees of freedom (the rotation translating a point along the 'meridians' of θ) is not shown in these diagrams, they display information only on one chosen section (e.g., the equatorial plane $\theta = \pi/2$, as we see on Figure 8). The figure plots several world lines of photons that are important for describing the properties of the Kerr geometry. The first thing to remember is that the closer the coordinates are to the horizon, the more they are twisted around the black hole. The world lines of photons falling into the black hole are mapped by straight lines. In Boyer-Lindquist coordinates (a rigid grid; see Sections 3.2 and 3.3), they would appear twisted. Here the coordinate lines are twisted just as the photon trajectories are, so that these trajectories appear as straight with respect to the coordinate lines (in fact, we chose the coordinate lines precisely to have them coincide with the trajectories of the falling photons). At the static limit r_1 [see (3.80) and (3.81)], the r, θ, $\tilde{\phi}$ = const world line is a null line tangent to the light cone. At $r < r_1$, all photons and particles necessarily participate in the rotational motion around the black hole, moving at $d\tilde{\phi}/d\tilde{t} > 0$. But they can escape from below the static limit to $r > r_1$.

At the horizon, all time-like and null world lines point into the black hole, except a single null line, unique for each point of the horizon, of an 'escaping' photon; this null line is tangent to the horizon. This family of world lines 'winds up' on the horizon (see Figure 8), always staying on it. In Kerr coordinates, the equation of these null geodesics is

$$r = r_+, \qquad \theta = \text{const}, \qquad \tilde\phi = \frac{a\widetilde V}{r_+^2 + a^2}. \qquad (3.96)$$

All other photons and particles have to continue falling into the black hole after they reach the horizon.

The Kerr metric is invariant under the transformation $t \to -t$, $\phi \to -\phi$ which transforms incoming light rays into outgoing ones; hence, this transformation can be performed in (3.93). If we also substitute $\widetilde V \to -\widetilde U$, $\tilde\phi \to -\tilde\phi_+$, then equations $\widetilde U = \text{const}$, $\tilde\phi_+ = \text{const}$ describe the family of outgoing light rays and the coordinate $\widetilde U$ at infinity coincides with the ordinary coordinate of retarded time. The Kerr metric is obtained in these coordinates from (3.94) by the transformation $\widetilde V = -\widetilde U$, $\tilde\phi = -\tilde\phi_+$.

In contrast to the Schwarzschild metric, here we do not consider the continuation of the Kerr metric into the region within the horizon.[16] The reason is as follows. In the collapse of a spherical body (generating a Schwarzschild black hole), the spacetime metric beyond the collapsing body is exactly Schwarzschild both inside and outside the black hole. In the collapse of a nonrotating slightly nonspherical body, the metric outside the black hole rapidly tends to the Schwarzschild metric as $t \to \infty$. The same property holds inside the Schwarzschild black hole. The inner region of the Schwarzschild metric thus describes the real 'insides' of a nonrotating black hole.

These arguments do not hold for the Kerr metric. First, when an arbitrary rotating body contracts and turns into a black hole, the metric outside the body cannot become stationary immediately (and, hence, cannot be a Kerr metric) because gravitational waves are emitted in the course of the collapse. This statement holds both for the region outside the horizon and for that inside it. Outside the horizon all derivations from the Kerr metric are radiated away via gravitational waves, as we described in our book Novikov and Frolov (1989), and the limiting metric at $t \to \infty$ is the Kerr solution. In the external spacetime, therefore, the Kerr metric describes the real rotating black hole.

In the region inside the horizon, however, the metric does not tend to the Kerr solution, neither immediately after the collapse nor at later stages. For this reason this solution does not describe (inside the horizon) the inner structure of real rotating black hole. Note that all the above discussed properties of the black hole spacetime are valid only if $M \geq |a|$. Otherwise the horizon vanishes from the solution, and it ceases to describe the black hole. Pathological features appear [Hawking and Ellis 1973] so that this solution may hardly relate to reality. From the physics standpoint, the formation of an object with $M < |a|$ requires the compression of a rotating body with such a high angular momentum that at $r \approx r_+$ the linear velocity of rotation inevitably exceeds the speed of light. Hereafter we always assume (for noncharged black hole) that $M \geq |a|$.

---

[16] The structure of the maximal analytic continuation of the Kerr-Newman metric is discussed in our book Novikov, Frolov 1989.

3.5 Celestial Mechanics of a Rotating Black Hole

Consider the motion of test particles along geodesics in the gravitational field of a rotating black hole. In the general case, the trajectories are fairly complicated because the field has no spherical symmetry. For detailed analysis, see Bardeen et al. (1972), Stewart and Walker (1973), Ruffini and Wheeler (1971), Misner, Thorne, and Wheeler (1973), Shapiro and Teukolsky (1983), and Dymnikova (1986). Important aspects of the gravitational capture of particles by a rotating black hole were treated by Dymnikova (1982) and Bičak and Stuchlik (1976). The references given above cite numerous original publications.

We consider the motion of test particles with respect to the 'absolute' space introduced in Section 3.2, that is, with respect to the rigid lattice of the chronometric reference frame described by the Boyer-Lindquist coordinates (see Section 3.3).

First integrals of motion are written in the form

$$\rho^2 \frac{dr}{d\lambda} = \{[E(r^2 + a^2) - L_z a]^2 - \Delta[m^2 r^2 + (L_z - aE)^2 + Q^*]\}^{1/2}, \qquad (3.97)$$

$$\rho^2 \frac{d\theta}{d\lambda} = \left\{Q^* - \cos^2\theta\left[a^2(m^2 - E^2) + \frac{L_z^2}{\sin^2\theta}\right]\right\}^{1/2}, \qquad (3.98)$$

$$\rho^2 \frac{d\phi}{d\lambda} = -\left(aE - \frac{L_z}{\sin^2\theta}\right) + \frac{a}{\Delta}\left[E(r^2 + a^2) - L_z a\right], \qquad (3.99)$$

$$\rho^2 \frac{dt}{d\lambda} = -a(aE\sin^2\theta - L_z) + \frac{r^2 + a^2}{\Delta}\left[E(r^2 + a^2) - L_z a\right]. \qquad (3.100)$$

Here m is the test particle mass, $\lambda = \tau/m$, where τ is the proper time of the particle, E is the constant energy of the test particle, L_z is the constant projection of the angular momentum of a particle on the rotation axis of the black hole, and Q^* is the integral of the motion found by Carter (1968):[17]

$$Q^* = p_\theta^2 + \cos^2\theta\left[a^2(m^2 - E^2) + \sin^{-2}\theta\, L_z^2\right], \qquad (3.101)$$

where p_θ is the θ component of the four-momentum of the test particle. The motion of an ultrarelativistic particle corresponds to the limit as $m \to 0$.

3.6 Motion of Particle in the Equatorial Plane

First we consider the characteristic features of the motion of particles in the equatorial plane of a rotating black hole. In this case the expressions for $dr/d\lambda$ and $d\phi/d\lambda$ can be written in the form [Shapiro and Teukolsky 1983)]

$$r^3 \left(\frac{dr}{d\lambda}\right)^2 = E^2(r^3 + a^2 r + 2Ma^2) - 4aMEL_z - (r - 2M)L_z^2 - m^2 r\Delta, \qquad (3.102)$$

---

[17] This integral of motion is implied by the existence of the Killing tensor field in the Kerr metric [Carter 1968, 1973, 1977, Walker and Penrose 1970].

$$\frac{d\phi}{d\lambda} = \frac{(r-2M)L_z + 2aME}{r\Delta}. \qquad (3.103)$$

These expressions are analogue of equations (2.58)–(2.59) for a Schwarzschild black hole. An analysis of the peculiarities of motion is performed in the same way as in Sections 2.10, 2.11. Thus, equating the right-hand side of equation (3.102) to zero and solving it for E, we obtain the 'effective' potential. The analysis of the circular motion can be performed in the same way as in Section 2.11. The extrema of the effective potential correspond to circular motion. In this case the expressions for E_{circ} and L_{circ} have the form (for $m=1$)

$$E_{\text{circ}} = \frac{r^2 - 2Mr \pm a\sqrt{Mr}}{r(r^2 - 3Mr \pm a\sqrt{Mr})^{1/2}}, \qquad (3.104)$$

$$L_{\text{circ}} = \pm \frac{\sqrt{Mr}\,(r^2 \mp 2a\sqrt{Mr} + a^2)}{r(r^2 - 3Mr \pm 2a\sqrt{Mr})^{1/2}}. \qquad (3.105)$$

The upper signs in these and subsequent formulas correspond to the same direction of rotation of the particle as that of the black hole, and the lower signs to the opposite sense of rotation, so that we always assume that $a \geq 0$.

The radius of the circular orbit, closest to the black hole, (the motion along it is at the speed of light) is

$$r_{\text{photon}} = 2M\left\{1 + \cos\left[\frac{2}{3}\arccos\left(\mp\frac{a}{M}\right)\right]\right\}. \qquad (3.106)$$

This orbit is unstable.

The unstable circular orbit on which $E_{\text{circ}} = m$ is given by the expression

$$r_{\text{bind}} = 2M \mp a + 2M^{1/2}(M \mp a)^{1/2}. \qquad (3.107)$$

These values of the radius are the minima of periastra of all parabolic orbits. If the orbit of a particle, which comes in the equatorial plane from infinity where its velocity $v_\infty \ll c$, passes by the black hole closer than r_{bind}, the particle is captured. The radius r_0 of the periastron of the parabolic orbit is determined by the parameter $\tilde{L}$ of the particle:

$$r_0 = M[\tilde{L}^2 + \sqrt{\tilde{L}^4 - (2\tilde{L} - a/M)^2}\,], \qquad (3.108)$$

where $|\tilde{L}| < 1 + \sqrt{1 \mp a/M}$. Finally the radius of the boundary circle separating stable circular orbits from unstable ones is given by the expression

$$r_{\text{bound}} = M\{3 + Z_2 \mp [(3 - Z_1)(3 + Z_1 + 2Z_2)]^{1/2}\}, \qquad (3.109)$$

where

$$Z_1 = 1 + (1 - a^2/M^2)^{1/3}[(1 + a/M)^{1/3} + (1 - a/M)^{1/3}],$$
$$Z_2 = (3a^2/M^2 + Z_1^2)^{1/2}.$$

Table 1 lists r_{photon}, r_{bind}, and r_{bound} for the black hole rotating at the limiting angular velocity, $a = M$, and gives a comparison with the case of $a = 0$ (in units

Table 1.

Orbit	$a=0$	$a=M$	
		$L>0$	$L<0$
r_{photon}	1.5	0.5	2.0
r_{bind}	2.0	0.5	2.92
r_{bound}	3.0	0.5	4.5

of $r_{\text{g}} = 2GM/c^2$). Note that as $a \to M$, the invariant distance from a point r to the horizon r_+, which is equal to

$$\int_{r_+}^{r} \frac{\rho dr'}{\Delta(r')^{1/2}},$$

diverges. As a result, this does not mean that all orbits coincide in this limit and lie at the horizon, although at $L > 0$ the radii r of all three orbits tend to the same limit r_+ [see Bardeen et al. 1972].

Finally, we will give the values of specific energy E/m, specific binding energy $(m-E)/m$, and specific angular momentum $|L|/mM$ of a test particle at the last stable orbit, r_{bound} (see Table 2).

Table 2.

Orbit	$a=0$	$a=M$	
		$L>0$	$L<0$
E/m	$\sqrt{8/9}$	$\sqrt{1/3}$	$\sqrt{25/27}$
$(m-E)/m$	0.0572	0.423	0.0377
$|L|/mM$	$2\sqrt{3}$	$2/\sqrt{3}$	$22/3\sqrt{3}$

Equation (3.102) shows that particle motion with negative E is possible in the neighborhood of a rotating black hole. Let us solve this equation for E:

$$E = \frac{2aML + [L^2 r^2 \Delta + m^2 r\Delta + r^3 (dr/d\lambda)^2]^{1/2}}{r^3 + a^2 r + 2Ma^2}. \tag{3.110}$$

The positive sign of the radical was chosen because it corresponds to the direction of the four-momentum of the particle into the future [Misner, Thorne, and Wheeler 1973]. The numerator in (3.110) is negative if $L < 0$ and the first term is of greater magnitude than the square root of the expression in brackets.

The second and third terms in brackets can be made arbitrary small ($m \to 0$ corresponds to the transition to an ultrarelativistic particle, and $dr/d\lambda \to 0$ is the transition to the motion in the azimuthal direction). Then E may become negative if we choose points inside the ergosphere, $r < r_1$. Additional constraints appear if $m \neq 0$ and $dr/d\lambda \neq 0$.

The expression (3.110) holds only for $\theta = \pi/2$. It can be shown that orbits with negative E are possible within the ergosphere for any $\theta \neq 0$. Orbits with $E < 0$ make it possible to organize processes that extract the 'rotational energy' of the black hole. Such processes were discovered by Penrose (1969).

3.7 Motion of Particles off the Equatorial Plane

Now we consider some forms of motion of test particles off the equatorial plane. First of all let us consider the motion of nonrelativistic particles moving at parabolic velocity ($v_\infty = 0$) and zero angular momentum ($L = 0$). Such particles fall at constant θ and are dragged into rotation around the black hole in the latitudinal direction at angular velocity (3.85), that is, at the angular velocity of locally 'nonrotating observers'. Therefore, these particles fall radially at each point in the reference frame of locally 'nonrotating observers'.

Another important case is the falling of ultrarelativistic particles (photons) which move at infinity at $d\theta/d\lambda = 0$ and $L_z = aE\sin^2\theta$. Equations (3.97)–(3.100) reduce for such particles to

$$\frac{dr}{d\lambda} = -E, \quad \frac{d\theta}{d\lambda} = 0, \quad \frac{d\phi}{d\lambda} = \frac{aE}{\Delta}, \quad \frac{dt}{d\lambda} = \frac{(r^2+a^2)E}{\Delta}. \quad (3.111)$$

The world lines of these photons were used in constructing the Kerr coordinate system (Section 3.4). In Figure 9 a trajectory of nonrelativistic particle moving in a bounded region in space off the equatorial plane of a rotating black hole is plotted. The orbit is not plane.

3.8 Peculiarities of the Gravitational Capture of Bodies by a Rotating Black Hole

By analogy to Section 2.9, consider the gravitational capture of particles by a rotating black hole [this topic is reviewed in Dymnikova 1986].

The impact parameter $b_\perp$ of capturing a nonrelativistic particle moving in the equatorial plane is given by the expression

$$b_\perp = \pm 2M\frac{1}{v_\infty}\left(1 + \sqrt{1 \mp \frac{a}{M}}\right). \quad (3.112)$$

The capture cross-section for particles falling perpendicularly to the rotation axis of the black hole with $a = M$ is plotted in Figure 10 [Young 1976]. In this case the cross-section area is

$$\sigma_\perp = 14.2\,\pi(1/v_\infty)^2 M^2. \quad (3.113)$$

The impact parameter of particles falling parallel to the rotation axis, $b_\parallel$, can be found in the following manner. Let us denote $\tilde{b}_\parallel = b_\parallel/M$, $\tilde{a} = a/M$. Then $\tilde{b}_\parallel$ is found as the solution of the equation

$$(1-\tilde{a})^2 q_0^4 + 4(5\tilde{a}^2 - 4)q_0^3 - 8\tilde{a}^2(6+\tilde{a}^2)q_0^2 - 48\tilde{a}^4 q_0 - 16\tilde{a}^6 = 0, \quad (3.114)$$

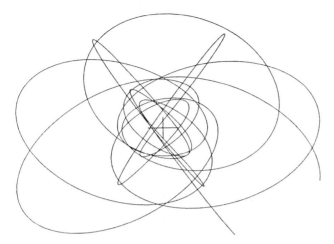

Fig. 9. (a) The plots of the trajectory of a particle around a black hole with $a = M$. Parameters of the motion are $E = 0.96m$, $L_z = 0.8mr_g$, $Q = 2.0m^2r_g^2$. Cross represents the black hole. Plot (a) shows the projection of the orbit on the XZ-plane. (Computations by P.Diener)

Fig. 9. (b) The same as Fig. 9a. Plot (b) is the yz-plane.

where $q_0 = v_\infty^2(\tilde{b}_\parallel^2 - \tilde{a}^2)$. If $\tilde{a} = 1$, then

$$\tilde{b}_\parallel = 3.85\left(\frac{1}{v_\infty}\right)M, \qquad \sigma_\parallel = 14.8\,\pi\left(\frac{1}{v_\infty}\right)^2 M^2. \qquad (3.115)$$

Consider now ultrarelativistic particles. The impact parameters of capture, $b_\perp$, for the motion in the equatorial plane are given by the following formulas:

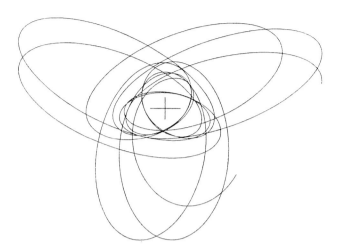

Fig. 9. (c) The same as Fig. 9a. Plot (c) is the xy-plane.

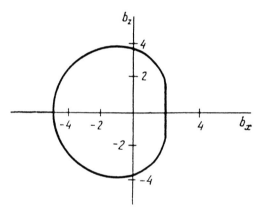

Fig. 10. The capture cross-section for particles moving with $v_\infty \to 0$ at right angles to the rotation axis of a black hole with $a = M$. Coordinate axes are marked off in units of $M(1/v_\infty)$

If the angular momentum is positive, then

$$\frac{b_\perp^+}{M} = 8\cos^3\left[\frac{1}{3}(\pi - \arccos\tilde{a})\right] + \tilde{a}. \qquad (3.116)$$

If the angular momentum is negative, then

$$\frac{b_\perp^-}{M} = -8\cos^3\left(\frac{1}{3}\arccos|\tilde{a}|\right) + \tilde{a}. \qquad (3.117)$$

In this case the cross-section for $\tilde{a} = 1$ is

$$\sigma_\perp = 24.3\,\pi M^2. \qquad (3.118)$$

For photons propagating parallel to the rotation axis of the black hole with $\tilde{a} = 1$, we have

$$\frac{b_{\|}}{M} = 2(1 + \sqrt{2}), \qquad \sigma_{\|} = 23.3\,\pi M^2. \tag{3.119}$$

A comparison of cross-sections given in this section with those of Section 2.12 demonstrates that a rotating black hole captures incident particles with lower efficiency that a nonrotating black hole of the same mass does. All these properties are important for the analysis of accretion of the interstellar medium onto a rotating black hole.

4 Electromagnetic Fields Near a Black Hole

4.1 Introduction

In this section we will consider the electric and magnetic fields near a black hole.

At first glance, black-hole electrodynamics is quite trivial. Indeed, it is known (see Novikov and Frolov 1989) that the electromagnetic field of a stationary black hole (of a given mass M) is determined unambiguously by its electric charge Q and rotation parameter a. If the charged black hole does not rotate, its electromagnetic field reduces to the radial electric field of the charge Q and is static. Any multipoles higher than the monopole are absent.

If the black hole rotates, the electromagnetic field has the form described in the section 4.8 of our book Novikov and Frolov (1989). The field is stationary but now the rotation of the black hole, first, induces a magnetic field and, second, distorts the geometry of space and generates the higher-order electric (and magnetic) moments of the fields. However, these higher-order moments are determined unambiguously by the quantities M, a, and Q, not being in any way independent, as we find in the case of ordinary bodies.

In astrophysics, the electric charge of a black hole (as of any other celestial body) cannot be high. The magnetic field must also be very weak: the dipole magnetic moment of a black hole is $\mu^* = Qa$.

There can be no other stationary electromagnetic field inherent in a black hole. In this sense, the electrodynamics of, say, radio pulsars is definitely much richer than that of the intrinsic fields of black holes. A radio pulsar is a rapidly rotating neutron star of approximately solar mass possessing a gigantic 'frozen-in' magnetic field about 10^{12} Gauss. Rotation induces high electric fields that 'rip' charges of the surface of the star, accelerate them to high energies. Energetic particles moving along curved magnetic field lines radiate hard curvature photons, which generate secondary e^+-e^--pairs in a strong magnetic field [Goldreich and Julian 1969, Sturrock 1971, Ruderman and Sutherland 1975, Arons 1983, Kardashev et al. 1984, Gurevich and Istomin 1985]. As the result, a complicated magnetosphere of the pulsar is created, and a host of other related phenomena are generated [Michel 1991, Beskin, Gurevich, and Istomin 1993].

Black holes have neither strong magnetic or electric fields [see Ginzburg 1964, Ginzburg and Ozernoi 1964)], nor a surface from which charges could be ejected. Complex electrodynamics processes are thus impossible. However, if a black hole is placed in an *external* electromagnetic field and charges in its surroundings can be produced, the situation changes dramatically and complex electrodynamics does appear. It is this aspect that we mean when black hole electrodynamics is discussed.

The case important for astrophysics applications is that of external magnetic (not electric) fields and rarefied plasma in which a black hole is embedded. A regular magnetic field in this system arises, for example, as a result of the cleaning of magnetic loops falling down into a black hole [Thorne, Price, and Macdonald 1986]. The regular magnetic field can also be generated in an accretion disk by the dynamo-action [Pudritz 1981ab, Camenzind 1990].

4.2 Maxwell's Equations in the Neighborhood of a Rotating Black Hole

The general problem of strong electromagnetic fields in the General Relativity is rather very complicated (Misner, Thorne and Wheeler 1973). We will consider electromagnetic fields against the background of a given metric, that is, we assume that these fields are not sufficiently strong for affecting the metric. As a rule, this condition is met in astrophysics.[18]

Electrodynamics equations in four-dimensional form, using the tensor $F_{\alpha\beta}$, can hardly suggest anything to our intuition. They are difficult enough to apply to even modestly difficult concrete problems of physics. Thorne and Macdonald (1982) [see also Macdonald and Thorne 1982, Thorne, Price, and Macdonald 1986] have rewritten these equations using the '3+1' split for the exterior spacetime of a rotating black hole (see Section 4.3). Their formalism operates with familiar concepts: field strength, charge density, electric current density, and so forth, 'absolute' space and unified 'time'. The equations of electrodynamics are written in a form similar to their form in the flat spacetime in the Lorentz reference frame. As a result, one cannot only use well-developed methods of solving electrodynamics but also 'rely' on conventional notions and the intuition stemming from the experience of solving such problems. Furthermore, the work of Thorne and Macdonald cited above makes use of the so-called 'membrane' interpretation of a black hole. The main point of this approach is that for a distant observer (who does not fall into the black hole), the boundary of a black hole can often be thought of as a thin membrane characterized by special electromagnetic, thermodynamic, and mechanical properties. This interpretation will be described in more detail in Section 4.4 [see also Thorne 1986, Price and Thorne 1986, Thorne, Price, and Macdonald 1986]. Of course, in reality there is no membrane. This concept needs very careful handling and one must constantly bear in mind that it is only conditional and convenient for solving certain problems. The methods outlined above make it relatively simple to apply black-hole electrodynamics to astrophysics, even when the astrophysicist is not a specialist in relativity. For a review of this field, see Thorne *et al.* (1986).

---

[18] The slow change in the parameters of the black hole as a result of electromagnetic processes is analyzed later (Sections 4.4 and 4.6).

This section presents the results of the papers of Thorne and Macdonald for axial symmetric black holes cited above. All physical quantities introduced below are three-dimensional vectors (or tensors) that are characterized by a position occupied in the 'absolute' three-dimensional space (outside the black hole) and in the 'absolute' unified 'time' t. These are quantities that a locally nonrotating observer measures by conventional instruments.[19]

We introduce the following notation for electrodynamic physical quantities measured by locally nonrotating observers: $\mathbf{E}$ is the electric field strength, $\mathbf{B}$ is the magnetic field strength, ρ_e is the electric charge density, and $\mathbf{j}$ is the electric current density. Denote ϖ the norm of a Killing vector $\xi^\mu_{(\phi)}$ reflecting the axial symmetry of spacetime

$$\varpi \equiv g_{\phi\phi}^{1/2} = A^{1/2}\sin\theta/\rho, \qquad (4.120)$$

and denote by $\mathbf{e}_{\hat{\phi}}$ a three-dimensional unit vector in the direction of the Killing vector $\xi^\mu_{(\phi)}$.

By using these notations Maxwell's equations can be written in the following form:[20]

$$\boldsymbol{\nabla}\mathbf{E} = 4\pi\rho_\mathbf{e}, \qquad (4.121)$$

$$\boldsymbol{\nabla}\mathbf{B} = \mathbf{0}, \qquad (4.122)$$

$$\boldsymbol{\nabla}\times(\alpha\,\mathbf{B}) = \frac{4\pi\alpha\mathbf{j}}{c} + \frac{1}{c}\left[\dot{\mathbf{E}} + \mathcal{L}_{\boldsymbol{\beta}}\mathbf{E}\right], \qquad (4.123)$$

$$\boldsymbol{\nabla}\times(\alpha\,\mathbf{E}) = -\frac{1}{c}\left[\dot{\mathbf{B}} + \mathcal{L}_{\boldsymbol{\beta}}\mathbf{B}\right]. \qquad (4.124)$$

Here

$$\boldsymbol{\beta} = \omega\varpi\,\mathbf{e}_{\hat{\phi}}, \qquad (4.125)$$

α is the lapse function given by (4.198), and ω is the angular velocity of rotation (in time t) of locally nonrotating observers [see (4.194)]. The notation $\mathcal{L}_{\boldsymbol{\beta}}$ is used for the Lie derivative of $\mathbf{E}$ or $\mathbf{B}$ along the vector $\boldsymbol{\beta}$, describing how these vectors vary with respect to the field of $\boldsymbol{\beta}$ e.g.,

$$\mathcal{L}_{\boldsymbol{\beta}}\mathbf{E} \equiv (\boldsymbol{\beta}\boldsymbol{\nabla})\,\mathbf{E} - (\mathbf{E}\boldsymbol{\nabla})\,\boldsymbol{\beta}. \qquad (4.126)$$

This derivative equals zero when the origin and end of the vector $\mathbf{E}$ are 'glued', in a displacement by $\boldsymbol{\beta}\,d\phi$, to the vectors $\boldsymbol{\beta}\,d\phi$. A dot denotes differentiation with respect to t and $\boldsymbol{\nabla}$ is the three-dimensional (covariant) gradient operator in the curved 'absolute' space.

Equations (4.121)–(4.122) have familiar form, while that of equations (4.123)–(4.124) is slightly unusual. The following differences are evident. A function α has appeared because the physical time flows differently at different points of

---

[19] We remind that a locally nonrotating observers are observers which possess zero angular momentum (see Section 4.3). They are also known as ZAMO's [Thorne, Price, and Macdonald 1986].

[20] In view of the applications where the electrodynamic formulas of this chapter are used in astrophysics, we write them in a conventional system of units, retaining c.

space while the equations are written in terms of the global 'time' t (recall that the acceleration of free fall, $\mathbf{F}$, is related to α in the reference frame of locally nonrotating observers by the formula $\mathbf{F} = -\mathbf{c}^2\, \nabla \ln \alpha$). Furthermore, the expressions in brackets in (4.123) and (4.124) are 'Lie-type' derivatives (with respect to time) for the set of locally nonrotating observers that move in the absolute space and for which $d\mathbf{x}/d\mathbf{t} = \boldsymbol{\beta}$. These expressions thus correspond to total derivative with respect to the time of $\mathbf{E}$ and $\mathbf{B}$, respectively, with the motion of locally nonrotating observers taken into account.

The electrodynamic equations become especially lucid and convenient for the analysis of specific problems if written in the integral form [see, e.g., Pikel'ner 1961]. Here we give only one of such integral expression (we will need it later) for the external space of the black hole, namely, Faraday's law:

$$\int_{\partial A^*(t)} \alpha \left(\mathbf{E} + \frac{1}{c} \mathbf{v} \times \mathbf{B} \right) d\mathbf{l} = -\frac{1}{c} \frac{d}{dt} \int_{A^*(t)} \mathbf{B}\, d\boldsymbol{\Sigma}. \qquad (4.127)$$

Here $d\Sigma$ is the vector of a surface element, with the vector length equal to the surface area of the element; $A^*(t)$ is a two-dimensional surface that does not intersect the horizon and is bounded by the curve $\partial A^*(t)$; $d\mathbf{l}$ is an element of $\partial A^*(t)$; and $\mathbf{v}$ is the physical velocity of $A^*(t)$ or $\partial A^*(t)$ relative to local nonrotating observers.

4.3 Stationary Electrodynamics

A rotating black hole and the space beyond are stationary and axially symmetric. The motion of matter around a black hole can very often be regarded, with high accuracy, as stationary and axisymmetric as well. It is then natural to assume that the electromagnetic field too has these properties.

In this section we assume that these conditions are satisfied.[21] Then the derivatives with respect to time t vanish and the Lie derivatives of vectors $\mathbf{E}$ and $\mathbf{B}$ take the form

$$\mathcal{L}_\beta \mathbf{E} = -(\mathbf{E}\, \nabla \omega)\, \varpi\, \mathbf{e}_{\hat{\phi}}, \qquad \mathcal{L}_\beta \mathbf{B} = -(\mathbf{B}\, \nabla \omega)\, \varpi\, \mathbf{e}_{\hat{\phi}}. \qquad (4.128)$$

It is found that under stationarity and axial symmetry, the directly measured values of $\mathbf{E}$, $\mathbf{B}$, ρ_e, and $\mathbf{j}$ are expressible in terms of three scalar functions that can be chosen in the following manner.

Let ∂A^* be a closed coordinate line for constant r and θ in the absolute three-space, and A^* be a two-dimensional surface, bounded by ∂A^*, that does not intersect the black hole. This surface necessarily intersects the axis of symmetry. We denote by $d\Sigma$ the surface element on A^* and choose its orientation so that at the axis of symmetry it is directed along z-axis (i.e., from a black hole in the northern hemisphere, and to the black hole in the southern one). Then three scalar functions mentioned above are:

---

[21] The case of nonstationary field is discussed, e.g., by Macdonald and Suen (1985), Thorne (1986), Park and Vishniak (1989, 1990).

1. total magnetic flux across ∂A^*:
$$\Psi \equiv \int_{A^*} \mathbf{B}\, d\Sigma\,; \tag{4.129}$$

2. total current inside the loop ∂A^* (taken with reversed sign):
$$I \equiv -\int_{A^*} \alpha \mathbf{j}\, d\Sigma\,; \tag{4.130}$$

3. electric potential
$$\mathcal{U}_0 \equiv -\alpha \Phi - \mathbf{A}\,\boldsymbol{\beta}. \tag{4.131}$$

In the latter relation Φ is the scalar and $\mathbf{A}$ is the vector potentials. The quantities I and Ψ depend only on the choice of the position of the loop ∂A^* but are independent of the shape of A^* (we assume the black hole to have zero magnetic charge). In other words Ψ, I, and $\mathcal{U}_0$ are the functions of r, θ-coordinates.

Before expressing $\mathbf{E}$ and $\mathbf{B}$ in terms of Ψ, I, and $\mathcal{U}_0$ we decompose the fields into poloidal (superscript P) and toroidal (superscript T) components that are perpendicular and parallel to the vector $\mathbf{e}_{\hat{\phi}}$, respectively:

$$\mathbf{E}^{\mathbf{T}} \equiv (\mathbf{E}\,\mathbf{e}_{\hat{\phi}})\,\mathbf{e}_{\hat{\phi}}\,, \qquad \mathbf{E}^{\mathbf{P}} = \mathbf{E} - \mathbf{E}^{\mathbf{T}}, \tag{4.132}$$

$$\mathbf{B}^{\mathbf{T}} \equiv (\mathbf{B}\,\mathbf{e}_{\hat{\phi}})\,\mathbf{e}_{\hat{\phi}}\,, \qquad \mathbf{B}^{\mathbf{P}} = \mathbf{B} - \mathbf{B}^{\mathbf{T}}. \tag{4.133}$$

The current density $\mathbf{j}$ is also decomposed into the poloidal ($\mathbf{j}^{\mathbf{P}}$) and toroidal ($\mathbf{j}^{\mathbf{T}}$) components.

Faraday's law (4.127) and the stationarity condition imply $\mathbf{E}^{\mathbf{T}} = \mathbf{0}$. Equation (4.122) and the condition of axial symmetry of $\mathbf{B}$ imply

$$\nabla \mathbf{B}^{\mathbf{T}} = \mathbf{0}\,, \qquad \nabla \mathbf{B}^{\mathbf{P}} = \mathbf{0}\,, \tag{4.134}$$

that is, the poloidal and toroidal magnetic lines of force can be treated independently (as not terminating anywhere).

Now we can give expressions for all electromagnetic quantities in terms of Ψ, I, and $\mathcal{U}_0$:

$$\mathbf{E}^{\mathbf{P}} = \alpha^{-1}\left(\nabla \mathcal{U}_0 + \frac{\omega}{2\pi c}\nabla \Psi\right), \tag{4.135}$$

$$\mathbf{E}^{\mathbf{T}} = \mathbf{0}\,, \tag{4.136}$$

$$\mathbf{B}^{\mathbf{P}} = \frac{\nabla \Psi \times \mathbf{e}_{\hat{\phi}}}{2\pi \varpi}\,, \tag{4.137}$$

$$\mathbf{B}^{\mathbf{T}} = -\frac{2I}{\alpha c \varpi}\,\mathbf{e}_{\hat{\phi}}\,, \tag{4.138}$$

$$\mathbf{j}^{\mathbf{P}} = \frac{\mathbf{e}_{\hat{\phi}} \times \nabla I}{2\pi \alpha \varpi}\,, \tag{4.139}$$

$$j^T \equiv \mathbf{j}^{\mathbf{T}}\mathbf{e}_{\hat{\phi}} = \frac{\varpi}{4\pi\alpha}\left\{-c\,\nabla\left[\frac{\alpha \nabla \Psi}{2\pi \varpi^2}\right] + \frac{1}{\alpha}\nabla\omega\left(\nabla \mathcal{U}_0 + \frac{\omega \nabla \Psi}{2\pi c}\right)\right\}, \tag{4.140}$$

$$\rho_e = \frac{1}{4\pi} \nabla \left[\frac{1}{\alpha} \left(\nabla \mathcal{U}_o + \frac{\omega \nabla \Psi}{2\pi c} \right) \right]. \tag{4.141}$$

Note that the last three equations can be treated as differential equations for determining $\Psi, I, \mathcal{U}_0$ (and, hence, **E** and **B** as well) provided the field sources $\mathbf{j}^P$, $\mathbf{j}^T$, and ρ_e are assumed to be given stationary and axisymmetric but otherwise arbitrary functions. Note also that in the stationary and axisymmetric case the current **j** must be prescribed in such a way that the condition $\nabla(\alpha \mathbf{j}) = \mathbf{0}$ (the charge conservation law) is satisfied, that is, $\alpha \mathbf{j}^P$ must be divergence-free.

Consider now the physical conditions in the plasma surrounding the black hole. In the case of the maximum importance for astrophysics, the conductivity of the plasma is so high that the electric field in the reference frame comoving with the plasma vanishes and the magnetic lines of force are 'frozen' into the plasma. In this case, the electric and magnetic fields in an arbitrary reference frame are perpendicular to each other (degenerate fields):

$$\mathbf{EB} = \mathbf{0}. \tag{4.142}$$

Note that this condition is only some approximation and generally a small longitudinal electric field is present. To solve problems concerning the configuration of fields, currents, and charge distributions, it is only necessary that the inequality

$$|\mathbf{EB}| \ll |\mathbf{E}^2 - \mathbf{B}^2|, \tag{4.143}$$

is satisfied instead of (4.142). Small deviations from exact equation (4.142) in the neighborhood of a black hole may prove to be essential for a number of astrophysical processes [e.g., see Kardashev *et al.* 1983, Beskin, Istomin, and Pariev 1992].

The equation (4.136) shows that the field **E** is purely poloidal. The condition (4.142) implies that **E** can be presented as the vector product of $\mathbf{B}^P$ by a vector $-\mathbf{v}^F/c$ which is a function of only r and θ and is parallel to $\mathbf{e}_{\hat{\phi}}$:

$$\mathbf{E} \equiv \mathbf{E}^P = -\frac{\mathbf{v}^F}{c} \times \mathbf{B}^P. \tag{4.144}$$

Recall that **E** and **B** are fields measured by locally nonrotating observers. It is clear that for $|\mathbf{v}^F| < c$ the equation (4.144) implies that the observer moving at a velocity $\mathbf{v}^F$ with respect to locally nonrotating observers measures only the magnetic field. The electric field is zero for this observer owing to the Lorentz transformation. Therefore, $\mathbf{v}^F$ can be interpreted as the linear velocity of the points of a magnetic lines of force with respect to locally nonrotating observers. The field **E** is completely induced by this motion. If the vector $\mathbf{v}^F$ is written in the form

$$\mathbf{v}^F = \left(\frac{\Omega^F - \omega}{\alpha} \right) \varpi \, \mathbf{e}_{\hat{\phi}}, \tag{4.145}$$

then Ω^F is the angular velocity of the points on the lines of force of the poloidal magnetic field in the 'absolute' space. The relations (4.144) and (4.145) remain also formally valid for $|\mathbf{v}^F| > c$.

The surface obtained when a magnetic line of force rotates around the symmetry axis is called the *magnetic surface*. The quantity Ψ is obviously constant on this surface. By substituting (4.144)–(4.145) into the Maxwell's equations (4.124) we get that Ω^F depends only on Ψ

$$\Omega^F = \Omega^F(\Psi). \tag{4.146}$$

It means that each line of force revolves around the black hole as a whole at the angular velocity Ω^F which is constant in t in 'absolute' space. Finally by comparing relations (4.144), (4.145), and (4.135) we obtain

$$\frac{d\mathcal{U}_0}{d\Psi} = -\frac{\Omega^F}{2\pi c}. \tag{4.147}$$

Thus $\mathcal{U}_0$ is also a function of Ψ and the equation (4.147) allows one to define it if $\Omega^F(\Psi)$ is known. In what follows we use $\Omega^F(\Psi)$ as the basic quantity instead of $\mathcal{U}_0$. Equations (4.140) for j^T and (4.141) for ρ_e get somewhat simplified

$$j^T = -\frac{\varpi}{8\pi^2 \alpha}\left[c\,\nabla\left(\frac{\alpha\,\nabla\Psi}{\varpi^2}\right) + \frac{1}{\alpha c}\left(\Omega^F - \omega\right)\nabla\Psi\,\nabla\omega\right], \tag{4.148}$$

$$\rho_e = -\frac{1}{8\pi^2 c}\nabla\left[\left(\frac{\Omega^F - \omega}{\alpha}\right)\nabla\Psi\right]. \tag{4.149}$$

We consider at first the simplest case in which inertial (and gravitational) forces acting on plasma are small in comparison with electromagnetic forces. The configuration of fields and currents is then such that currents in the reference frame comoving with the plasma are parallel to the magnetic lines of force and no Lorentz force acts on moving charges. Such fields are known as *force-free fields*. In an arbitrary reference frame the condition of existence of a force-free field is

$$\rho_e \mathbf{E} + \frac{1}{c}\mathbf{j} \times \mathbf{B} = \mathbf{0}. \tag{4.150}$$

This equation implies that $I = I(\Psi)$. In order to exclude any possible misunderstanding we stress that usually more than one magnetic surface correspond to a given value of the flux Ψ. That is why the current I as well as the other integrals of motion, which will be introduced later and which are constant on the magnetic surfaces, generally are not single-valued functions of Ψ. However localy the relations similar to $I = I(\Psi)$ have a well defined meaning.

Note that the condition (4.150) is definitely violated somewhere in the exterior space of the black hole. Indeed, the outer magnetic field in normal conditions survives in the space around the black hole because the ends of magnetic lines of force are 'frozen' into the sufficiently dense massive plasma that exists somewhat further away and that has 'transported' the magnetic field to the black hole. Condition (4.142) is met in this plasma but condition (4.150) is not. The gravitation (and inertia) keeps this plasma in the vicinity of the black hole, together with the magnetic field 'frozen' into it. The lines of force of the field pass from the dense plasma into the region of much more rarefied plasma where condition (4.150) is satisfied. Some of these lines of force go by the black hole

Black Holes 291

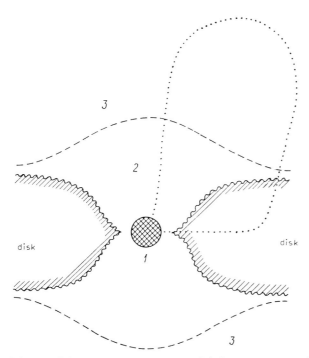

Fig. 11. Schematic representation of disk accretion to a black hole: 1—rotating black hole, 2—region of force-free field, 3—'acceleration region' where conditions (4.142) and (4.150) are violated. The dashed curve is the boundary between the regions 2 and 3. The dotted curve is an example of electric current line

and some go through it. This is the case in, for example, the model of widely discussed disk accretion on a black hole (Figure 11).

If condition (4.150) was not violated somewhere and the dense plasma did not counteract the spreading pressure of the magnetic field, this pressure would drive the outward motion of the lines of force together with the rarefied plasma.

Condition (4.142) is likely to be violated far from the black hole (region 3 in Figure 11), where the magnetic field is sufficiently weak while the inertial forces become relatively large (see the next section).

The most important and salient fact is that in the force-free approximation Ψ, I, and $\mathcal{U}_0$ are not any more arbitrary and independent, hence, ρ_e, j^T, and the divergence less part of αj^P are equally not arbitrary and independent as they would be if stationarity and axial symmetry were the only constraints. All arbitrariness in choosing them has now been eliminated. The necessary (and sufficient) condition for a force-free field to exist is that Ψ satisfies the equation called the *stream equation*:

$$\nabla \left\{ \frac{\alpha}{\varpi^2} \left[1 - \frac{(\Omega^F - \omega)^2 \varpi^2}{\alpha^2 c^2} \right] \nabla \Psi \right\} +$$

$$+ \frac{(\Omega^F - \omega)}{\alpha c^2} \frac{d\Omega^F}{d\Psi} (\boldsymbol{\nabla}\Psi)^2 + \frac{16\pi^2}{\alpha c^2 \varpi^2} I \frac{dI}{d\Psi} = 0. \tag{4.151}$$

The equation (4.151) follows directly from the force-free condition (4.150). Indeed, the force-free condition, after substituting into it the expressions for the charge density ρ_e [(4.149)] and the electric current $\mathbf{j}$ [(4.139) and (4.148)], coincides (up to the common factor $\boldsymbol{\nabla}\Psi$) with (4.151).

The stream equation (4.151) is a nonlinear elliptic equation for the potential Ψ in which the current $I(\Psi)$ and the angular velocity $\Omega^F(\Psi)$ are to be specified independently. If Ψ, $\Omega^F(\Psi)$, and $I(\Psi)$ are chosen so as to satisfy (4.151), then $\mathbf{E}$ for a region with a force-free field is found from (4.144) after substituting into it (4.145) and (4.137); $\mathbf{B}$ is found from (4.137) and (4.138), $\mathbf{j}^P$ from (4.139), and $\mathbf{j}^T$ and ρ_e from (4.148) and (4.149), respectively.

In a more general case when the mass of the particles cannot be neglected and the force-free condition (4.150) is not satisfied, a more general approach based on the one-liquid magneto-hydrodynamics (MHD) approximation can be applied. The applications of this approach in the strong gravitational field of black holes were considered by Phinney (1983a, b), Camenzind (1986a, b, 1987), Punsly and Coronity (1990a,b), Takahashi et al. (1990), Punsly (1991), Nitta, Takahashi, and Tomimatsu (1991), Hirotani et al. (1992), Beskin and Pariev[22] (1993). In this case the Maxwell's equations are to be supplemented by the continuity equation and the equation of state.

We assume that the number of particles is conserved and the continuity equation reads

$$\boldsymbol{\nabla}(\alpha n \mathbf{u}) = \mathbf{0}. \tag{4.152}$$

where $\mathbf{u} = \gamma \mathbf{v}/c$, $\gamma = (1 - v^2/c^2)^{-1/2}$, $\mathbf{v}$ is a three-velocity of matter, and n is the matter density in the comoving reference frame. In the MHD-approximation the 'frozen-in condition' (4.142) takes the form

$$\mathbf{E} + (\mathbf{v}/c) \times \mathbf{B} = \mathbf{0}. \tag{4.153}$$

This equation and (4.122) allow one to write the general solution of the continuity equation (4.152)

$$\mathbf{u} = \frac{\eta}{\alpha n} \mathbf{B} + \frac{\gamma}{\alpha} \frac{(\Omega^F - \omega)\varpi}{c} \mathbf{e}_{\hat{\phi}}. \tag{4.154}$$

The parameter η is the 'integration constant' depending on Ψ, $\eta = \eta(\Psi)$. In other words the value of η is constant on the magnetic surface. The relation (4.154) in particular shows that the poloidal component $\mathbf{u}^P$ of $\mathbf{u}$ is

$$\mathbf{u}^P = \frac{\eta}{\alpha n} \mathbf{B}^P. \tag{4.155}$$

To specify the equation of state it is convenient to use the pressure P and the entropy per a particle s as thermodynamical variables. The corresponding thermodynamical potential is the specific enthalpy μ

$$\mu = \mu(P, s) = (\rho_m + P)/n, \tag{4.156}$$

---

[22] The consideration below in this section is based on the results presented in this last paper.

where ρ_m is the energy density of matter. The first law of thermodynamics implies [see e.g., Landau and Lifshitz 1959]

$$d\mu = \frac{1}{n} dP + T\, ds. \tag{4.157}$$

The matter density n and the temperature T are defined as

$$n = n(P, s) = \left[\left(\frac{\partial \mu}{\partial P}\right)_s\right]^{-1}, \tag{4.158}$$

$$T = T(P, s) = +\left(\frac{\partial \mu}{\partial s}\right)_P. \tag{4.159}$$

The above relations allow one to express μ, T, and P as functions of n and s

$$\mu = \mu(n, s), \qquad T = T(n, s), \qquad P = P(n, s). \tag{4.160}$$

We also make an additional assumption that the matter flow is isentropic

$$(\mathbf{u}\nabla) s = 0. \tag{4.161}$$

In the axisymmetric case equations (4.155) and (4.161) yield

$$s = s(\Psi). \tag{4.162}$$

In the MHD-approximation the current I fails to be constant on the magnetic surface. Nevertheless the following two combinations $E(\Psi)$ and $L(\Psi)$ containing I are constant on the magnetic surface

$$E(\Psi) = \frac{\Omega^F I}{2\pi} + \mu c^2 \eta (\alpha \gamma + \frac{\omega \varpi}{c} u_{\hat{\phi}}), \tag{4.163}$$

$$L(\Psi) = \frac{I}{2\pi} + \mu c \eta \varpi\, u_{\hat{\phi}}. \tag{4.164}$$

Here $u_{\hat{\phi}} = \mathbf{u}\mathbf{e}_{\hat{\phi}}$. These relations reflect the conservation of the total energy and the ϕ-component of the momentum in the system. The first term on the right-hand side of (4.163) and (4.164) corresponds to the contribution of the electromagnetic field, while the second ones are connected with the contribution of particles. Thus in the one-liquid MHD approximation there exist five quantities which are constant on the magnetic surface: $\Omega^F(\Psi)$, $E(\Psi)$, $L(\Psi)$, $\eta(\Psi)$, and $s(\Psi)$. We call these quantities integrals of motion.

The next step is to show that for a known poloidal field $\mathbf{B}^\mathbf{P}$ (i.e., for known potential Ψ) and for given integrals of motion one can reconstruct the toroidal magnetic field $\mathbf{B}^\mathbf{T}$, the matter density n, and velocity $\mathbf{v}$. For this purpose we use the conservation laws (4.163) and (4.164) which, together with the ϕ-component of (4.154), allows one to express the electric current I, the Lorentz factor γ, and $u_{\hat{\phi}}$ as follows

$$\frac{I}{2\pi} = \frac{\alpha^2 L - (\Omega^F - \omega)(E - \omega L)\varpi^2/c^2}{\alpha^2 - (\Omega^F - \omega)^2 (\varpi^2/c^2) - \mathcal{M}^2}, \tag{4.165}$$

$$\gamma = \frac{1}{\alpha\mu c^2\eta} \frac{\alpha^2(E - \Omega^F L) - \mathcal{M}^2(E - \omega L)}{\alpha^2 - (\Omega^F - \omega)^2 (\varpi^2/c^2) - \mathcal{M}^2}, \qquad (4.166)$$

$$u_{\hat{\phi}} = \frac{1}{\mu c\eta\varpi} \frac{(E - \Omega^F L)(\Omega^F - \omega)\varpi^2/c^2 - \alpha^2 L\mathcal{M}^2}{\alpha^2 - (\Omega^F - \omega)^2 (\varpi^2/c^2) - \mathcal{M}^2}, \qquad (4.167)$$

Here we denote

$$\mathcal{M}^2 \equiv \frac{4\pi\mu\eta^2}{n}. \qquad (4.168)$$

We emphasize that $\mathcal{M}$ remains finite at the horizon, provided the field and matter fluxes are regular there.[23] According to the equation (4.160) $\mu = \mu(n, s)$ and, hence, the relation (4.168) allows one to express n (and, hence, μ) as a function of η, s, and $\mathcal{M}$. It means that besides the integrals of motion only one additional quantity ($\mathcal{M}$) enters the expressions for I, γ, and $u_{\hat{\phi}}$.

We show now that if $\Psi = \Psi(r, \theta)$ is known then one can define $\mathcal{M}$ in terms of integrals of motions. For this purpose one can use the relation $\gamma^2 - \mathbf{u}^2 = 1$. After substitution of expressions (4.166) for γ, (4.167) for $u_{\hat{\phi}}$, and (4.155), (4.137) for u^P into this relation one gets

$$\frac{K}{(\alpha\mu c\eta\varpi\mathcal{A})^2} - \frac{\mathcal{M}^4(\nabla\Psi)^2}{64\,\pi^4(\alpha\mu\eta\varpi)^2} = 1, \qquad (4.169)$$

where

$$\mathcal{A} = \alpha^2 - (\Omega^F - \omega)^2 (\varpi^2/c^2) - \mathcal{M}^2, \qquad (4.170)$$

and

$$K = \alpha^2(\varpi^2/c^2)(E - \Omega^F L)^2(\mathcal{A} - \mathcal{M}^2) + \mathcal{M}^4[(\varpi^2/c^2)(E - \omega L)^2 - \alpha^2 L^2]. \quad (4.171)$$

The equation (4.169) defines $\mathcal{M}$ as the function of $(\nabla\Psi)^2$ and integrals of motion

$$\mathcal{M} = \mathcal{M}[(\nabla\Psi)^2, E, L, \Omega^F, \eta, s\,]. \qquad (4.172)$$

We remind that

$$(\nabla\Psi)^2 = 4\pi^2\varpi^2\,(\mathbf{B^P})^2. \qquad (4.173)$$

Thus for the known poloidal field $\mathbf{B^P}$ the relations (4.160), (4.165)–(4.168), and (4.172) define the toroidal magnetic field $\mathbf{B^T}$, the matter density n, and velocity $\mathbf{v}$.

The stream equation (4.151) defining the magnetic flux Ψ can be also generalized for the MHD-approximation. In this approximation the force-free equation (4.150), defining the structure of the poloidal magnetic field, is to be changed by the following equation, which is the poloidal component of the covariant conservation law

$$n[(\mathbf{u}\nabla)(\mu\mathbf{u})]^\mathbf{P} + \frac{n}{\alpha}\mu\gamma[\frac{\varpi\,\mathbf{u}_{\hat{\phi}}}{c}\nabla\omega + \gamma\nabla\alpha] = -\nabla P + \rho_\mathbf{e}\,\mathbf{E} + [(\mathbf{j}/\mathbf{c}) \times \mathbf{B}]^\mathbf{P}, \quad (4.174)$$

---

[23] By introducing the Alfven velocity $u_A = B^P(4\pi n\mu)^{-1/2}$ we can also write $\mathcal{M}$ in the form $\mathcal{M} = \alpha u^P/u_A$. Thus $\mathcal{M}$ is (up to the factor α) the Mach number calculated for the poloidal velocity u^P with respect to the Alfven velocity u_A.

where P is the pressure, n is the density, and $\mathbf{u} = \gamma\mathbf{v}$. As a result, the stream equation takes the form

$$\mathcal{A}\left[\nabla\left(\frac{1}{\alpha\varpi^2}\nabla\Psi\right) + \frac{1}{\mathcal{D}\alpha\varpi^2(\nabla\Psi)^2}\nabla^i\Psi\nabla^j\Psi\nabla_i\nabla_j\Psi\right]$$

$$+ \frac{1}{\alpha\varpi^2}\nabla'\mathcal{A}\nabla\Psi - \frac{\mathcal{A}}{2\mathcal{D}\alpha\varpi^2(\nabla\Psi)^2}\nabla'F\nabla\Psi + \frac{(\Omega^F - \omega)}{\alpha c^2}\frac{d\Omega^F}{d\Psi}(\nabla\Psi)^2 \quad (4.175)$$

$$+ \frac{32\pi^4}{\alpha\varpi^2 c^2 \mathcal{M}^2}\frac{\partial}{\partial\Psi}\left(\frac{G}{\mathcal{A}}\right) - 16\pi^3\alpha\mu n \frac{1}{\eta}\frac{d\eta}{d\Psi} - 16\pi^3\alpha nT\frac{ds}{d\Psi} = 0.$$

Here

$$D = \frac{1}{\mathcal{M}^2}\left[\mathcal{A} + \frac{16\pi^2 I^2}{c^2(\nabla\Psi)^2} - \frac{\mathcal{A}}{(u^P)^2}\frac{a_s^2}{c^2 - a_s^2}\right], \quad (4.176)$$

$$F = \frac{64\pi^4}{\mathcal{M}^4}\left(\frac{K}{c^2\mathcal{A}^2} - \alpha^2\varpi^2\mu^2\eta^2\right), \quad (4.177)$$

$$G = \alpha^2(E - \Omega^F L)^2(\varpi^2/c^2) + \alpha^2\mathcal{M}^2 L^2 - \mathcal{M}^2(E - \omega L)^2(\varpi^2/c^2), \quad (4.178)$$

$$(u^P)^2 = \frac{\mathcal{M}^4(\nabla\Psi)^2}{64\pi^4(\alpha\mu\eta\varpi)^2}, \quad (4.179)$$

and a_s is a sound velocity

$$a_s^2 = c^2\frac{1}{\mu}\left(\frac{\partial P}{\partial n}\right)_s. \quad (4.180)$$

In the equation (4.175) the quantities μ, n, T, and a_s are to be expressed in terms of $\mathcal{M}$ and integrals of motions. The gradient ∇' denotes the action of ∇ under the condition that $\mathcal{M}$ is fixed. The derivative $\partial/\partial\Psi$ in the expression $\partial/\partial\Psi(G/\mathcal{A})$ is acting only on the integrals of motions $\Omega^F(\Psi)$, $E(\Psi)$, $L(\Psi)$, $\eta(\Psi)$, while other variables in a $G/\mathcal{A}$ are considered as constants. Finally, in the obtained relation the quantity $\mathcal{M}$ must be expressed by means of (4.172). (For more details, see [Nitta, Takahashi, and Tomimatsu 1991, Beskin and Pariev 1993]).

The stream equation (4.175) is the desired equation for the poloidal field, which contains only the magnetic flux Ψ and five integrals of motion depending on it. When the mass of the particle can be neglected ($\mu = 0$) this equation reduces to the force-free stream equation (4.151).

An MHD-flow, described by the equation (4.175), is characterized by the following singular points, in which the matter velocity becomes equal to the velocity of the electromagnetic waves. These points are:

1. *Alfven points* defined by the condition $\mathcal{A} = 0$. In these points the total velocity of matter u is equal to the Alfven velocity $u_A = B(4\pi\mu n)^{-1/2}$.
2. *Fast and slow magneto-sonic points* defined by the condition $D = 0$. In these points the matter velocity is equal to the velocity of the fast or slow MHD-wave.

The stream equation (4.175) is elliptic for $D > 0$ and is hyperbolic for $D < 0$ and unlike the force-free equation (4.151) it is of mixed type. In the case when the density of matter n remains finite at the horizon of a black hole one has

$$D(r_+) = -1. \quad (4.181)$$

If in the region where plasma is generated its velocity is less than the fast magneto-sonic velocity (and hence $D > 0$) then it is evident that between this region and the horizon there exists the fast magneto-sonic singular point. It should be stressed that the singular points described above are well known in solar physics [Weber and Davis 1967, Mestel 1968, Michel 1969 and Sakurai 1985] and in pulsar physics [Okamoto 1978, Kennel, Fujimura, and Okamoto 1983, Bogovalov 1989, Takahashi 1991]. In the case of black holes, besides similar singular points connected with ejection of matter there is a new additional family of singular points related with the accretion of matter in a strong gravitational field.

To conclude this section we emphasize that the above consideration does not take into account a number of processes which might be important for real systems. The plasma interaction with the radiation field, the radiation of the plasma itself and electron-positron pair creation in the magnetic field of a black hole are among them [Novikov and Thorne 1973, Shapiro and Teukolsky 1983, Begelman, Blandford, and Rees 1984, Blandford 1990].

4.4 Boundary Conditions at the Event Horizon

Black-hole electrodynamics treats only processes outside the event horizon. Generally speaking, the solution of electrodynamic equations requires that boundary conditions far from the black hole are to be supplemented with boundary conditions on its surface. Formally, this resembles the situation in pulsar electrodynamics where boundary conditions on the surface of the neutron star must also be specified.

Nevertheless, the two situations are principally different. In contrast to a neutron star a black hole has no material surface that differs from the surrounding space. For the black holes the role of the boundary conditions are played by the obvious physical requirements that the region of spacetime lying inside the black hole cannot affect the processes outside it and that all physical observable measured at the horizon in the freely-falling reference frame are to be finite.

The event horizon is generated by null geodesics which are characteristics of Maxwell's equations. For this reason the regularity conditions allow an attractive mathematical formulation [Znajek 1978 and Damour 1978]. Namely, it was found that the corresponding boundary conditions can be written in a very clear form if one assumes that the surface of the black hole has a *fictitious* surface electric charge density σ^H that compensates for the flux of electric field across the surface, and a *fictitious* surface electric current $\mathbf{i}^H$ that closes tangent components of the magnetic fields at the horizon. This interpretation is used in the *membrane formalism* [Thorne 1986, Thorne et al. 1986].

Black Holes

We remind that in the '3+1'-formalism the horizon is a two-dimensional surface of infinite gravitational red-shift, $\alpha = 0$. The red-shifted gravitational acceleration $\alpha \mathbf{F} \equiv -\mathbf{c}^2 \alpha \nabla \ln \alpha$ remains finite at the horizon

$$(\sigma \mathbf{F})_\mathbf{H} = -\kappa \mathbf{n}, \tag{4.182}$$

where $\mathbf{n}$ is a unit vector pointing orthogonally out of the horizon, and κ is the surface gravity. In calculations near the horizon it is convenient to introduce a coordinate system (α, λ, ϕ), where λ is a proper distance along the horizon from the north pole toward the equator. In this coordinates the metric of the absolute 3-space near the horizon takes the form

$$ds^2 = (c^2/\kappa)^2 \, d\alpha^2 + d\lambda^2 + \varpi^2 \, d\phi^2, \tag{4.183}$$

and the unit vectors along the 'toroidal' (ϕ), 'poloidal' (λ), and 'normal' (α) directions are

$$e_{\hat{\phi}}^\mu \frac{\partial}{\partial x^\mu} = \varpi^{-1} \frac{\partial}{\partial \phi}, \quad e_{\hat{\lambda}}^\mu \frac{\partial}{\partial x^\mu} = \frac{\partial}{\partial \lambda}, \quad n^\mu \frac{\partial}{\partial x^\mu} = \frac{\kappa}{c^2} \frac{\partial}{\partial \alpha}. \tag{4.184}$$

Macdonald and Thorne (1982) formulated the conditions at the horizon as follows:

1. Gauss' law

$$\mathbf{E} \mathbf{n} \equiv \mathbf{E}_\perp \to 4\pi \, \sigma^\mathbf{H}. \tag{4.185}$$

2. Charge conservation law

$$\alpha \, \mathbf{jn} \to -\frac{\partial \sigma^\mathbf{H}}{\partial \mathbf{t}} - {}^{(2)}\nabla \mathbf{i}^\mathbf{H}. \tag{4.186}$$

3. Ampere's law

$$\alpha \mathbf{B}_\| \to \mathbf{B}^\mathbf{H} \equiv \left(\frac{4\pi}{c}\right) \mathbf{i}^\mathbf{H} \times \mathbf{n}. \tag{4.187}$$

4. Ohm's law

$$\alpha \mathbf{E}_\| \to \mathbf{E}^\mathbf{H} \equiv \mathbf{R}^\mathbf{H} \mathbf{i}^\mathbf{H}. \tag{4.188}$$

In these relations the symbol $\to$ indicates the approach to the black-hole horizon along the trajectory of a freely falling observer, ${}^{(2)}\nabla$ is the two-dimensional divergence at the horizon, and $\mathbf{B}_\|$ and $\mathbf{E}_\|$ are the magnetic and electric field components tangent to the horizon. $R^H \equiv 4\pi/c$ is the effective surface resistance of the event horizon ($R^H = 377 \, \text{Ohm}$). The lapse function α in conditions (4.186)—(4.188) reflects the slowdown in the flow of physical time for locally nonrotating observers in the neighborhood of the black hole.

The values of $\mathbf{E}^\mathbf{H}$ and $\mathbf{B}^\mathbf{H}$ at the horizon are finite, and $\alpha \to 0$. Hence, taking into account the conditions given above, we arrive at the following properties of fields at the horizon:

$$\mathbf{E}_\perp \text{ and } \mathbf{B}_\perp \text{ are finite at the horizon,} \tag{4.189}$$

$|\mathbf{E}_\parallel|$ and $|\mathbf{B}_\parallel|$ generally diverge at the horizon as $1/\alpha$, \hfill (4.190)

$$|\mathbf{E}_\parallel - \mathbf{n} \times \mathbf{B}_\parallel| \propto \alpha \to 0 \quad \text{at the horizon.} \quad (4.191)$$

Condition (4.191) signifies that for locally nonrotating observers, the electromagnetic field at the horizon acquires (in the general case) the characteristics of an electromagnetic wave sinking into the black hole at infinite blue shift.

In the presence of matter one needs also to require that the matter density n in a freely-falling reference frame remains finite at the horizon. It is interesting to note that in the MHD-approximation the latter condition automatically implies the condition (4.191).

The conditions listed above make it possible to imagine quite clearly how electromagnetic processes will effect the properties of the black hole, slowly varying its parameters (the change is slow because we have assumed from the beginning that the electromagnetic field is relatively weak and the plasma is rare enough; see Section 4.2).

The change in the energy Mc^2 and angular momentum J of a black hole are equal to the total flux of the energy and angular momentum across the horizon, respectively. All quantities are considered in the global time t.

The electromagnetic field contribution to the change of the black hole's parameters can be written as

$$\frac{d(Mc^2)}{dt} = \int_H \{\Omega^H[\sigma^H\,\mathbf{E}^H + (\mathbf{i}^H/c) \times \mathbf{B}_\perp)]\,\varpi\,\mathbf{e}_{\hat{\phi}} + \mathbf{E}^H\,\mathbf{i}^H\}\,d\Sigma^H, \quad (4.192)$$

$$\frac{dJ}{dt} = \int_H (\sigma^H\,\mathbf{E}^H + (\mathbf{i}^H/c) \times \mathbf{B}_\perp)\,\varpi\,\mathbf{e}_{\hat{\phi}}\,d\Sigma^H. \quad (4.193)$$

Here $d\Sigma^H$ is an element of the horizon area. The first term in the braces in (4.192) describes the change in the rotational energy of the black hole and the second term gives the change due to the 'heating' of the black hole by the surface current.

We thus find that the boundary conditions at the event horizon, described in this section, make it possible to model a black hole for electrodynamic problems in the exterior space as an imaginary sphere with specific electromagnetic properties. The sphere can carry surface charge and electric currents. This visually clear picture helps greatly in solving specific problems. As we already mentioned, this approach is called the membrane formalism.

We wish to emphasize again that there are no real sphere, no charges, and no current at the black-hole boundary. Note also that the fields $\mathbf{E}$ and $\mathbf{B}$ that a locally nonrotating observer measures at the event horizon, differ drastically from $\mathbf{E}^H$ and $\mathbf{B}^H$ that appear in the boundary conditions: the reason is the factor α in the definition of $\mathbf{E}^H$ and $\mathbf{B}^H$ [see (4.187) and (4.188)]. This factor appears (recall the formulation of Maxwell's equations) as a result of using the 'global' time t.

Note here the following important aspects. In the '3+1' split of the black hole spacetime, the $t =$ const surfaces behave as shown in Figure 12.

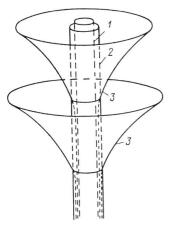

Fig. 12. '3+1' split of spacetime close to the black hole horizon: 1—black hole event horizon, 2—'stretched' horizon, 3—const sections

As we approach the horizon, these surfaces recede into distant past in parameter V (2.36) or in time T of a freely falling observer [T is the time of reference frame (2.28) for the Schwarzschild metric or the time in a similar reference frame for the Kerr metric].[24] Hence, as we approach the horizon along this section, it contains the values of the electromagnetic field that correspond to the distant-past history. This fact is of no importance if a stationary problem is considered. But if we are interested in the evolution of the fields, it becomes important and may lead to serious inconveniences. It was therefore suggested that the concept of 'stretched' horizon should be introduced. This horizon is defined as the surface (membrane) that lies in the immediate vicinity and outside of the horizon but is time-like, in contrast to the event horizon (see Figure 12). The exact position of the 'stretched' horizon is not specified; it is chosen to suit a concrete problem.

The boundary conditions are fixed in this approach on the 'stretched' horizon. The entire distant past history of the fields at $t = $ const close to the true horizon is then cut off and ignored. A detailed theory of 'stretched' horizon and the relevant bibliography can be found in Price and Thorne (1986), and in the monograph by Thorne *et al.* (1986).

Let us return to the true horizon. As in the preceding section, we introduce 'special' physical conditions. First we assume that the problem is stationary and axisymmetric. The fictitious surface current $\mathbf{i}^{\mathbf{H}}$ and electric field $\mathbf{E}^{\mathbf{H}}$ at the horizon are then completely poloidal while the magnetic field $\mathbf{B}^{\mathbf{H}}$ is toroidal. The values of these quantities at the horizon read

$$\mathbf{i}^{\mathbf{H}} = \frac{\mathbf{I}^{\mathbf{H}}}{2\pi\,\varpi_{\mathbf{H}}}\,\mathbf{e}_{\hat{\lambda}}\,, \tag{4.194}$$

---

[24] Note that the metric distance from any point in the section $t = $ const to the event horizon is nevertheless finite (excepting the case of $a = M$). This effect occurs because the null geodesic tends to align parallel to this section as the horizon approaches.

$$\mathbf{E^H} = \frac{2I^H}{c\,\varpi_H}\,\mathbf{e_{\hat{\lambda}}}, \qquad (4.195)$$

$$\mathbf{B^H} = -\frac{I^H}{2\pi\,\varpi_H}\,\mathbf{e_{\hat{\phi}}}, \qquad (4.196)$$

where I^H and ϖ_H are the values of I and ϖ at the horizon. Furthermore, the poloidal magnetic field measured by nonrotating observers intersects the event horizon at right angles. Recall that the toroidal component diverges at the horizon.

In the particular case of the MHD-approximation the conditions (4.195) and (4.196) imply

$$\frac{(E - \Omega^H L)}{\mathcal{M}_H^2 + (\Omega^F - \Omega^H)^2\,\varpi_H^2/c^2} = -\frac{c^2}{8\pi^2\,\rho_H\,\varpi_H}\left(\frac{\partial\Psi}{\partial\theta}\right)_H. \qquad (4.197)$$

For regular flow ($\mathcal{M}(r_+) \neq 0$) this condition follows directly from the $\alpha \to 0$-limit of algebraic constraint (4.169). Moreover the solutions of the equation (4.175) without any nonphysical singularity satisfy automatically the boundary condition (4.197) at the event horizon. According to the definitions (4.135) and (4.137) the electric field $\mathbf{E^H}$ at the horizon is directly expressible in terms of $\mathbf{B}_\perp$:

$$\mathbf{E^H} = -\frac{1}{c}(\Omega^F - \Omega^H)\,\varpi\,\mathbf{e_{\hat{\phi}}} \times \mathbf{B}_\perp. \qquad (4.198)$$

In the MHD-approximation the change of the energy and angular momentum of a black hole can be simply expressed in terms of integrals of motion E and L taken at the surface of a black hole:

$$\frac{d(Mc^2)}{dt} = -\frac{1}{c}\int_0^\pi E|_H \left(\frac{d\Psi}{d\theta}\right)_H d\theta, \qquad (4.199)$$

$$\frac{dJ}{dt} = -\frac{1}{c}\int_0^\pi L|_H \left(\frac{d\Psi}{d\theta}\right)_H d\theta. \qquad (4.200)$$

On the black hole surface the quantities $E|_H$ and $L|_H$ are well defined (single valued) functions of θ and hence the integrals have well defined meaning.

In the force-free approximation the following 'principle of least action' is found to hold [Macdonald and Thorne 1982]. The lines of the poloidal magnetic field that intersect the horizon have a distribution that ensures extremal total surface energy $\mathcal{E}$ of the tangential electromagnetic field at the horizon. The expression for $\mathcal{E}$ has the form

$$\mathcal{E} = \frac{1}{8\pi}\int_H [\,(\mathbf{B^H})^2 + (\mathbf{E^H})^2\,]\,d\Sigma^H, \qquad (4.201)$$

where the integration is carried out over the horizon.

Now the boundary condition (4.197) can be rewritten in the form

$$4\pi\,I(\Psi) = [\,\Omega^H - \Omega^F(\Psi)\,]\,\frac{\varpi_H}{\rho_H}\left(\frac{\partial\Psi}{\partial\theta}\right)_H. \qquad (4.202)$$

Then equations (4.199) and (4.200) take the form

$$\frac{d(Mc^2)}{dt} = \int_H \frac{\Omega^F(\Omega^F - \Omega^H)}{4\pi c} \frac{A^H \sin^2\theta}{\rho_H^2} (B_\perp)^2 d\Sigma^H, \qquad (4.203)$$

$$\frac{dJ}{dt} = \int_H \frac{(\Omega^F - \Omega^H)}{4\pi c} \frac{A^H \sin^2\theta}{\rho_H^2} (B_\perp)^2 d\Sigma^H. \qquad (4.204)$$

The angular momentum and energy lost by the black hole are transferred along the lines of force of the poloidal field in the force-free region to those 'regions' where condition (4.150) is violated.

Note that if $\Omega^F = 0$ (i.e., if the magnetic lines of force are, say, frozen into plasma far from the black hole and this plasma *does not participate* in the rotation around the hole), then $dM = 0$, that is, the total mass of the black hole is conserved and $dJ < 0$; in other words, the rotation of the black hole slows down. The entire energy of rotation transforms into the mass of the black hole (the so-called irreducible mass; see our book Novikov and Frolov 1989) so that nothing escapes.

If the parameters of the black hole are fixed, its angular velocity Ω^F is determined by the boundary condition far from the black hole in the external plasma. The situation in realistic astrophysical conditions will be discussed in Section 4.5.

4.5 Electromagnetic Fields in Vacuum

Before beginning the description of the magnetosphere of a rotating black hole [it is formed via the accretion of magnetized gas (see the next section)], we illustrate the above analysis by the solutions of the following problems on electromagnetic fields in vacuum:

1. electric charge in the vacuum in the Schwarzschild metric [Copson 1928, Linet 1976, Hanni and Ruffini 1973];
2. magnetic field in the vacuum in the Kerr metric, uniform at infinity [Wald 1974b, Thorne and Macdonald 1982, King and Lasota 1977].

We begin with problem (1). Let a point-like charge q be at rest in the Schwarzschild coordinates at $r = b$, $\theta = 0$. The problem reduces to solving system (4.139)–(4.141) with δ-function for ρ_e and $j^P = j^T = 0$. Conditions (4.139) and (4.140) are satisfied when $\Psi = 0$ and $I = 0$. Expressions (4.137) and (4.138) then imply that the external magnetic field is absent. External currents also vanish. Hence [see expressions (4.186) and (4.187)], the surface current at the horizon is also zero, $\mathbf{i}^H = \mathbf{0}$. As follows from condition (4.191), at the horizon $\mathbf{E}_\| \to \mathbf{0}$, so that electric lines of force intersect the horizon at right angles. The total flux of $\mathbf{E}$ across the horizon is zero (the black hole is not charged). With these boundary conditions, $\mathcal{U}_0$ is found from the solution (4.141) which for $\omega = 0$ is of the form

$$\nabla(\alpha^{-1}\nabla\mathcal{U}_0) = 4\pi\rho_e. \qquad (4.205)$$

By solving this equation for a point-like charge q located at the symmetry axis at a point $r = b$ and using the equation (4.135) one finds the following expression for $\mathbf{E}^P$:

$$\mathbf{E}^P = \frac{q}{br^2}\left\{M\left(1 - \frac{b - M + M\cos\theta}{R}\right) + \right.$$
$$+ \frac{r[(r-M)(b-M) - M^2\cos\theta]}{R^3}\left. [r - M - (b-M)\cos\theta]\right\}\mathbf{e}_{\hat{r}} +$$
$$+ \frac{q(b - 2M)(1 - 2M/r)^{1/2}\sin\theta}{R^3}\mathbf{e}_{\hat{\theta}}, \qquad (4.206)$$

where $\mathbf{e}_{\hat{r}}$ and $\mathbf{e}_{\hat{\theta}}$ are unit vectors along the directions of r and θ, respectively, and

$$R \equiv [(r-M)^2 + (b-M)^2 - M^2 - 2(r-M)(b-M)\cos\theta + M^2\cos^2\theta]^{1/2}. \qquad (4.207)$$

(Throughout this section, with the exception of the final formulas, we set $G = 1$, $c = 1$)

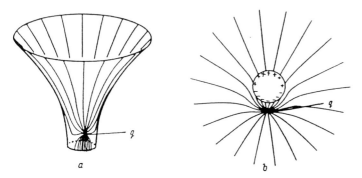

Fig. 13. The electric-field lines of force of a test charge q at rest in the Schwarzschild metric, in a $\phi = $ const section: (a) lines of force on a curved surface whose geometry coincides with the section $\phi = $ const of the Schwarzschild metric; (b) the same lines projected on a plane ('top view'). The distribution of the fictitious surface charge σ^H is shown on the horizon. The charge q is assumed to be positive

The pattern of electric lines of force is shown in Figure 13. The charge surface density at the boundary of the black hole follows from (4.185):

$$\sigma^H = \frac{q[M(1 + \cos^2\theta) - 2(b - M)\cos\theta]}{8\pi b[b - M(1 + \cos\theta)]^2}. \qquad (4.208)$$

Let us bring the charge closer to the horizon ($b \to 2M$). At a distance $r \gg b - 2M$ from the horizon, the lines of force become practically radial, and the field strength tends to q/r^2. With the exception of a narrow region close to the horizon, the general picture is almost the same as for a charge placed at the center of the black hole.

Now we will give, without detailed justification, the solution of problem (2). A rotating black hole is placed in a magnetic field B_0, uniform at infinity. In the Kerr metric, the magnetic field is given by the expression

$$\mathbf{B} = \frac{B_0}{2A^{1/2}\sin\theta}\left[\frac{\partial X}{\partial \theta}\mathbf{e}_{\hat{r}} - \Delta^{1/2}\frac{\partial X}{\partial r}\mathbf{e}_{\hat{\theta}}\right], \qquad (4.209)$$

where $X \equiv (A - 4a^2 Mr)(\sin^2\theta)/\rho^2$ and A is given by (4.129)

The electric field induced by the rotation of the black hole is proportional to a:

$$\mathbf{E} = -\frac{B_0\, a\, A^{1/2}}{\rho^2}\left\{\left[\frac{\partial(\alpha^2)}{\partial r} + \frac{M\sin^2\theta}{\rho^2}(A - 4a^2 Mr)\frac{\partial}{\partial r}\left(\frac{r}{A}\right)\right]\mathbf{e}_{\hat{r}} + \right.$$
$$\left. + \Delta^{-1/2}\left[\frac{\partial(\alpha^2)}{\partial \theta} + \frac{Mr\sin^2\theta}{\rho^2}(A - 4a^2 Mr)\frac{\partial}{\partial \theta}\left(\frac{1}{A}\right)\right]\mathbf{e}_{\hat{\theta}}\right\}. \qquad (4.210)$$

As in problem (1), $\mathbf{E}^H$, $\mathbf{B}^H$, and $\mathbf{i}^H$ are absent.[25] Formulas (4.193), (4.192) imply that the angular momentum $\mathbf{J}$ of a black hole and its mass M remain invariant. King and Lasota (1977) demonstrated that the angular momentum changes if the magnetic field is tilted at an angle to the black hole axis. Their result is as follows.

Let a magnetic field $\mathbf{B}$, uniform at infinity, be tilted at an angle to the direction of the angular momentum $\mathbf{J}$. Decompose $\mathbf{J}$ into a component $\mathbf{J}_\parallel$ parallel to the field and a component $\mathbf{J}_\perp$ perpendicular to it. The following formulas give $\mathbf{J}$ as a function of time:

$$\mathbf{J}_\parallel = \text{const}, \qquad (4.211)$$

$$\mathbf{J}_\perp = \mathbf{J}_\perp(t=0)\exp(-t/\tau), \qquad (4.212)$$

where

$$\tau = \frac{3c^5}{2GMB^2} = 10^{36}\,\text{years}\left(\frac{M}{M_\odot}\right)^{-1}\left(\frac{B}{10^{-5}\,\text{Gauss}}\right)^{-2}. \qquad (4.213)$$

The component $\mathbf{J}_\perp$ of the black hole angular momentum is thus completely lost with time. (But the time of desintegration is fantastically long!) The rotational energy connected with $\mathbf{J}_\perp$ thus transforms in the static magnetic field into the 'irreducible' mass of the black hole while the component $\mathbf{J}_\parallel$ remains unaltered.

The final state of the black hole corresponds to Hawking's theorem that states that the stationary state must be axially symmetric. Press (1972) points

---

[25] If we calculate the components of the field $E_\perp$ and $B_\perp$ at the event horizon (see Section 4.4), we find that both are proportional to $r_+ - M$, where $r_+ = M + (M^2 - a^2)^{1/2}$. Hence, we obtain for a black hole rotating at maximum angular velocity, with $a_{\max} = M$, that $E_\perp = 0$ and $B_\perp = 0$ at the horizon, that is, the lines of force of the axisymmetric field do not pass through the black hole (the case of asymmetric field is treated later). Problems in black-hole electrodynamics were also treated by Léaute and Linet (1976, 1982), Misra (1977), Gal'tsov and Petukhov (1978), Linet (1976, 1977a, b, 1979), Demianski and Novikov (1982), Bičak and Dvorak (1980).

out that if the external magnetic field (or any other field) is not axisymmetric, the black hole finally loses its angular momentum (by Hawking's theorem). If the field **B** varies smoothly on a scale much greater than the black hole size, **J** can again be decomposed into $\mathbf{J}_\parallel$ and $\mathbf{J}_\perp$ with respect to the field direction in the black hole neighborhood. By the order of the magnitude, the decrease of $\mathbf{J}_\perp$ is again determined by formula (4.212), while $\mathbf{J}_\parallel$ decreases according to the formula

$$\frac{d\mathbf{J}_\parallel}{dt} \approx -\frac{\mathbf{J}_\parallel}{\tau} O\left(\frac{r_g}{R}\right), \tag{4.214}$$

where R is the scale of nonuniformity of the field.

We have mentioned in the note 25 to page 303 that if $a = a_{\max}$, an axisymmetric magnetic field does not intersect the black hole horizon. Bičak (1983, 1985) proved that if a magnetic field B_0, uniform at infinity, is tilted with respect to the rotation axis, then the flux through one half of the horizon[26] for the field component $B_{0\perp}$ is maximal when $a = a_{\max}$; it equals

$$\Psi^H_{\max} \approx 2.25\, B_{0\perp}\, \pi M^2. \tag{4.215}$$

Finally, consider a *nonrotating* black hole placed in a strong magnetic field B_0, uniform at infinity [Bičak 1983]. Let the field be so strong that its self-gravitation has to be taken into account. It is then found that for a black hole of fixed mass M, there is a critical field $B_{0,\mathrm{cr}}$ at which the flux Ψ^H through one half of the event horizon is maximal:

$$B_{0,\mathrm{cr}} = c^4\, G^{-3/2}\, M^{-1} = 2.5 \times 10^{13}\, Gauss \times \left(\frac{10^6 M_\odot}{M}\right), \tag{4.216}$$

$$\Psi^H_{\max,1} = 2\pi\, G^{1/2}\, M. \tag{4.217}$$

The flux across the horizon cannot be greater than $\Psi^H_{\max,1}$.

About a rotating electrically charged black hole in an external magnitude field, see Dokuchayev (1987).

4.6 Magnetosphere of a Black Hole

The problems discussed in the preceding section illustrate some important properties of electric and magnetic fields in the neighborhood of a black hole. These problems can hardly be used, however, for the description of the actual electrodynamic processes expected to take place in astrophysical conditions. We have already mentioned that the reason for this is the nonvacuum nature of the fields in the vicinity of black holes. The space is always filled with rarefied plasma (so that the fields becomes force-free) or even more complicated situations arise.

---

[26] Of course, the total flux of magnetic field across any closed surface, including the horizon, equals zero (see Section 4.2). In speaking about the flux through a black hole, one considers the incoming (or the outgoing) lines of force.

Recall that the fields in the neighborhood of rotating magnetized neutron stars (pulsars) cannot be considered as vacuum fields either. A complicated pulsar magnetosphere is formed. For more details, see e.g. the books of Michel (1991) and Beskin, Gurevich, and Istomin (1993).

By analogy to pulsars, the region of magnetized plasma around a black hole is called the black-hole *magnetosphere*. The complexity and diversity of processes in this region stand in the way of developing a complete magnetosphere theory. In fact, we still lack an acceptably complete theory of a pulsar's magnetosphere, despite the great expenditure of effort and time that has been put into it.

We will not discuss all aspects of the theory of the black-hole magnetosphere here. We will only look at the important side of the electrodynamic processes that occur in the black-hole neighborhood and are black-hole-specific. Correspondingly, we restrict the representation to the simplest model. Only the processes caused by the black hole itself are covered. (For discussion of other problems, such as the motion of plasma in the accreted gas disk formed around the black hole and the problem of jets formation, which are not discussed in the book, see e.g., Shapiro and Teukolsky (1983), Blandford and Payne (1982), Heyvaerts and Norman (1989), Li, Chiuen, and Begelman (1992), Pelletier and Pudritz (1992).)

We choose the stationary axisymmetric model of the force-free magnetosphere of a black hole inducing disc accretion of magnetized gas (the magnetosphere schematically shown in Figure 11). This model was analyzed by Blandford (1976), Blandford and Znajek (1977), Macdonald and Thorne (1982), and Phinney (1983ab); see also Lovelace (1976), Lovelace et al. (1979), Thorne and Blandford (1982), Rees et al. (1982), Macdonald (1984), Okamoto (1992), Beskin, Istomin, and Pariev (1992).

For the condition of force-free field existence, (4.150), to be satisfied in the neighborhood of a black hole, it is necessary to have a rarefied plasma with electric currents flowing along magnetic lines of force. Charges for these currents crossing the black hole must be constantly replenished because they sink into the hole and obviously, charges cannot flow back out.[27] Mechanisms of free charge generation thus have to exist in the neighborhood of the black hole. Such mechanisms were analyzed by Blandford and Znajek (1977), Kardashev et al. (1983), and Beskin, Istomin, and Pariev (1992). We remark, without going into the details, that these mechanisms require a small component of the electric field, parallel to the magnetic field. This component is so small that inequality (4.143) is not violated.

Two examples of magnetospheres of a black hole are shown in Figures 14 and 15. They correspond to the solutions of the force-free equation (4.151) for a slowly rotating black hole ($\Omega^H r_+/c \ll 1$). Arrows indicate the longitudinal electric current $\mathbf{j}^\mathrm{P}$. Dashed lines separate regions with positive and negative charge density ρ_e.

---

[27] Obviously, the total charge of the black hole in the stationary solution cannot be changed by this process because the total numbers of charges of opposite signs, sinking into the black hole, are equal to each other.

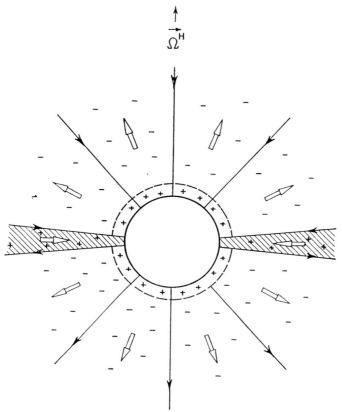

Fig. 14. The model of a black hole's magnetosphere considered by Blandford and Znajek (1977). Arrows indicate the longitudinal electric current $\mathbf{j}^P$. Dashed lines separate regions with positive and negative charge density ρ_e

The solution presented in Figure 14 corresponds to the following choice of integrals of motion [Blandford and Znajek 1977]

$$\Omega^F(\Psi) = \Omega^H/2, \qquad (4.218)$$

$$I(\Psi) = \pm \frac{\Omega^F}{4\pi}\left(2\Psi - \Psi^2/\Psi_0\right), \qquad (4.219)$$

where the sign $+$ $(-)$ is to be taken for the upper (lower) hemisphere. It can be shown that for a slowly rotating black hole, up to terms which remain small everywhere outside the horizon, the solution to the equation (4.151) is of the form

$$\Psi = \Psi_0\left(1 - |\cos\theta|\right). \qquad (4.220)$$

Thus $\Psi_0 = \Psi_{\max}^H$ is the maximal magnetic flux through the horizon. Such split-monopole magnetic field can be generated by toroidal currents flowing in the accretion disk provided that the latter (formally) extend up to the horizon. For this solution the influence of the rotating charged plasma is totally compensated

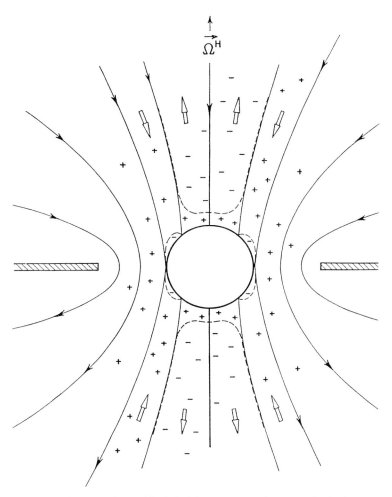

Fig. 15. The model of a black hole's magnetosphere in which the magnetic field near the black hole is close to homogeneous [Beskin, Istomin, and Pariev 1992]. In this case the closure of electric current takes place at the magnetic field lines crossing the horizon at $\theta \sim \pi/2$

by the longitudinal current I. The charge density ρ_e (4.149) changes the sign at $r = 2^{1/3} r_+$. The magnetic field is always (even at the distances lying outside the *light cylinder* $R_L = c/\Omega^F$) larger than the electric one. The validity of the solution (4.220) (under the condition (4.219)) at large distances ($r \gg R_L$) was shown by Michel (1973) for the case of a pulsar magnetosphere. It is interesting to emphasize that the expression (4.220) is the exact vacuum solution of Maxwell's equations (4.123) and (4.137) for a non-rotating black hole.

The solution presented in Figure 15 corresponds to the case when the magnetic field near the black hole is close to the homogeneous one, while it is monopole-like at large distances [Beskin, Istomin, and Pariev 1992]. Such a magnetic field can be generated in the disc with finite inner radius. The longitudinal current closure takes place not inside the disk (like in the Blandford-Znajek solution) but at the magnetic field lines crossing the horizon at $\theta \sim \pi/2$. It can be shown that the electric current obeys the following condition at the horizon

$$\int_H \mathbf{j} \, d\Sigma^{\mathbf{H}} = \mathbf{0}, \qquad (4.221)$$

and, hence, the electric charge of the black hole is not changed (see also Okamoto (1992)). The surface separating the regions with positive and negative charge densities (shown by dashed lines) has a more complicated form. One can argue that the domain where plasma is generated inside the magnetosphere lies near this surface [Beskin, Istomin, and Pariev 1992].

We discuss now the efficiency of the power-generation process near a rotating, magnetized black hole. Consider a thin tube of the lines of force that pass through the black hole. This tube rotates around the black hole at a constant angular velocity Ω^F (see Section 4.3). In the force-free approximation formula (4.203) shows that, the rotational energy of the black hole is extracted at the rate

$$\mathcal{P} = -\frac{d(Mc^2)}{dt} = \int_H \frac{\Omega^F(\Omega^H - \Omega^F)}{4\pi c} \frac{A^H \sin^2\theta}{\rho_H^2} B_\perp^2 \, d\Sigma^H. \qquad (4.222)$$

This energy is transferred along magnetic lines of force into region 3 (see Figure 11) where the force-free condition is violated, energy is pumped into accelerated particles, and so forth.

In this region, the particles exert back-reaction on the line of force, owing to their inertia, and, thus, determine Ω^F. If inertia is large, the angular velocity Ω^F is small ($\Omega^F \ll \Omega^H$); in the limit $\Omega^F \to 0$. The power $\mathcal{P}$ of the above 'engine' is quite low, as follows from (4.222). Otherwise, i.e. when the inertia of the particles in the region 3 is low, $\Omega^F \to \Omega^H$ and (4.222) again gives low power. The power is maximal when $\Omega^F = \Omega^H/2$.

Macdonald and Thorne (1982) demonstrated that this condition is very likely to be implemented in the described model. Their arguments run as follows. The angular velocity of nonrotating observers far from the black hole can be assumed zero, and the velocity $\mathbf{v^F}$ of the points of the lines of force far from the symmetry axis (with respect to nonrotating observers and, hence, in the absolute space, because $\omega = 0$) is much greater than the speed of light:

$$|\mathbf{v^F}| = \Omega^F \varpi \gg c. \qquad (4.223)$$

Charged particles cannot move at a velocity greater than c. Staying on a line of force, however, and sliding along it outward (see Figure 16a), they can have velocities less than $\mathbf{v^F}$. Figure 16b shows a segment of a line of force in the plane defined by the vectors $\mathbf{B^P}$ and $\mathbf{B^T}$. There is an optimal velocity of sliding along the line of force (which, in turn, moves at a velocity $\mathbf{v^F}$), such that the total velocity of the particle in the absolute space is minimal (this is clear from the

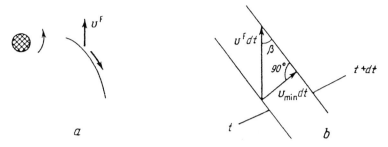

Fig. 16. (a) Scheme of motion of a charged particle along a magnetic field line of force rotating around a black hole. (b) The position of a segment of a magnetic line of force in the plane of the vectors $\mathbf{B^P}$ and $\mathbf{B^T}$ at the moments t and $t + dt$. If the velocity vector of a particle that slides outward in the absolute space along a line of force is perpendicular to this line, the velocity of the particle is minimal

Figure 16b). The angle β is the angle between the direction of the line of force and the direction of the ϕ coordinate; hence, β is found from the relation

$$\sin \beta = \frac{|\mathbf{B^P}|}{\sqrt{(\mathbf{B^P})^2 + (\mathbf{B^T})^2}}. \tag{4.224}$$

Using (4.224), we find $|\mathbf{v}_{\min}|$:

$$|\mathbf{v}_{\min}| = \frac{|\mathbf{v^F}|}{\sqrt{1 + (\mathbf{B^T})^2/(\mathbf{B^P})^2}}. \tag{4.225}$$

It is found that the condition $|\mathbf{v}_{\min}| \approx c$ is equivalent to the maximal energy condition $\Omega^F \approx \Omega^H/2$. Indeed, let us write for $|\mathbf{v^F}|$, $|\mathbf{B^T}|$ and $|\mathbf{B^P}|$ the following expression valid *far from both the black hole and the symmetry axis*. For $|\mathbf{v^F}|$ we find

$$|\mathbf{v^F}| = \Omega^F \varpi. \tag{4.226}$$

Making use of formulas (4.138), (4.195), (4.198), and definition (4.129), we obtain

$$|\mathbf{B^T}| \approx \frac{(\Omega^H - \Omega^F)\Psi}{\pi c \varpi}. \tag{4.227}$$

Finally, formula (4.137) and relation $|\nabla \Psi| \approx 2\Psi/\varpi$ yield

$$|\mathbf{B^P}| \approx \frac{\Psi}{\pi c \varpi^2}. \tag{4.228}$$

Substituting (4.226)–(4.228) into (4.225), we find

$$|\mathbf{v}_{\min}| \approx \frac{c\,\Omega^F}{\Omega^H - \Omega^F}. \tag{4.229}$$

The last formula implies that $|\mathbf{v}_{\min}| \approx c$ when $\Omega^F = \Omega^H/2$. If $\Omega^F \ll \Omega^H/2$ and $|\mathbf{v}_{\min}| \ll c$, then the inertia of particles in the region 3 is small and Ω^F

increases until the velocity $|\mathbf{v}_{\min}|$ grows close to c. If $|\mathbf{v}_{\min}| \gg \mathbf{c}$, then particles cannot stay long on the lines of force and their back-reaction on the field reduces Ω^F until we obtain $|\mathbf{v}_{\min}| \approx \mathbf{c}$.[28]

These are the reasons why it is likely that $\Omega^F \approx \Omega^H/2$ and the rate of extraction of rotational energy from the black hole (4.222) is almost optimal.

By the order of magnitude, the power of the 'electric engine' outlined above is

$$\mathcal{P} \approx \left(10^{39} \frac{erg}{s}\right) \left(\frac{M}{10^6 M_\odot}\right)^2 \left(\frac{a}{a_{\max}}\right)^2 \left(\frac{B}{10^4\, Gauss}\right)^2. \tag{4.230}$$

Here B is the magnetic field strength in the neighborhood of the black hole. Sometimes this electric engine is described in terms of electrical engineering [Blandford 1979, Znajek 1978, Damour 1978, Macdonald and Thorne 1972, Thorne and Blandford 1982, Phinney 1983a]. We will give the expression for quantities at the black hole horizon using this notation.

Equipotential curves at the horizon are the lines of constant θ, since the field $\mathbf{E}^H$ is meridional [see (4.195)]. Hence, the potential difference between two equipotential lines (marked by 1 and 2) are [see also (4.198)]

$$\Delta U^H = \int_1^2 \mathbf{E}^H \mathbf{dl} = (\Omega^H - \Omega^F)\frac{\Delta \Psi}{2\pi \mathbf{c}} \approx$$

$$\approx (10^{17}\, \mathrm{V}) \left(\frac{M}{10^6 M_\odot}\right) \left(\frac{B}{10^4\, Gauss}\right) \left(\frac{a}{a_{\max}}\right). \tag{4.231}$$

where $d\mathbf{l}$ is an element of distance along a meridian on the black hole surface, and $\Delta \Psi$ is the difference between the values of Ψ on the equipotential 1 and 2. The approximate equality in (4.231) is written for the condition $\Omega^F \approx \Omega^H/2$, maximal Ω^H, and the equipotentials 2 and 1 corresponding to the equatorial and polar regions, respectively.

On the other hand, ΔU^H can be written in terms of the surface current $\mathbf{i}^H$ and resistance:

$$\Delta U^H = R^H \mathbf{i}^H \Delta \mathbf{l}, \tag{4.232}$$

where $\Delta \mathbf{l}$ is the distance along the meridian between the equipotentials 2 and 1. Substituting expression (4.194) for $\mathbf{i}^H$, we obtain

$$\Delta U^H = \frac{I\, R^H\, |\Delta \mathbf{l}|}{2\pi\, \varpi_H} = I\, \Delta Z^H, \tag{4.233}$$

where

$$\Delta Z^H \equiv \frac{R^H\, |\Delta \mathbf{l}|}{2\pi\, \varpi_H} \tag{4.234}$$

is the total resistance between the equipotential lines 2 and 1 (if the equipotentials 2 and 1 correspond to the equator and to $\theta \approx \pi/4$, the integration of (4.234) yields $\Delta Z^H \approx 30\,\mathrm{Ohm}$).

---

[28] A similar analysis of slipping of particles along the magnetic lines of force at the black hole horizon shows that the condition $|v_{\min}| \approx c$ corresponds to boundary conditions (4.195) and (4.198).

Formulas (4.231) and (4.234) permit the conclusion that in this model the rotating black hole acts as a battery with e.m.f. in the order of

$$(10^{17}\,\text{V})\left(\frac{M}{10^6\,M_\odot}\right)\left(\frac{B}{10^4\,Gauss}\right) \qquad (4.235)$$

and internal resistance of about 30 Ohm.

This mechanism (and a number of its versions) has been employed in numerous papers for the explanation of the activity of the nuclei of galaxies and quasars [see, for example, Ruffini and Wilson 1975, Blandford 1976, Blandford and Znajek 1977, Blandford and Rees 1978, Blandford 1979, Lovelace et al. 1979, Kardashev et al. 1983, Rees 1982, Phinney 1983ab, Begelman, Blandford, and Rees 1984, Novikov and Stern 1986, Camenzind 1986ab, 1987, Punsly and Coronity 1990ab, Takahashi et al. 1990, Punsly 1991, Nitta, Takahashi, and Tomimatsu 1991, Hirotani et al. 1992, Beskin, Istomin, and Pariev 1992, Okamoto (1992), Beskin and Pariev 1993].

5 Some Aspects of Physics of Black Holes, Wormholes, and Time Machines

The branch of physics that is now referred to as black-hole physics was born and actually took shape as a full-blooded scientific discipline during the past three decades at the junction of the theory of gravitation, astrophysics, and classical and quantum field theory. Profound links were found between black-hole theory and such seemingly very distant fields as thermodynamics, information theory, and quantum theory. By now, a fairly detailed understanding has been achieved of the properties of the black holes, their possible astrophysical manifestations, and the specifics of the various physical processes involved.

For a review of the black holes physics see the book by Novikov and Frolov (1989). In this short section I will outline three key problems which have principle interest, and which are in a focus of attention. The review of all these problems is given by Frolov and Novikov (1995); and references may be found in this book. The first is a problem of the entropy of black holes. A black hole considered as a part of a thermodynamical system possesses the Bekenstein-Hawking entropy (in natural units) $S = A/4l_p^2$, where A is the area of a black hole surface and l_p is the Planck length.

Two questions arise. The first – why is the entropy so big, and the second – why is it universal, and does not depend on the number and characteristics of the fields ? It was proposed to explain the dynamical origin of the entropy of a black hole by identifying its dynamical degrees of freedom with the physical modes propagating in the black hole interior. The universality of the entropy is connected with the fact that in the state of thermal equilibrium the parameters of the internal dynamical degrees of freedom of a black hole depend on the temperature of the system in the universal way.

The second key problem is connected with the question: Can one see what happens inside a black hole? The standard answer to this question is no. Recently it was demonstrated that such a possibility exists and, in principle, one could search a black hole's interior practically without disturbing the metric describing a black hole. The gedanken experiments was discussed when the black hole's interior could be studied by means of a traversable wormhole, and it was shown that many habitual concepts concerning the properties of black holes (including the impossibility of extracting energy and information from the black holes) were to be taken with care.

The last key problem is connected with the possibility of creating, in principle, time machines (closed timelike curves), allowing one to travel into the past. Two crucial questions are discussed. The first: do the laws of physics prevent the time machines from ever forming? The second: whether and how the laws of physics deal with close timelike curves. What are the conclusion of these discussions? I believe that the following quotation from Thorne (1993) describes the modern situation: "These studies are giving us glimpses of how closed timelike curves influence physics; but whether those glimpses are teaching us something deep and important or we are just playing fun mental games, is far from clear."

6 Observational Appearance of the Black Holes in the Universe

6.1 Black Holes in the Interstellar Medium

What is the observational appearance of black holes in the Universe?

The most important physical process which leads to observable manifestations of a black hole's presence is gas accretion [Zeldovich, 1964, Salpeter 1964]. If a black hole is in a gaseous nebula the gas will be falling in its gravitational field.

Assume that a black hole is at rest with respect to the gas and is a Schwarzschild black hole. We'll discuss adiabatic accretion first. In the stationary accretion pattern the amount of matter falling upon the black hole per unit time, $\dot{M}$ and other basic parameters are determined by the gas properties and the gravitational field at large distances greatly exceeding the gravitational radius r_g where the gravitational field may be treated as a Newtonian one. Properties of the flux close to the black hole will be considered later.

The problem in the Newtonian theory was solved by Bondi (1952). Pioneering studies of radial accretion by astrophysical black holes were performed by Shvartsman (1971), Shapiro (1973a,b). The general scenario is described by Novikov and Thorne (1973), Shapiro and Teukolsky (1983). The gas flow is governed by two equations:
Conservation of mass

$$4\pi r^2 \rho u = \dot{M} = const \; with \; respect \; to \; r, \qquad (6.236)$$

and the Euler equation

$$u\frac{du}{dr} = -\frac{1}{\rho}\frac{dP}{dr} - \frac{GM_h}{r^2} \; . \qquad (6.237)$$

Here M_h is the mass of the hole, P and ρ are pressure and mass density of the gas, u is its radial velocity, $\dot{M}$ is the rate at which the hole accretes mass. We assume that ρ and P are related by the adiabatic law:

$$P = K\rho^\Gamma , \qquad (6.238)$$

K and Γ are constants. The adiabatic sound velocity is given by $a = (\Gamma P/\rho)^{1/2}$. For the conditions in the accretion flux $\Gamma \approx 1.4$. The solution of the system (6.236)-(6.238) determines the parameters of the flow. The general picture of accretion is as follows. The crucial role is played by the "sonic radius":

$$r_s \approx \frac{GM_h}{a_\infty^2} , \qquad (6.239)$$

where a_∞ is the sound velocity at infinity. At $r > r_s$ gravitation of the hole does not change the parameters of the gas at infinity and the velocity of the gas is much smaller than a_∞. At a distance of the order of r_s the velocity u of the gas flow increases and becomes equal to the local sound velocity a. The local value of a at this r is greater but still of the order of a_∞. At distances $r < r_s$, the velocity $u \gg a_\infty$, and the gas is nearly in free-fall. In this region

$$\begin{aligned}\rho &\approx 0.66\rho_\infty (r_s/r)^{3/2} \\ T &\approx 0.76 T_\infty (r_s/r)^{(3/2)(\Gamma-1)}\end{aligned} , \qquad (6.240)$$

(coefficients correspond to $\Gamma \approx 1.4$ near the sonic radius). Here ρ_∞ and T_∞ are the mass density and the temperature at infinity. The rate of accretion is

$$\dot{M} \approx 4\pi r_s^2 \rho_\infty a_\infty \approx (10^{11} g/sec) \left(\frac{M_h}{M_\odot}\right)^2 \left(\frac{\rho_\infty}{10^{-24} g/cm^3}\right) \left(\frac{T_\infty}{10^4 K}\right)^{-3/2} . \qquad (6.241)$$

So far we discussed the adiabatic hydrodynamic accretion. Now we shall consider various processes which can complicate the picture.

Thermal bremsstrahlung for accretion on a stellar black hole is many orders of magnitudes less than the luminosity of ordinary stars and is negligible. However, the influence of interstellar magnetic fields on accretion and radiation of infalling gas can be significant. Estimates [see Bisnovaty-Kogan 1989] show that the total luminosity of the accreting black hole due to synchrotron radiation of electrons in the magnetic fields which are frozen into the falling gas is

$$\begin{aligned} L_{synch} &\approx (3 \times 10^{31} \tfrac{ergs}{sec}) \left(\tfrac{M_h}{M_\odot}\right)^3 \left(\tfrac{\rho_\infty}{10^{-24} g/cm^3}\right)^2 \left(\tfrac{T_\infty}{10^4 K}\right)^{-3}, & a < 1, \\ L_{synch} &\approx (3 \times 10^{31} \tfrac{ergs}{sec}) \left(\tfrac{M_h}{M_\odot}\right)^2 \left(\tfrac{\rho_\infty}{10^{-24} g/cm^3}\right) \left(\tfrac{T_\infty}{10^4 K}\right)^{-3/2}, & a \gg 1, \end{aligned} \qquad (6.242)$$

where $a = \left(\frac{M_h}{M_\odot}\right)^{1/2} \left(\frac{T_\infty}{10^4 K}\right)^{-3/4} \left(\frac{\rho_\infty}{10^{-24} g/cm^3}\right)$. Most radiation comes from the high-temperature, strong field region near the black hole. The radiation has the broad maximum at $\nu \sim 7 \times 10^{14}$Hz for typical conditions. The full spectrum of the radiation is depicted in the book by Bisnovaty-Kogan (1989).

An important measure of any accretion luminosity of a black hole is provided by the Eddington critical luminosity

$$L_E = 4\pi GM_h\mu m_p c/\sigma_T = (1.3 \times 10^{38} ergs/sec)\, \mu \left(\frac{M_h}{M_\odot}\right). \qquad (6.243)$$

Here μ is the molecular weight per electron, m_p is the rest mass of the proton, and σ_T is the Thomson cross section. L_E is the luminosity at which the radiation pressure just balances the gravitational force of mass M_h for a fully ionized plasma. Eddington (1926) derived this value for the discussion of stellar equilibrium, Zeldovich and Novikov (1964) introduced it for the discussion of black holes in quasars and accretion.

If $L \ll L_E$ gravity dominates and photon pressure is not important. Comparing expressions (6.242) and (6.243) demonstrates that $L \ll L_E$ for the case under consideration.

Some aspects of spherical symmetrical accretion were recently reanalyzed by Park (1990ab). The influence of the radiation from inner regions on the flux far away is discussed by Ostriker et al. (1976), Cowie, Ostriker and Stark (1978), Wandel, Yahil and Milgrom (1984), Park (1990ab).

Near the boundary of the black hole the gas is practically in the state of free-fall. The moment of crossing the horizon is in no way peculiar, as seen by an observer falling together with the gas. Typical conditions are $\rho \simeq (6 \times 10^{-12} g/cm^3)\left(\frac{r}{r_g}\right)^{-3/2}$, $T \simeq (10^{12} K)\left(\frac{r_g}{r}\right)$. However the radiation of some infalling gas element which reaches infinity will be sharply cut off as the gas element crosses the horizon.

Review of works on possible pair creation and possible solutions with shocks close to event horizon was done by Lamb (1991). Some important details of radial accretion are discussed in the papers by Krolik and London (1983), Colpi, Marashi and Treves (1984), Schultz and Price (1985). In the case of a rotating black hole the dragging of inertial frames in its vicinity is very essential. This dragging swings the accretion gas into orbital rotation around the black hole [Bardeen-Peterson 1975 effect see, Thorne et al. 1986].

Spherically symmetric accretion onto a Schwarzschild black hole has probably only an academic interest as a testbed for theoretical ideas and it is of little relevance to interpreting the observational data. More realistic is the probe on a hole that moves with respect to interstellar gas. Once again the main processes which are responsible for the rate of accretion take place far away from the boundary of a black hole and one can use the Newtonian approximation. The corresponding problem was analyzed by Hoyle and Littleton (1939), Bondi and Hoyle (1944), Bondi (1952), Salpeter (1964). For a review of more recent work see Novikov and Thorne (1973) and Bisnovaty-Kogan (1989).

When a black hole moves through the interstellar medium with a speed v much larger than the sound speed a_∞ a shock front must develop around the black hole.

If the remaining kinetic energy of a gas element behind the shock front exceeds the potential energy in the gravitational field of the black hole, then this element will escape the pull of the hole. Otherwise it will be captured by the black hole.

The accretion rate is

$$\dot{M} \approx \frac{4\pi G^2 M_h^2 \rho_\infty}{v^3} \approx \left(10^{11} g/sec\right) \left(\frac{M_h}{M_\odot}\right)^2 \left(\frac{\rho_\infty}{10^{-24} g/cm^3}\right) \left(\frac{v}{10\frac{km}{sec}}\right)^{-3}. \tag{6.244}$$

The accretion rate for the supersonic motion, and the rate for a hole at rest are identical (in order of magnitude) except for the replacement of sound speed a_∞ by speed of the hole v.

In conclusion of the discussion of accretion onto the isolated stellar black holes we note that the luminosity of these objects is low and it is rather difficult to distinguish their observational appearances from some other astrophysical objects at great distances.

So far no one such object has been identified.

6.2 Disk Accretion

For the purpose of finding out and investigating black holes, two specific cases of accretion are of a particular importance: accretion in binary systems and accretion onto supermassive black holes which probably reside at the centers of galaxies.

In both cases the accreting gas has angular momentum $\tilde{L} \gg r_g c$. As a result the gas elements go into Keplerian orbits around the hole, forming a disk or a torus around it. The inner edge of the disk is in the region of the last stable circular orbit. Close to a rotating black hole the Lense-Thiring precession drags the gas around it in the equatorial plane [see Bardeen and Petterson 1975]. Viscosity plays a crucial role in the accretion. It removes angular momentum from each gas element, permitting it to gradually spiral inward toward the black hole. As the gas reaches the inner edge of the disk it then spirals down the hole practically in the state of free fall. At the same time the viscosity heats the gas, causing it to radiate. The sources of viscosity are probably turbulence in the gas disk and random magnetic fields. Unfortunately, we are no nearer to having a good physical understanding of the effective viscosity.

Large-scale magnetic fields can also play an important role in the physics of accretion.

The properties of the accreting disk are determined by the rate of gas accretion. It is worth noting the following. In the problem of the spherical accretion that was discussed in the section 6.1, the rate of accretion was the result of the solution of the equations governing the gas flow. In the case of disk accretion the rate is an independent (external) parameter and is determined by the evolution of a binary system or the conditions in a galactic nucleus. As a result the rate of accretion can be much higher than in the spherical case, and observational manifestations are much more prominent.

The useful measure of the accretion rate is a so called "critical accretion rate":

$$\dot{M}_E = L_E c^{-2}, \tag{6.245}$$

where L_E is given by (6.243). We shall use the dimensionless expression $\dot{m} \equiv \dot{M}/\dot{M}_E$. Excluding the innermost parts of the disk, relativistic effects are not important in physical processes of accretion.

Lynden-Bell (1969) was the first to propose the model of gaseous disk accretion onto black hole. Shakura (1972), Pringle and Rees (1972), Shakura and Sunyaev (1973) built Newtonian models of accretion disk. At last, Novikov and Thorne (1973) gave the theory of disk accretion in the framework of General Relativity.

The first models were rather simple. They focused on the case of moderate rate of accretion $\dot{m} < 1$. Subsequently the theory with $\dot{m} \sim 1$ and $\dot{m} \gg 1$ were developed. They take into account various complex processes in radiative plasma and various types of instability. The review is given by Lamb (1991). However these processes have no direct relation (or at least almost no relation) with specific properties of space-time in the vicinity of a black hole. Thus, we will not describe them here in detail. We only give some theoretical estimates of observational properties of accreting disks around black holes in binary systems and in galactic nuclei.

The source of luminosity of disk accretion is gravitational energy which is released when gas elements spiral down in the disk. Most of the gravitational energy is released and most of the luminosity is emitted from the inner parts of the disk. The total energy radiated by the gas element must be equal to the gravitational binding energy of the element when it is at the last stable circular orbit. This energy for the mass m_* is

$$\begin{aligned} E_{bind} &\approx 0.057 m_* c^2 \quad for\ nonrotating\ hole, \\ E_{bind} &\approx 0.42 m_* c^2 \quad for\ maximally\ rotating\ hole. \end{aligned} \tag{6.246}$$

The total luminosity of the disk is

$$\begin{aligned} L &\approx 0.057 \cdot \dot{M}c^2 \approx \left(3 \times 10^{36} \tfrac{erg}{s}\right)\left(\tfrac{\dot{M}}{10^{-9} M_\odot/yr}\right) \quad for\ nonrotating\ hole \\ L &\approx 0.42 \cdot \dot{M}c^2 \approx \left(3 \times 10^{37} \tfrac{erg}{s}\right)\left(\tfrac{\dot{M}}{10^{-9} M_\odot/yr}\right) \quad for\ maximally\ rotating\ hole \end{aligned} \tag{6.247}$$

Here $\dot{M}$ is an accretion rate. This parameter is an arbitrary external parameter, which is determined by the source of gas (for example by flux of the gas from the upper atmosphere of the companion star in the binary system). We normalized $\dot{M}$ to the value $\dot{M}_0 = 10^{-9} M_\odot$/year because it is probably the typical rate at which a normal star is dumping gas onto its companion black hole (see 6.3). The ratio $\dot{M}_0/\dot{M}_{crit} \equiv \dot{m}_0 \approx 0.1 \left(\frac{M_h}{M_\odot}\right)^{-1} \leq 1$ for cases of interest (see subsequent section 6.3). As estimates showed [see Lamb 1991], under this condition probably a geometrically thin disk (with heights $h \ll r$) might be formed. This is so called the "standard disk model" [Shakura and Sunyaev 1973, Novikov and Thorne 1973]. In this model the electron and ion temperatures are equal, the disk is effectively optically thick. Temperatures of the gas in the inner parts of the disk reach $T \approx 10^7 \div 10^8 \mathrm{K}$. In this region electron scattering opacity modifies the emitted spectrum so it is no longer blackbody. The total spectrum of the disk radiation has the power law $F \sim \omega^{1/3}$ with the exponential cut off at high frequencies (Shakura and Sunyaev, 1973). The innermost regions of such "standard" disks are probably unstable.

In the model proposed by Thorne and Price (1975) the ions in the inner region are hot ($T_i \approx 10^{11}\mathrm{K}$) but the electrons are considerably cooler ($T_e \approx 10^9\mathrm{K}$). This inner disk is thicker than in the "standard" model and produces most of the X-ray emission.

Further development of the theory of disk accretion led to more complex models. Abramovich et al., (1988, 1989) have demonstrated that when the luminosity reaches the critical one (specifically at $2 < \dot{m} < 80$) radiation pressure in the inner parts of the disk dominates the gas pressure and the disk is thermally and viscously unstable. For $\dot{m} > 80$ the essential part of the energy of plasma is lost by advection into the black hole horizon. This process stabilizes the gas flow against perturbation.

For high mass accretion rate, when $\dot{m} \geq 10$, the height of the accretion disk becomes comparable to the radius. In this case the inner edge of the disk can be closer to the black hole than the marginally stable circular orbit because of the essential pressure gradient.

In conclusion we note that in some models of disk accretion the pair production can be important. The review of these aspects is given by Lamb (1991).

For new development see, for example, Björnsson and Svensson (1991, 1992) and Artemova et al. (1994). Many aspects of the physics of accretion disks, including the development of instabilities were discussed in the following works: Mineshige and Wheeler (1989), Wheeler, Soon-Wook Kim and Moscoco (1993), Moscoco and Wheeler (1994) and Wheeler (1994).

6.3 Black Holes in Stellar Binary Systems

Probably the best evidence that black holes exist comes from studies of X-ray binaries. Galactic X-ray sources were discovered by Giacconi et al. (1962). Hayakawa and Matsuoko (1964) pointed out that X-rays might be produced by accretion of gas in close binary systems. However, they discussed the accretion into the atmosphere of a normal companion star, rather than onto a compact companion. Novikov and Zeldovich (1966) were the first to point out that the accretion onto compact relativistic objects (neutron stars and black holes) in binary systems should produce X-rays. They inferred also that Sco X-1 which had been just discovered, might be a neutron star in a state of accretion. After that the observational data were analyzed by Shklovsky (1967). Models for X-ray sources have been discussed in some detail by Prendergast and Burbidge (1968). They argued that the gas flow forms a disk around the compact object, with approximately a Keplerian velocity distribution.

A new era started in December 1970 when the X-ray satellite "UHURU" was launched. This satellite has provided much new data about the sources [see

Giacconi et al. 1972]. For the history of the early period of the investigations of the observational evidence for black holes in stellar binary systems, see in Novikov (1974). In this section we summarize the observational evidence for black holes in stellar binary systems which, at present, appear to be the best documented cases (for review see McClintock 1992, Cowley 1992).

The argument is as follows:
1) The X-ray-emitting object in a binary system is very compact, and therefore cannot be an ordinary star, thus it is either a neutron star or a black hole. This argument comes mainly from the analysis of the features of X-ray emission.
2) The analysis of the observational data to determine the orbital motion in the binary system makes it possible to obtain the mass of the compact object. Here the most important are data on the observed velocity of the optical companion star. Note that the Newtonian gravitational theory is always enough for the analysis. The technique of weighing of stars in binaries is well known in astronomy.

If the mass of the compact component is greater than the maximal possible mass of neutron stars $M_0 \approx 3 M_\odot$ (see section 1.2), then it is a black hole.

It is worth noting that the evidence is somewhat indirect because it does not confront us with the specific relativistic effects that occur near black holes and which are peculiar to black holes alone. However, it is the best that modern astronomy could propose so far. In spite of these circumstances we believe that this logic of arguments is reliable enough.

According to the generally accepted interpretation only three presently known systems have the necessary observational confirmation and it is strongly believed that the compact X-ray emitting companions of the systems are black holes. Some characteristics of the three leading black hole candidates are summarized in Table 3.

Table 3. Black-Hole Candidates in Binary Systems (mainly from McClintock,1992)

Property	Cyg X-1	LMC X-3	A0620-00
X-ray luminosity (ergs s^{-1})	$2 \cdot 10^{37}$	3×10^{38}	10^{38}
Spectral type of the optic companion	O9.7Iab	B3V	K5V
Distance (Kpc)	2.5(?)	55	1(?)
Orbital eccentricity	0.00 ± 0.01	0.13 ± 0.05	0.01 ± 0.01
Orbital period (days)	5.6	1.7	0.32
Plausible masses ($M_\odot$)			
a) Optical companion	33	6	0.5
b) Compact companion	16	9	5

Table 4. Estimates of the Minimum Mass (in $M_\odot$) for the Compact Objects

Cyg X-1	Ref	LMC X-3	Ref.	A0620-00	Ref.
3	(1,2)	6	(5)	3.2	(11)
7	(3)	2.5	(6)	2.90 ± 0.8	(12)
$3.4\left(\frac{d}{2Kpc}\right)^2$	(4)*)	2.3	(7)	4.5	(12)
		6	(7)	3.30 ± 0.95	(13)
		3	(8,9)	6.6	(13)
		4	(10)	3.1 ± 0.2	(14)
				3.82 ± 0.24	(15)
				6	(16)

*) d-distance from the Solar System

1. Webster and Murdin (1972)
2. Bolton (1972a,b; 1975)
3. Gies and Bolton (1986)
4. Paczynski (1974)
5. Paczynski (1983)
6. Mazeh et al. (1986)
7. Cowley et al. (1983)
8. Kuiper et al. (1988)
9. Bochkarev et al. (1988)
10. Treves et al. (1990)
11. McClintock (1988)
12. McClintock and Remillard (1990)
13. Johnston et al. (1989)
14. Johnston and Kulkarni (1990)
15. Haswell and Shafter (1990)
16. Cowley et al. (1983)

The plausible masses of the compact objects in these systems are essentially larger than $M_0 \approx 3M_\odot$. In Table 4 there are estimates of the minimum masses of these three candidates. These estimates were obtained by various methods and have different reliability. In most cases even the minimum masses are greater than M_0. Thus, these three objects are strong black hole candidates.

Using data from Table 3 and the expressions (6.243), (6.245), (6.247), one can estimate $\dot{m} \equiv \dot{M}/\dot{M}_E$ (see 6.2). In all cases $\dot{m} < 1$ or $\dot{m} \approx 1$.

Several other systems have been suggested as black hole candidates [the review is given by Cowley 1992, Cominsky 1994 and Tutukov and Cherepashchuk 1993]. Three systems: LMC X-1 (Mass of compact companion $M_c = 2.91\pm 0.08$), Nova Muskae 1991 ($M_c = 3.1 \pm 0.4$) and V404 Cyg ($M_c = 6.26 \pm 0.31$), see Cominsky (1994), could be considered as good candidates. The total number of systems that are frequently mentioned as possible candidates for black holes is about 20. All seriously discussed candidates are X-ray sources in binary systems. Some of them are persistent, other are transient. For example, Cyg X-1 and LMC X-1 are persistent, while A0620-00 is transient.

During the more than 20 years elapsed since the discovery of the first black hole candidate Cyg X-1 only a few new candidates have been added. This is in contrast to the rapid increase of the number of identified neutron stars. At present many hundreds of neutron stars have been identified in the Galaxy. About 100 of them are in binary systems (Lamb, 1991). One might conclude that black holes in binary systems are exceedingly rare objects. Probably that is not true, and the small number of identified black hole candidates is related with the specific conditions which are necessary for their observable manifestation (summary see Cowley, 1992). Inoue (1991) estimates the total number of only soft X-ray transient black-hole candidates in the Galaxy to be $(1-3) \cdot 10^3$. Thus such systems may be as frequent as neutron star binaries. For the estimate of the formation rate for black-hole binaries see van den Heuvel and Habets (1984).

At present we know only one persistent X-ray source in the Galaxy and one in another galaxy, the LMC, which are strong candidates for black holes (Cyg X-1 and LMC X-3 correspondingly). The probable explanation was provided by van den Heuvel (1983). He estimates that the evolutionary stage when a black hole binary continuously radiates X-rays may last only 10^4 years. We can detect it during this short period only. Thus the population of black-hole binaries may be much larger than we can presently see.

6.4 Black Holes in Galactic Centers

Since the middle of this century astronomers have increasingly often come across violent or even catastrophic processes associated with galaxies. These processes are accompanied by powerful releases of energy and are fast not only by astronomic but also by earthly standards. They may last a few days or even minutes. Most such processes occur in the central area of galaxies, the galactic nuclei.

About one percent of all galactic nuclei eject radio-emitting plasma and gas clouds, and are themselves powerful sources of radiation in the radio, infrared, gamma, and especially, the "hard" (short wavelength) X-ray regions of the spectrum. The full luminosity of the nucleus is in some cases $L \approx 10^{47} \frac{erg}{s}$, millions of times the luminosity of the nuclei of more stable galaxies, such as ours. These objects have been termed active galactic nuclei (AGN). Practically all the energy of activity and of the giant jets released by galaxies originates from the centers of their nuclei.

Quasars are a class in themselves among AGN. Their total energy release is hundreds of times greater than the combined radiation of all the stars in a big galaxy. At the same time the average linear dimensions of the proper radiating areas are small: a mere one-hundred-millionth part of the linear size of a galaxy.

Quasars are the most powerful energy sources registered in the Universe to date. What processes are responsible for the extraordinary outbursts of energy from AGN and quasars ? For review of the problem see Blandford (1987) and Wallinder (1993).

Learning about the nature of these objects involves measuring their sizes and masses. This is not easy at all. The center of the emitting areas of AGN and quasars are so small that a telescope view reveals them just as point sources of light. The job has been made much simpler by the fact that the brightness of the quasar which has the number 3C 273 is not constant, and sometimes it changes very fast, in just a week. After that, even faster changes (as fast as few hours or even less) were detected in other galactic nuclei. From those changes one could get the dimensions of the central parts of the nuclei. The conclusion was that they were no more than a few light-hours in diameter, that is, were comparable to the solar system in size.

In spite of the rather small linear dimensions of quasars and many galactic nuclei, their masses turned out to be enormous. For the first time they were estimated by Zeldovich and Novikov (1964). It was done using formula (6.243). For quasistatic objects luminosity cannot be greater than L_E. The comparison of the observed luminosity with the expression (6.243) gives the estimate of the lower limit of the central mass. In some quasars this limit is $M \approx (10 - 10^3) \times 10^6 M_\odot$. These estimates are supported by data on the velocities of stars within the galactic nuclei and gas clouds accelerated in the gravitational fields of the center of the nuclei. We will discuss this in the next section.

Great mass but small linear dimensions prompt the guess that it could be a black hole. Zeldovich and Novikov (1964) and Salpeter (1964) suggested that the centers of quasars and AGN could harbor supermassive black holes accounting for all extraordinary qualities of these objects.

It is interesting to note that the estimates of masses of black holes in AGN – $M \approx 10^7 M_\odot$ and more – are close to the estimates of masses of supermassive invisible "stars" in the Universe (black holes in our terminology) which were done at the end of the 18-th century by Mitchell and Laplace. They speculated on supermassive stars, which would generate a gravitational field strong enough to trap light rays, so they would be invisible.

In the modern history of the idea of supermassive black holes the crucial step was done by Linden-Bell (1969). He derived and applied a theory of thin accretion disks in orbit around massive black holes. Now it is generally accepted that in AGN there are supermassive black holes with accretion gas (and maybe also dust) disks. One of the most important facts derived from observations, and especially radio observations, is the existence of directed jets from the nuclei of some active galaxies. Sometimes there is evidence that radio components move with ultrarelativistic velocity away from the nucleus. The observation of an axis of ejection strongly suggests that there is some stable compact gyroscope, probably a rotating black hole. In some cases one can observe evidence that there is also precession of such a gyroscope. Probably the essential role in physics of processes in the centers of AGN is played by black hole electrodynamics which we discussed in Chapter 4.

The review of physics of AGN is given by Lamb (1991) and Svensson (1994).

In the model of a supermassive black hole with an accretion disk for AGN one requires sources of fuel – gas or dust. The following sources of fueling were discussed: gas from a nearby galactic companion as a result of the interaction of the host galaxy and the companion, interstellar gas of the host galaxy, disruption of stars by high velocity collision in the vicinity of a black hole, disruption of stars by the tidal field of a black hole and some others (see Shlosman, Begelman and Franc (1990)). An excellent review of all problems of supermassive black holes in galactic centers including the problem of their origin and evolution, is given by Rees (1990a) and Haehnelt and Rees (1993).

Clearly, the processes taking place in quasars and other nuclei are still a mystery in many respects. But the suggestion that we are witnessing the work of a supermassive black hole with an accretion disk seems rather plausible.

6.5 Dynamical Evidence for Black Holes in Galaxy Nuclei

So far we considered supermassive black holes as the most probable explanation of the nuclei of some galaxies. Is there more conclusive evidence of the presence of black holes?

First of all, massive black holes should not only be in active galactic nuclei but also in the centers of "normal" galaxies including nearly galaxies and our Milky Way (Rees 1990a). They are quiescent because they are now starved of fuel (gas). Observations show that activity of galactic nuclei was more common in the past. Thus, "dead quasars" (massive black holes without fuel) should be common now.

How could they be detected? It was pointed out that black holes produce cuspy potentials and hence they should produce cuspy density distributions of the stars in the central regions of the galaxies. Some authors argued that the brightness profiles of the central regions of particular galaxies imply that they contain black holes. J. Kormendy (1993), emphasized that arguments based only on surface brightness profiles are inconclusive. The point is that a high central number density of stars on the core with small radius can be the consequence of the dissipation, and cuspy profiles can be a result of anisotropy of the velocity dispersion of stars. Thus these properties are not sufficient to establish the presence of a black hole.

The reliable way to detect black holes in galactic nuclei is analogous to the case of black holes in binaries. Namely, one needs to give the proof that there is a big dark mass in a small volume, and that it can be nothing else but a black hole.

In order to give such a proof we need stellar kinematics as well as surface photometry of galactic nuclei.

The mass M inside radius r is given by the formula (Kormendy 1993):

$$M(r) = \frac{v^2 r}{G} + \frac{\sigma_r^2 r}{G}\left[-\frac{d\ln I}{d\ln r} - \frac{d\ln \sigma_r^2}{d\ln r} - \left(1 - \frac{\sigma_\theta^2}{\sigma_r^2}\right) - \left(1 - \frac{\sigma_\phi^2}{\sigma_r^2}\right)\right], \quad (6.248)$$

where I is brightness, v is the rotation velocity, σ_r, σ_θ, σ_ϕ are the radial, and two tangential components of the velocity dispersion. These values must be obtained from observations after a few special corrections. More complex formulae are used for more sophisticated models.

Now we can consider the mass-to-light ration M/L (in solar units) as a function of radius. This ratio is well known for different types of stellar populations. As a rule this ratio is between 1 and 10 for elliptical galaxies and globular clusters (old stellar population dominates there). If for some galaxy the ratio M/L is almost constant at rather large radius r (and has a "normal" value between 1 and 10) but starting from some small r rises rapidly toward the center to values much larger than 10, then there is evidence for a central dark object, probably a black hole.

As an example consider the galaxy NGC 3115 which is at a distance of 9.2Mpc (Kormendy and Richstone 1992). For this galaxy $M/L \approx 4$ and almost constant over a large radius range at radii $r > 4''$ (in angular units). This value is exactly normal for a bulge of this type of galaxies. At radius $r < 2''$ the ratio M/L rises rapidly up to $M/L \approx 40$. If this is due to a central dark mass added to a stellar distribution with constant M/L, then $M_{b.h.} = 10^{9.2 \pm 0.5} M_\odot$.

Is it possible to give another explanation of the large mass-to-light ratio in the central region of a galaxy? We cannot exclude the possibility that a galaxy contains a central compact cluster of dim stars. But it is unlikely. The central density of stars in the galaxy NGC 3115 is not peculiar, it is the same as in the centers of globular clusters. The direct observational data (spectra and colors) of this galaxy do not give any evidence of a dramatic population gradient near the center. Thus, the most plausible conclusion is that there is a central massive black hole.

Unfortunately, it is most difficult to detect massive black holes in giant elliptical galaxies with active nuclei, where we are almost sure black hole must exist because we observe their active manifestation (Kormendy 1993).

The reason for that is due to a fundamental difference between giant elliptical galaxies (the nuclei of some of them are among the most extreme examples of AGN) and dwarf elliptical galaxies and spiral galaxies. Dwarf ellipticals rotate rapidly and stellar velocity dispersions are nearly isotropic. Giant elliptical galaxies do not rotate significantly and they have anisotropic velocities. It is not so easy to model these dispersions. This leads to uncertainties in equation (6.248).

Giant elliptical galaxies have large cores and shallow brightness profiles, and therefore projected spectra are dominated by light from large radii, where a black hole has no effect.

The type of technique described above was used for black hole search in galactic nuclei. Another possibility is the observation of rotational velocities of gas in the very vicinity of a galactic center. So far (the beginning of 1995) black hole detections have been reported for the following galaxies (for a review see Kormendy 1993, see also van der Marel *et al.* 1994, Miyoshi *et al.* 1995): M32, M31, NGC 3377, NGC 4594, Milky Way, NGC3115, and NGC 4258. Special investigations were performed in the case of the galaxy M87 (see Dressler 1989 for a review of earlier works and Lauer *et al.* 1992 and Ford *et al.* 1994, Harms *et al.* 1994 for Hubble Space Telescope observations). This is a giant elliptical galaxy with an active nucleus and a jet from the center. At present there is secure stellar-dynamical evidence for a black hole with mass $M \approx 3 \cdot 10^9 M_\odot$ in this galaxy. The presence of a black hole in M87 is especially important for our understanding of the nature of the central regions of the galaxies since in this case we also observe the activity of the "central engine".

In conclusion we give the table with estimates of masses of black holes in the nuclei of some galaxies (Kormendy 1993, van der Marel *et al.* 1994, van der Marel 1995, Miyoshi *et al.* 1995).

Progress in this field is very fast and in the nearest future our knowledge about evidence of supermassive black holes in galactic nuclei will be more profound. We also want to mention the possibility of the formation of binary supermassive black hole systems in the process of merging of galaxies (see for example Rees (1990b), Polnarev and Rees (1994)). The radiation of gravitational waves

Table 5. Mass estimates of black holes in galactic nuclei

Galaxy	Mass of black hole (in $M_\odot$)
M31	2×10^7
M32	$(2-5) \times 10^6$
Milky Way	3×10^6
NGC 4594	10^9
NGC 3115	10^9
NGC 3377	10^8
M87	2.4×10^9
NGC 4258(M106)	3.6×10^7

of such a system leads to a decay of the orbits of black holes. Eventually they coalesce. The final asymetric blast of gravitational radiation may eject the merged hole from the merged galaxy.

7 Primordial Black Holes

Let us now consider the possibility of existence of primordial black holes. The smaller the mass of matter, the greater the density to which it must be compressed in order to create a black hole. Powerful pressure develops at high densities, counteracting the compression. As a result, black holes of mass $M \ll M_\odot$ cannot form in the contemporary Universe. However, the density of matter at the beginning of expansion in the Universe was enormously high. Zel'dovich and Novikov (1967a, 1967b), and then Hawking (1971) hypothesized that black holes could have been produced at the early stages of the cosmological expansion of the Universe. Such black holes are known as primordial. Very special conditions are needed for primordial black holes to form. Lifshitz (1946) proved that small perturbations in a homogeneous isotropic hot Universe (with the equation of state of the matter $p = \varepsilon/3$) cannot produce appreciable inhomogeneities. A hot Universe is stable under small perturbations [see Bisnovaty-Kogan et al. 1980]. Large deviations from homogeneity must exist from the very beginning in the metric describing the Universe (i.e., the gravitational field had to be strongly inhomogeneous) even though the spatial distribution of matter density close to the beginning of the cosmological expansion was very uniform. When the quantity $l = ct$, where t is the time elapsed since the Big Bang, grows in the course of expansion to a value of the order of the linear size of an inhomogeneity of the metric, the possibility appears for the formation of a black hole of the mass contained in the volume l^3 at the time t. The formation of black holes with masses substantially smaller than stellar masses was thus possible, provided that such holes were created at a sufficiently early stage (see below).

Primordial black holes are of special interest because Hawking's quantum evaporation is important for small-mass black holes, while only primordial black holes can have such masses. (Note that quantum evaporation of massive and even supermassive black holes may be essential for the distant future of the Universe.)

First of all, the following two questions arise:

1. How large must the deviations from the metric of a homogeneous isotropic model of the Universe be for black holes to be born?
2. What is the behavior of the accretion of the surrounding hot matter to the created hole and how does the accretion change the mass of the hole?

The second question arises because of a remark made in the pioneer paper of Zel'dovich and Novikov: If a stationary flux of gas to the black hole builds up, the black-hole mass grows at a catastrophically fast rate. But if such a stationary accretion does not build up immediately after the black hole is formed, accretion is quite negligible at later stages because the density of the surrounding gas in the expanding Universe falls off very rapidly.

Both questions can be answered via numerical modeling. The required computation for the case of spherical symmetry were carried out by Nadezhin *et al.* (1977, 1978) and Novikov and Polnarev (1980).

The main results of these computations are as follows. The dimensionless amplitude of metric perturbations, δg_β^α, necessary for the formation of a black hole, is about 0.75–0.9. The uncertainty of the result reflects the dependence on the perturbation profile. Recall that the amplitude of the metric perturbation is independent of time as long as $l = ct$ remains much less than the linear size of the perturbed region. If δg_β^α is less than 0.75–0.9, the created density perturbations transform into acoustic waves after $l = ct$ increases to about the size of the perturbation.

This answers the first of the questions formulated above.

As for the second question, the computations show that the black-hole mass at the moment of formation incorporates 10 % to 15 % of the mass within the scale $l = ct$. This means that the accretion of the gas to the newborn black hole cannot become catastrophic. Computations confirm that the gas falling into the black hole from the surrounding space only slightly increases its mass.

If in the early history of the Universe there were periods when pressure was reduced for a while, then all pressure effects were not important. Formation of primordial black holes (PBHs) under these conditions was discussed in the papers of Khlopov and Polnarev (1980); Polnarev and Khlopov (1981).

Other exotic formation mechanisms include cosmic phase transitions (Kodama *et al.* 1982; Hawking *et al.* 1982; Kardashev and Novikov 1983, Naselsky and Polnarev 1985; Hsu 1990) or collapse of loops of cosmic strings (Hawking 1989; Polnarev and Zembowicz 1991, Polnarev 1994). The review of the problem is given by Carr (1992).

The discovery by Hawking that black holes can evaporate by thermal emission made the study of PBH formation and evaporation of considerable astrophysical interest, because the evaporation effects of small black holes are potentially observable. Their absence from observational searches therefore enables powerful limits to be placed upon the structure of the very early universe (Zeldovich and

Novikov 1967ab: Carr and Hawking 1974; Novikov et al. 1979; Carr 1983; Carr et al. 1994).

When we consider black holes of solar-mass size or supermassive black holes, Hawking radiation is quite negligible, but for small enough PBHs, it becomes the controlling influence in the black hole's evolution and very important for their possible observational manifestations. Using the formulae for Hawking radiation (see, for example, Chapter 12 of Novikov and Frolov 1989) it is easy to conclude that PBHs of $M \approx 5 \times 10^{14}$g or less will indeed have evaporated entirely in the 10^{10} years or so in the Universe's history. PBHs a little more massive than this initially will still be evaporating in the modern Universe. A rate of their evaporation is large enough that the stream of energetic particles and radiation they emit can be turned into a strong observational limit on their presence.

Searches for PBHs attempt to detect a diffuse photon (or another particle) background from a distribution of PBHs or to search directly for the final emission stage of individual black holes. Using the theoretical spectra of particles and radiation emitting by evaporating black holes of different masses (see Chapter 12 of Novikov and Frolov 1989), one can calculate the theoretical background of photons and other particles produced by a distribution of PBHs emitting over the lifetime of the Universe. The level of this background depends on the integrated density of PBHs with initial masses in the considered range.

A comparison of the theoretical estimates with the observational cosmic ray and γ-ray background place an upper limit on the integrated density of PBHs with initial masses in this range. According to estimates of MacGibbon and Carr 1991, this limit corresponds to $\approx 10^{-6}$ of the integrated mass density of the visible matter in the Universe (matter in the visible galaxies). The comparison of the theory with other observational data gives weaker limits (for a review see Halzen et al. 1991; Carr 1992, Coyne 1993).

The search for high energy gamma-ray bursts as direct manifestation of the final emission of the evaporating (exploding) individual PBHs has continued for more than 20 years. No positive evidence for the existence of PBHs has been reported (see Cline and Hong 1992, 1994).

A population of PBHs whose influence is small today may have been more important in the earlier epochs of the evolution of the Universe. Radiation from PBHs could perturb the usual picture of cosmological nucleosynthesis, distort the microwave background and could have produced too much entropy in relation to the matter density of the Universe. As we mentioned above, limits on the density of PBHs, now or at earlier times, can be used to provide information on the homogeneity and isotropy of the very early Universe, when they were formed. For a review see Novikov et al. (1979), Carr et al. (1994).

The final state of the black hole evaporation is still unclear (see Chapter 17 of Novikov and Frolov 1989). There is a possibility that the endpoint of the black hole evaporation is a stable relic. The possible role of such relics in cosmology was first discussed by MacGibbon (1987), for the recent review see Barrow et al. (1992).

In this chapter we have reviewed the current search for evidence for black holes in the Universe. Our conclusion is the following. Now, at the beginning of

1995, we are almost 100 percent sure that black holes of stellar masses exist in the binary systems.

Probably we have to repeat the same about the supermassive black holes in the centers of many galaxies. However, so far there is not any evidence for the existence of PBHs in the Universe.

Acknowledgements

This work is supported in part by Danish Natural Science Research Council through grant N9401635 and in part by Danmarks Grundforskningsfond through its support for the establishment of the Theoretical Astrophysical Center. The author is greatful to V.Beskin for help in the work on the Section 4 and the coauthors of my works on the problem of black holes.

References

Abramowicz, M.A., B. Czerny, J.P. Lasota and E. Szuzkievicz: 1988, *Astrophys. J.* **332**, 646.
Abramowicz, M.A., Szuskiewicz and Wallinder: 1989, in *Theory of Accretion Discs*, eds. F. Meyer, W.J. Duschl, J. Frank and E. Meyer-Hofmeister, NATO ASI Series, Vol. **290**, (Kluwer), p.141.
Alcook, C., E. Farhi, and A.V. Olinto: 1986, *Astrophys. J.* **310**, 261.
Alpar, M.A.: 1987, *Phys. Rev. Lett.* **58**, 2152.
Arons, I.: 1983, *Astrophys. J.* **266**, 215.
Artemova, J., G. Bisnovatyi-Kogan, G. Björnsson, I. Novikov: 1994, *Proceeding of Vulcano Workshop, in press*.
Bahcall, S., B.W. Lynn, and S.B. Selipsky: 1990, *Astrophys. J.* **362**, 251.
Bardeen, J.M., and J.A. Petterson: 1975, *Astrophys. J.* (Lett.), **195**, L65.
Bardeen, J.M., W.H. Press, and S.A. Teukolsky: 1972, *Astrophys. J.* **178**, 347.
Barrow J.D., E.J. Copeland, and A.R. Liddle: 1992, *phys. Rev. D.* **46**, 645.
Baym, G., and C.J. Pethick: 1979, *Ann.Rev.Astr.Ap.* **17**, 415.
Begelman, M.C., R.D.Blandford, and M.I.Rees: 1984, *Rev. Mod. Phys.* **56**, 255.
Beskin, V.S., A.V.Gurevich, and Ya.N.Istomin: 1993, *Physics of Pulsar Magnetosphere*, Cambridge University Press, Cambridge.
Beskin, V.S., Ya.N.Istomin, and V.I.Pariev: 1992, in *Extragalactic radio sources – from Beams to Jets*. Proc. 7th A.I.P.Meeting ed. J.Roland, H.Sol, and G.Pelletier. Cambridge University Press, Cambridge.
Beskin, V.S., and V.I.Pariev: 1993, *Sov. Phys. Uspekhi* (in press)
Bičak, J.: 1983, in B. Bertotti, F. de Felice, A. Pascolini (eds.), *10th Int. Conf. on General Relativity and Gravitation*, Padova, **2**, p. 688.
Bičak, J.: 1985, *Mon. Not. RAS* **183**, 1000.
Bičak, J. and L. Dvorak: 1980, *Phys. Rev.* **D22**, 2933.
Bičak, J. and Z. Stuchlik: 1976, *Mon. Not. RAS* **175**, 381.
Birkhoff, G.D.: 1923, *Relativity and Modern Physics*, Harvard Univ. Press.
Bisnovaty-Kogan G.S.: 1989, Physical Problems of the Theory of Stellar Evolution, Nauka, Moscow.
Bisnovaty-Kogan G.S., V.N. Lukash, and I.D. Novikov: 1980, in *Proc of the Fifth European Regional Meeting in Astronomy, Liege, Belgium*, G. 1.1.

Björnsson G., R. Svensson: 1991, *Astrophys. J.* **371**, L69.
Björnsson G.,R. Svensson: 1992, *Astrophys. J.* **394**, 500.
Blandford, R.D.: 1976, *Mon. Not. RAS* **176**, 465.
Blandford, R.D.: 1979, in C. Hazard and S. Mitton (eds.),*Active Galactic Nuclei*, Cambridge Univ. Press, p. 241.
Blandford, R.D.: 1987, in "300 Years of Gravitation" ed. by S.W. Hawking and W. Israel, p.199, Cambridge University Press.
Blandford, R.D.: 1990, in *Active Galactic Nuclei,* ed. R.D.Blandford, H. Netzer, and L. Woltier. Spring Verlag, Berlin.
Blandford, R.D. and D.G.Payne: 1982, *Mon. Nor RAS* **199**, 883.
Blandford, R.D. and M.J. Rees: 1978, *Phys. Sci.* **17**, 265.
Blandford, R.D. and R.L. Znajek: 1977, *Mon. Not. RAS* **179**, 433.
Bochkarev, N.G., R.A. Sunyaev, T.S. Kruzina, A.M. Cherepashchuk, N.I. Shakura: 1988, *Sov. Astron. J.* **32**, 405.
Bogovalov S.V.: 1989, *Soviet Astron Lett.* **15**, 469.
Bogorodsky, A.F.: 1962, *Einstein's Field Equations and Their Application to Astronomy*, Kiev Univ.
Bolton, C.T.: 1972a, *Nature* **235**, 271.
Bolton, C.T.: 1972b, *Nature Phys. Sci.* **240**, 124.
Bolton, C.T.: 1975, *Astrophys. J.* **200**, 269.
Bondi, H.: 1952, *Mon. Not. RAS* **112**, 195
Bondi, H. and F. Hoyle: 1944, *Mon. Not. RAS* **104**, 273.
Boyer, R.H. and R.W. Lindquist: 1967, *J. Math. Phys.* **8**, 265.
Camenzind, M: 1986a, *Astron. Astrophys.* **156,** 137.
Camenzind, M: 1986b, *Astron. Astrophys.* **162,** 32.
Camenzind, M: 1987, *Astron. Astrophys.* **184**, 341.
Camenzind, M: 1986a, *Rev. Mod. Astron.* **3,** 234.
Carr, B.J.: 1992, *Report on Zeldovich's meeting*, Moscow.
Carr, B.J.: 1983, in M.A. Markov and P.C. West (eds.), *Quantum Gravity*, Plenum Press, N.Y., p.337
Carr, B.J., J.H. Gilbert, J.E. Lidsey: 1994 *Phys. Rev. D.* **48**, 1000.
Carr, B.J. and S.W. Hawking: 1974, *Mon. Not. R. Astron. Soc.*, **168**, 399.
Carter, B.: 1968, *Phys. Rev.* **174**, 1559.
Carter, B.: 1973, *General Theory of Stationary Black Holes*, in C. DeWitt and B.S. DeWitt (eds.), *Black Holes*, Gordon and Breach, New York.
Carter, B.: 1977, *Phys. Rev.* **D16**, 3395.
Colpi, M., L. Marachi, and P. Treves: 1984, *Astrophys. J.* **280**, 319
Cominsky, L.: 1994, In *Proceeding of the Snowmass Workshop, in press.*
Copson, E.T.: 1928, *Proc. Roy. Soc.* **A118**, 184.
Cowie, L.L., J.P. Ostriker, and A.A. Stark: 1978, *Astrophys. J.* **226**, 1041.
Cowley, A.P., D. Crampton, J.B. Hutchings, R. Remillard, J.E. Penfold: 1983, *Astrophys. J.* **272**, 118.
Cowley, A.P., P.C. Schmidtke, D. Crampton and J.B. Hutchings: 1990, *Astrophys. J.* **350**, 288.
Cowley, A.P.: 1992, *Ann. Rev. Astron. Astrophys.* **30**, 287.
Coyne, D.G.: 1993 in the book: *International symposium on black holes, membranes, wormholes and superstrings*; World Scientific Publishing Co., p. 159.
Damour, T.: 1978, *Phys. Rev.* **D18**, 3598.
Davis, M., R. Ruffini, and R. Price: 1971, *Phys. Rev. Lett.* **27**, 1466.
Demianski, M. and I.D. Novikov: 1982, *GRG* **14**, 439.

Dokuchayev, V.I.: 1987, *Zh. Eksp. Teor. Fiz.* **92**, 1921.
Dressler, A.: 1989, in IAU Symposium 134, Active Galactic Nuclei, ed. D.E.Osterbrock and J.S.Miller (Kluwer. Dordrecht), p.217.
Dymnikova, I.G.: 1982, Preprint FTI-95, Leningrad.
Dymnikova, I.G.: 1986, *Usp. Fiz. Nauk* **148**, 393.
Eddington, A.S.: 1924, *Nature* **113**, 192.
Eddington, A.S.: 1926, *The Internal Constitution of the Stars* (Cambridge: Cambridge University Press).
Finkelstein, D.: 1958, *Phys. Rev.* **110**, 965.
Ford, H.C. *et al.*: 1994, *Astrophys. J* **435**, L27.
Friedman, J.L., J.R. Isper: 1987, *Astrophys. J.* **314**, 594.
Frolov, V.P. I.D. Novikov: 1995, *Physics of Black Holes; New Methods and Perspectives*, under preparation.
Fronsdal, C.: 1959, *Phys. Rev.* **116**, 778.
Gal'tsov, D.V. and V.I. Petukhov: 1978, *Zh. Eksp. Teor. Fiz.* **74**, 801.
Giacconi, R., S. Murray, H. Gursky, E. Kellogg, E. Schreier, H. Tananbaum: 1972, *Astrophys. J.* **178**, 281.
Giacconi, R., H. Gursky, F.R. Paolini and B.B. Rossi: 1962, *Phys. Rev. Lett.* **9**, 439.
Gies, D.R. and C.T. Bolton: 1986, *Astrophys. J.* **304**, 371.
Ginzburg, V.L.: 1964, *Dokl. Akad. Nauk USSR* **156**, 43.
Ginzburg, V.L. and L.M. Ozernoi: 1964, *Zh. Eksp. Teor. Fiz.* **47**, 1030.
Goldreich, P. and W.H. Julian: 1969, *Astrophys. J.* **157**, 869.
Gurevich, A.V. and Ya.N. Istomin: 1985, *Zh. Eksp. Teor. Fiz.* **89**, 3.
Haehnelt, M., and M. Rees: 1993, MNRAS, **263**, 168.
Hagihara, Y.: 1931, *Jap. J. Astr. Geoph.* **86**, 67.
Halzen, F., E. Zas, J.H. MacGibbon, and T.C. Weekes: 1991, *Nature,* **353**, 807.
Hanni, R.S. and R. Ruffini: 1973, *Phys. Rev.* **D8**, 3259.
Harms R.J. *et al.*: 1994, *Astrophys. J* **435**, L35 .
Haswell, C.A., E/L/ Robinson, K. Horne, R. Stiening & T.M.C. Abbott: 1993, *Astrophys. J.*, **411**, 802.
Haswell, C.A. and A.W. Shafter: 1990, *Astrophys. J. Lett.* **359**, L47.
Hawking, S.W.: 1971, *Mon. Not. RAS* **152**, 75.
Hawking, S.W.: 1989, *Phys. Lett. B.*, **231**, 237.
Hawking, S.W. and G.F. Ellis: 1973, *The Large-Scale Structure of Spacetime*, Cambridge Univ. Press.
Hawking, S.W., I. Moss and J. Stewart: 1982, *Phys. Rev. D.*, **26**, 2681.
Hayakawa, S. and M. Matsuoko: 1964, *Progr. Theor. Phys. Suppl.* **30**, 204.
van den Heuvel, E.P.J.: 1983, in *Accretion-Driven Stellar X-ray Sources*, ed. W.H.G. Lewin, E.P.J. van den Heuvel, p.303. Cambridge University Press
van den Heuvel, E.P.J. and G.M.H.J. Habets: 1984, *Nature* **309**, 598.
Heyvaerts, J. and C. Norman: 1989, *Astrophys, J.* **347**, 1055.
Hirotani, K. *et al.*: 1992, *Astrophys, J.* **386**, 455.
Hoyle, F. and W.A. Fowler: 1963, *Nature* **197**, 533.
Hoyle, F., W.A. Fowler, G. Burbidge, and E.M. Burbidge: 1964 *Astrophys. J.* **139**, 909.
Hoyle, F., and R.A. Lyttleton: 1939, *Proc. Camb. Phil. Soc.*, **35**, 405.
Hsu, S.D.U.: 1990, *Phys. Lett. B.*, **251**, 343.
Israel, W.: 1987, in *"300 Years of Gravitation"* ed. by S.W. Hawking and W. Israel, p.199, Cambridge University Press.
Inoue, H.: 1991, *28th Yamula Conf. The Frontiers of X-ray Astronomy* ed. K.Koyama.
Johnston, H.M., S.R. Kulkarni and J.B. Oke: 1989, *Astrophys. J.* **345**, 492.

Johnston, H.M. and S.R. Kulkarni: 1990, in *Accretion-Powered Compact Binaries*, Cambridge, Cambridge University Press.

Kardashev, N.S. and I.D. Novikov: 1983, in the book G.O. Abell and G. Chincarini (eds.) *Early Evolution of the Universe and Its Present Structure*, 463-468.

Kardashev, N.S., I.D. Novikov, A.G. Polnarev, and B.E. Stern: 1983, *Astron. Zh.* **60**, 209: in M. Burns et al. (eds.), *Positron-Electron Pairs in Astrophysics*, Goddard Space Flight Center.

Kardashev, N.S., I.G. Mitrofanov, and I.D. Novikov: 1984, *Astron. Zh.* **61**, 1113.

Kennel, C.F., F.S. Fujimura, and I.Okamoto: 1983 *Geophys. Astrophys. Fluid Dyn.* **26**, 147.

Khlopov, M.Yu. and A.G. Polnarev: 1980, *Phys. Lett. B.*, **97**, 383.

King, A.R., and J.P. Lasota: 1977, *Astr. Astrophys.* **58**, 175.

Kippenhahn, R. A.Weigert: 1990, *Stellar Structure and Evolution*, Springer-Verlag.

Kodama, H., M. Sasaki and K. Sato: 1982, *Prog. Theor. Phys.*, **68**, 1979.

Kormendy, J.: 1993, in The Nearest Active Galaxies, ed J.E. Beckman, H. Netzer, and L. Colina (Consejo Superior de Investigaciones Scientificas, Madrid)

Kormendy, J. and D. Richstone: 1992, *Astrophys. J* **393**, 559.

Krolik, J.H. and R.A. London: 1983, *Astrophys. J.* **267**, 371.

Kruskal, M.D.: 1960, *Phys. Rev.* **119**, 1743.

Kuiper, L., J. van Paradijs, M. van der Klis: 1988 *Astron. Astrophys.* **203**, 79.

Lamb, F.K: 1991, in *"Frontiers of Stellar Evolution"*, ed. by D.L.Lambert (Astronomical Society of the Pacific), p.299.

Landau, L.D. and E.M. Lifshitz: 1959, *Fluid Mechanics*, Pergamon Press, London.

Landau, L.D. and E.M. Lifshitz: 1975, *The Classical Theory of Fields*, Pergamon Press, Oxford.

Lauer, T.R.: 1992, AJ, **103** 703.

Lèaute, B. and B. Linet: 1976, *Phys. Lett.* **A58**, 5.

Lèaute, B. and B. Linet: 1982, *J. Phys.* **A15**, 1821.

Lemaitre, G: 1933, *Ann. Soc. Bruxelles* **A53**, 51.

Li, Zh.-Y., T.Chiuen, and M.C.Begelman: 1992, *Astrophys.J.* **394,** 459.

Lifshitz, E.M.: 1946, *Zh. Eksp. Teor. Fiz.* **16**, 587.

Linet, B.: 1976, *J. Phys.* **A9**, 1081.

Linet, B.: 1977a, *C. R. Acad. Sc. Paris* **A284**, 215.

Linet, B.: 1977b, *C. R. Acad. Sc. Paris* **A284**, 1167.

Linet, B.: 1979, *J. Phys.* **A12**, 839.

Lovelace, R.V.E.: 1976, *Nature* **262**, 649.

Lovelace, R.V.E., J. MacAuslan, and M. Burns: 1979, in *Proceedings of La Jolla Institute Workshop on Particle Acceleration Mechanism in Astrophysics*, American Inst. of Phys., N.Y.

Lipunov V.M.: 1987, *Astrophysics of Neutron Stars*, Nauka, Moscow.

Lynden-Bell, D.: 1969, *Nature* **233**, 690.

Macdonald, D.A.: 1984, *Mon. Not. RAS* **211**, 313.

Macdonald, D.A. and K.S. Thorne: 1982, *Mon. Not. RAS* **198**, 345.

Macdonald, D.A. and W.M. Suen: 1985, *Phys. Rev.* **D32**, 848.

MacGibbon, J.H.: 1987, *Nature* **329**, 308.

MacGibbon, J.H. and B.J. Carr: 1991, *Astrophys. J.* **371**, 447.

Madsen, K.: 1993, Proc. 2nd Int. Conf. on Physics and Astrophysics of Quark-Gluon Plasma, World Scientific.

van der Marel, R.P.: 1995, in *Highlights of Astronomy* **10**, Proc. of the XXII General Assembly of the IAU, The Hague, August 1994, ed. J. Bergeron, Kluwer Academic Publishers

van der Marel, R.P., N.W. Evans, H.W. Rix, S.D.M. White, P.T. de Zeeuw: 1994, MNRAS, in press.

Masevich, A.G. and A.V. Tutukov: 1988, *Stellar evolution: theory and observations*, Nauka, Moscow.

Mazeh, T., J. van Paradijs, E.P.J. van den Heuvel and G.J. Savonije: 1986, *Astron. Astrophys.* **157**, 113.

McClintock, J.E.: 1988, in *Supermassive Black Holes*, ed. Minas Kafatos, Cambridge University Press, p.1.

McClintock, J.E.: 1992, in *X-ray Binaries and Recycled Pulsars*, eds. E.P.J. van den Heuvel and S.A.Rappoport, p.27.

McClintock, J.E. and R.A. Remillard: 1990, *Bull. Am. Astron. Soc.* **21**, 1206.

Mestel, L.: 1968, , *Mon. Not. RAS* **138**, 359.

Michel, F.C.: 1969, *Astrophys.J.* **158**, 727.

Michel, F.C.: 1973, *Astrophys.J.* **180**, 207.

Michel, F.C.: 1991, *Theory of Neutron Star Magnetospheres*. University of Chicago Press. Chicago.

Mineshige, S. and J.C. Wheeler: 1989, *Astrophys. J.* **343**. 241.

Misner, C.W., K.S. Thorne, and J.A. Wheeler: 1973, *Gravitation*, Freeman, San Francisco.

Misra, R.M.: 1977, *Progr. Theor. Phys.* **58**, 1205.

Moscoso, M.D. and J.C. Wheeler: 1994, in *"Interacting Binary Stars"*, ed. by A.W. Shafter, Astronomical society of the Pacific, Conference Series, v. 56, p.100

Miyoshi, M., J. Moran, J. Herrnstein, L. Greenhill, N. Nakai, P. Diamond, M. Inoue: 1995, Nature **373**, 127.

Nadejin, D.K., I.D. Novikov, and A.G. Polnarev: 1977, in *Abstracts of contributed papers GR8*, Waterloo, Canada, p.382.

Nadejin, D.K., I.D. Novikov, and A.G. Polnarev: 1978, *Astr. Zh.*, **55**, 216.

Naselsky, P.D. and A.G. Polnarev: 1985, *Astr. Zh.* **29**, 487.

Ne'eman, Y.: 1965, *Astrophys. J.* **141**, 1303.

Nitta, S., M. Takahashi, and A. Tomimatsu: 1991, *Phys. Rev. Ser. D* **44**, 2295.

Novikov, I.D.: 1961, *Astron. Zh.* **38**, 564.

Novikov, I.D.: 1962a, *Vestnik MGU* Ser. III, No. **5**, 90.

Novikov, I.D.: 1962b, *Vestnik MGU* Ser. III, No. **6**, 61.

Novikov, I.D.: 1963, *Astron. Zh.* **40**, 772.

Novikov, I.D.: 1964a, *Soobshch. GAISH* No. **132**, 3, 43.

Novikov, I.D.: 1964b, *Astron. Zh.* **41**, 1075.

Novikov, I.D.: 1974, in *Astrophysics and Gravitation*, Proceedings of the 16th Solvay Conf. on Physics, 1973, edition de l'Universite de Bruxelles, p.317.

Novikov, I.D.: 1995a, *Black Holes, Wormholes, and Time Machines*, to be published in the Proceedings of the Conference "Birth of the Universe and Fundamental Physics", Rome, May, 1994.

Novikov, I.D.: 1995b, *Black Hole Astrophysics* to be published in the Proceedings of the Symposium: Particle and Nuclear Astrophysics and Cosmology in the Next Millennium, Snowmass, USA, 1994.

Novikov, I.D.: 1995c, *Physics and Astrophysics of Black Holes* Preprint TAC, 1995-002, to be published in the Proceedings of the Conference on "Key Problems in Astronomy" Tenerife, January 1995.

Novikov, I.D., V.P. Frolov: 1989, *Physics of Black Holes* Kluwer Academic Publishers.
Novikov, I.D. and L.M. Ozernoy: 1964, Preprint FIAN, A-17.
Novikov,I.D. and A.G. Polnarev: 1980, *Astr. Zh.*, **57**, 250.
Novikov,I.D., A.G. Polnarev, A.A. Starobinsky and Ya.B. Zeldovich: 1979, *Astron. and Astrophys.* **80**, 104.
Novikov, I.D. K.S. Thorne: 1973, in *"Black holes, Les Astres Occuls"*, ed. by C.Dewitt and B.S.DeWitt, Gordon and Breach Science Publishers, p.343.
Novikov, I.D. and B.E. Stern: 1986, in G. Giuricini *et al.* (eds.), *Structure and Evolution of Active Galactic Nuclei*, D. Reidel, Dordrecht, p. 149.
Novikov, I.D. and Ya.B. Zeldovich: 1966, *Nuovo Cim. Suppl.* **4**, 810, addendum 2.
Okamoto, I.: 1978, *Mon. Not. RAS* **185**, 69.
Okamoto, I.: 1992, *Mon. Not. RAS* **254**, 192.
Oppenheimer, J.R. and H.Snyder: 1939, *Phys. Rev.* **56**, 455.
Oppenheimer, J.R. and G. Volkoff: 1939, *Phys. Rev.* **55**, 374.
Ostriker, J.P., R. McCray, R. Weaver, and A. Yahil: 1976, *Astrophys. J. Lett.* **208**, L61.
Paczynski, B: 1974, *Astron. Astrophys.* **34**, 161.
Paczynski, B.: 1983, *Astrophys. J. Lett.* **273**, L81.
Park, M.G.: 1990a, *Astrophys. J.* **354**, 64.
Park, M.G.: 1990b, *Astrophys. J.* **354**, 83.
Park, S.I., and E.L. Vishniak: 1989, *Astrophys. J.* **337**, 78.
Park, S.I., and E.L. Vishniak: 1990, *Astrophys. J.* **353**, 1.
Pelletier, G. and R. Pudritz: 1992, *Astrophys. J.* **394**, 117.
Penrose, R.: 1969, *Rev. Nuovo Cim.* **1**, 252.
Phinney, S.: 1983a. in *Astrophysical Jets.* Proc. of International Workshop ed. A. Ferrari and A. Pacholczyk, D. Reidel Publishing Company, Dordrecht.
Phinney, S.: 1983b, PhD Thesis, Cambridge.
Pikel'ner, S.B.: 1961, *Foundations of Cosmical Electrodynamics*, Nauka, Moscow (1964, NASA).
Polnarev, A.G. and M.Yu. Khlopov: 1981, *Astr. Zh*, **58**, 706.
Polnarev, A.G., M. Rees: 1994, Astron Astrophys. **283**, 301.
Polnarev A.G.: 1994, *Astron. and Astrophys. Transactions*, **5**, 35.
Polnarev, A.G., and R. Zembowicz: 1991, *Phys. Rev. D.* **43**, 1106.
Prendergast, K.H. and G.R. Burbidge: 1968, *Astrophys. J. Lett.* **151**, L83.
Press, W.H.: 1972, *Astrophys. J.* **175**, 243.
Price, R.H. and K.S. Thorne: 1986, *Phys. Rev.* **D33**, 915.
Pringle, J.E. and M.J. Rees: 1972, *Astr. Astrophys.* **21**, 1.
Pudritz, R.: 1981a, *Mon. Not RAS* **195**, 881.
Pudritz, R.: 1981b, *Mon. Not RAS* **195**, 887.
Punsly, B.: 1991, *Astrophys. J.* **372**, 424.
Punsly, B. and F.C. Coroniti: 1990a, *Astrophys. J.* **350**, 518.
Punsly, B. and F.C. Coroniti: 1990b, *Astrophys. J.* **354**, 583.
Rees, M.J.: 1982, in G.R. Reigler and R.D. Blandford (eds.), *Proceedings of AIP Conference, 'The Galactic Center'*, Caltech, N.Y., p. 166.
Rees, M.: 1990a, Science **247**, N4944, 16 February, p.817.
Rees, M.: 1990b, Scientific American, November, 26.
Rees, M.J., M. Begelman, R. Blandford, and E. Phinney: 1982, *Nature* **295**, 17.
Ruderman, M.A.: 1989, Talk at the Workshop on Neutron Stars, Trieste.
Ruderman, M. and P.G. Sutherland: 1975, *Astrophys. J.* **196**, 51.
Ruffini, R. and J.A. Wheeler: 1971, in *Proceedings of the Conference on Space Physics*, European Space Research Organization, Paris.

Ruffini, R. and J.R. Wilson: 1975, *Phys. Rev.* **D12**, 2959.
Rylov, Yu.A.: 1961, *Zh. Eksp. Teor. Fiz.* **40**, 1955.
Sakurai, T.: 1985, *Astron. Astrophys.* **152**. 121.
Salpeter, E.: 1964, *Astrophys. J.* **140**, 796.
Schild, H. and A. Maeder: 1985, *Astron. Astrophys.* **143**, L7.
Schvartsman. V.F.: 1971, *Astr. Zh.* **15**, 377.
Schwarzschild, K.: 1916, *Sitzber. Deut. Akad. Wiss.* Berlin, K1. Math.-Phys. Tech., s. 189.
Schwarzschild, M.: 1958, Structure and Evolution of the Stars (Princeton, N.J.: Princeton University Press).
Schultz, A.L. and R.H. Price: 1985, *Astrophys. J.* **291**, 1.
Shakura, N.I.: 1972, *Astr. Zh.* **16**, 756.
Shakura, N.I. and R.A. Sunyaev, 1973, : 1972, *Astron. Astrophys.* **24**, 337.
Shapiro, S.L.: 1973a, *Astrophys. J.* **180**, 531.
Shapiro, S.L.: 1973a, *Astrophys. J.* **185**, 69.
Shapiro, S.L. and S.A. Teukolsky: 1983, Black Holes, White Dwarfs, and Neutron Stars. The Physics of Compact Objects (New York: John Wiley).
Shklovsky, I.S.: 1967,*Astrophys. J. Lett.* **148**, L1.
Shlosman, I., M. Begelman, J. Frank: 1990, Nature **345**, 679.
Stewart, J. and M. Walker: 1973, in *Springer Tracts in Modern Physics*, **69**, Springer-Verlag, Heidelberg, p. 69.
Sturrock, P.: 1971, *Astroph6s. J.* **164**, 529.
Svensson, R.: 1994 *Astrophys. J* Supplement Series, **92**, 585.
Synge, J.L.: 1950, *Proc. Roy. Irish. Acad.* **A53**, 83.
Szekeres, G.: 1960, *Publ. Mat. Debrecent* **7**, 285.
Takahachi M.: 1991. *Publ. Astron. Soc. Japan.* **43**. 569.
Takahachi M. *et al.*: 1990. *Astrophys. J.* **386**. 455.
Thirring, H. and J. Lense: 1918, *Phys. Z.* **19**, 156.
Thorne, K.S.: 1985, *Black Holes: The Membrane Viewpoint*, Preprint, Caltech, GRP-031.
Thorne, K.S.: 1986, in S.L. Shapiro and S.A. Teukolsky (eds.), *Highlights of Modern Astrophysics*, Wiley, N.Y.
Thorne, K.S.: 1993, Black Holes and Time Warps, Einstein's Outrageous Legacy, W.W.Norton and Company.
Thorne, K.S. and R.D. Blandford: 1982, in D. Heeschen and C. Wade (eds.), *Extragalactic Radio Sources*, p. 255.
Thorne, K.S. and D.A. Macdonald: 1982, *Mon. Not. RAS* **198**, 339.
Thorne, K.S. and R.H. Price: 1975, *Astrophys. J. Lett.* **195**, L101.
Thorne, K.S., R.H. Price, and D.A. Macdonald: 1986, *Black Holes: The Membrane Paradigm*, Yale Univ. Press, New Haven.
Tolman, R.G.: 1934, *Proc. Nat. Acad. Sci. US* **20**, 169.
Treves, A., T. Belloni, R.H.D. Corbet, K. Ebisawa, R. Falomo *et al.*: 1990, *Astrophys. J.* **364**, 266.
Tutukov, A.V. and A.M. Cherepashchuk: 1993, *Astr. Zh.*, **70**, 307.
Vladimirov, Yu.S.: 1982, *Frames of Reference in Gravitation Theory*, Energoizdat, Moscow.
Wald, R.M.: 1974b, *Phys. Rev.* **D10**, 1680.
Walker, M. and R. Penrose: 1970, *Commun. Math. Phys.* **18**, 265.
Wallinder, F.H.: 1993, *Comments on Astrophys.*, **16**, 331.
Wandell, A., A. Yahil, and M. Milgrom: 1984, *Astrophys. J.* **282**, 53.

Weber, E.J., and L. Davis: 1967, *Astrophys. J.* **148**, 217.
Webster, B.L. and P. Murdin: 1972, *Nature* **235**, 37-38.
Wheeler, J.C. Soon-Wook Kim and M. Moscoso: 1993, in *"Cataclysmic Variables and Related Physics"*, 2nd Technion Haifa Conference, Eilat, Israel. Annals of the Israel Physical Society, **10**, p.180.
Wheeler, J.C.: 1994, *Proceeding of Daejon Korea Workshop on Supernovae and the Evolution of the Galaxy, K.-T. kim ed., (in press)*
Young, P.Y.: 1976, *Phys. Rev.* **D14**, 12.
Zeldovich, Ya.B.: 1964, *Dokl. Akad. Nauk USSR* **155**, 67.
Zeldovich, Ya.B. and I.D. Novikov: 1964, *Dokl. Akad. Nauk USSR* **158**, 811.
Zeldovich, Ya.B. and I.D. Novikov: 1964, *Sov. Phys. Dokl.* **158**, 811.
Zeldovich, Ya.B. and I.D. Novikov: 1967a, *Astr. Zh.* **10**, 602.
Zeldovich, Ya.B. and I.D. Novikov: 1967b, Relativistic Astrophysics, Nauka, Moscow.
Zeldovich, Ya.B. and I.D. Novikov: 1971, *Relativistic Astrophysics* vol.1, *Stars and Relativity*, University of Chicago Press.
Zel'manov, A.L.: 1956, *Dokl. Akad. Nauk USSR* **107**, 815.
Zerilli, F.G.: 1970, *Phys. Rev. D*, **24**. 2141.
Znajek, R.L.: 1978, *Mon. Nat. RAS* **185**, 833.
Zwicky, F.: 1958, *Hdb. d. Phys.* **51**, 766.

Index

3C 273   320
47 Tuc   83
4U1626-67   214

A0620-00   318, 319
Abrikosov vortices   208, 215
Accretion disk   187
Adiabatic index   147–149
Age of the galactic disk   58–61, 82, 84
AGN   320, 321, 323
Alfvén radius   185–187
Alfvén waves   229
Alfvén points   295
Alfvén velocity   294, 295
Ampere's law   297
Asymptotic giant branch, AGB
– core mass–luminosity relation   3, 5, 37
– evolution along the   3, 32
– mass loss   4, 36
– pulsation driven mass loss   37
– thermal pulses   32–36

B-P plane   178, 212
Bardeen-Peterson effect   314
Beaming factor   172
Bekenstein-Hawking entropy   311
Birkhoff's theorem   241
Black holes
– angular velocity of rotation   273
– critical circular orbit   261
– disk accretion   291, 315–317
– in nuclei of galaxies   311, 320–321
– magnetosphere   301, 305–308
– membrane interpretation   285, 296, 298, 299
– number in the Galaxy   238
– of stellar masses   237
– primordial   237, 324–327
– rate of accretion onto   313–316
– rate of evaporation   326
– rate of formation   239

– standard accretion disk model   316, 317
– stellar progenitors   238
– supermassive   237, 315, 321–323, 325–327
Bohr radius   122
Bose condensation   159
Bose-Einstein condensate   157, 159
Boyle's law   170
Braking timescale   187
Breaking index   138
Brunt-Väisälä or buoyancy frequency   66, 67

Chandrasekhar limit   25, 104, 108, 147, 149, 238
Charge conservation law   289, 297
Chemical equilibrium   21
Chemical potential   21
Cluster age   83, 84
Co-rotation radius   178, 180, 186, 187
Coherence length   207, 209
Conductive diffusion coefficient   28–29
Conductive opacity   29
Convection   88–90
Cooper pairs of electrons   160
Coordinates
– Boyer-Lindquist   267, 276, 278
– curvature coordinates   239, 247
– Eddington–Finkelstein   249, 252, 276
– Kerr coordinate system   275, 276
– Lemaitre coordinates   246, 247, 249, 254, 255
– local Cartesian   241
– Schwarzschild coordinates   241, 246, 301
COS B   204
Cosmic rays   108, 109, 120, 159
Coulomb energy   30
Coulomb scattering cross section   29
Cowling approximation   65

Crab nebula
- historical facts   108–109
- luminosity   133, 139
- magnetic field   109, 135, 137, 138
- radio emission   108, 139, 175

Cracking timescale   229
Critical equilibrium period   187, 197, 202
Crystallization in white dwarfs   30, 31, 54–57, 59
Curvature radiation   113, 125, 126, 132, 204, 205
Cyclotron frequency   188
Cyg X-1   318–320

Debye cooling   31
Debye cooling phase   55, 57, 59
Debye temperature   31, 55
Degeneracy parameter   27–28
Diffusion
- coefficients   44
- equation   44
- ordinary   45, 48
- thermal   45

Diffusive equilibrium   45
Dispersion relation   66
Distribution of particle energies   23, 26
Dynamo process   206

Eddington accretion rate   200
Eddington critical luminosity   314
Eddington luminosity   187, 188
Effective potential   259, 261, 279
Effects
- of Coulomb interactions   30

Einstein observatory   167
Electron conduction   52, 53
Electron-positron cascade   122, 124, 127
Electrostatic energy   101
Energy density   24
Equation of state
- Baym-Bethe-Pethic equation   154
- below neutron drip   144
- complete degeneracy   24, 102
- Harrison-Wheeler equation   151, 155
- ideal gas   98
- Oppenheimer-Volkoff equation   151

Ergosphere   268, 271–273, 281
Escape velocity   260, 262
Event horizon   271, 273, 275, 296, 299, 300, 304, 314
- effective surface resistance   297

Faraday rotation   111
Faraday's law   287, 288
Fermi energy   23, 26, 28
Fermi momentum   23
Fermi–Dirac statistics   27, 170
Fermi-Dirac integrals   28
Ferromagnetic iron lattice   140
Free fall
- along a geodesic   244
- in Schwarzschild gravitational field   241

γ-ray burst   210, 229–230
γ-ray sources   204–205
g-modes   66, 67, 70, 89
- frequencies of nonradial   64
- kinetic energy of nonradial   69, 81
- magnetic or Zeeman splitting   70
- nonradial   61, 62, 78, 79
- period spacing   67–69, 81
- period spectrum   67
- periods   66
- rotational splitting   70
- secular period changes   79, 81

Gauss' law   297
Geodesic   244, 248, 249, 251, 252, 256–258, 264, 277, 278, 296, 299
Glitches   128, 130, 161, 218–223
- post-glitch recovery timescale   128, 130, 161, 219
Globular cluster   58, 83, 84
Goldreich-Julian model   134, 135
Gravitational capture   259, 260, 263–264, 278, 281
- capture cross section   264
- critical angle   264
- impact parameter   262–264, 281, 282
Gravitational field
- Schwarzschild   240
- spherically symmetric   239, 247, 248, 258
Gravitational radiation   182, 264–266, 324
Gravitational radius   242, 245, 246, 248, 256, 266, 312
Gravitational settling   45, 48
Gravitational shift   244

Hansen relation   68, 69
Hawking quantum process   237, 324, 325
Hawking radiation   326
Hawking's theorem   303
Helium flash   168

Index   337

Helium shell flash   34
Her X-1   230
Homologous contraction   147
Hubble Space Telescope   83–85
Hulse-Taylor pulsar   *see* PSR 1913+16, Hulse-Taylor pulsar
Hydrostatic equilibrium   98

Inverse beta decay   140
Inverse Compton effect   204
Inverse Compton scattering   132, 133

Killing vector field   268
Kramers opacity law   53, 87

Lamb, acoustic frequencies   66, 67
Lapse function   286, 297
Larmor radius   122
Lattice energy   143
Lense-Thiring precession   315
Leo group   85
Light cylinder radius   119, 121, 130, 133, 185
LMC X-1   319
LMC X-3   318–320
London penetration depth   207, 209
Lorentz factor   126

M31   85, 323, 324
M32   323, 324
M4   83, 84
M81   85
M87   323, 324
Mach number   294
Magellanic Clouds   85
Magnetic breaking   200
Magnetic dipole radiation   121, 138, 194
Magnetic dipole radiation model   121, 126
Magnetic pressure   70
Magneto-sonic points   295, 296
Magneto-sonic velocity   296
Magnus force   221, 227
Mass formula   143, 144
Mass loss   169
Mass-radius relation   25
– neutron stars   153
– white dwarfs   25–105
Maxwell–Boltzmann distribution   27
Mean-free-path
– of electrons   29
Meissner effect   207
Merging of galaxies   323

Metric
– Kerr   267, 274, 277, 278, 299, 301, 303
– Kerr-Newman   277
– Schwarzschild   252, 265, 277, 299, 301, 302
Milky Way   322–324
Mode trapping   68, 69, 80, 81

Neutrino cooling   41–43, 54–58, 80
Neutrino emission   163, 166
Neutrino luminosity   158, 159, 166
Neutron drip   141, 142, 154
Neutron fluid   152
Neutron stars
– as a unipolar inductor or a homopolar generator   117
– birth rate of   171
– cooling of   158, 163–167
– dynamical stability   147
– in a binary system   185, 186, 211, 213, 217, 218
– magnetic field   121, 140, 161, 202, 205–218, 230
– mass-radius relation   150, 153
– masses   108, 140, 155–157
– maximum mass   154–155
– mean density   140
– minimum mass   151, 154–155
– moment of inertia   140
– radius   140
– rotating   162, 185, 218
– rotation period   140
– runaway   185, 193
– surface temperature   140
– X-ray emitting   155, 167, 171
Neutronisation   140
NGC 3115   322–324
NGC 3377   323, 324
NGC 4258(M106)   323, 324
NGC 4594   323, 324
Nova Muskae 1991   319
Nuclear equilibrium   140
Nucleus
– Coulomb energy   143
– liquid drop model   143
– surface energy   143
Number density   22

Ohm's law   297
Onsagar-Feynman quantized vortices   163, 165, 208, 214–216
Oppenheimer-Volkoff limit   238
– effect of rotation on   238

338   Index

Oppenheimer-Volkoff's mass   108
Order parameter   159

p–modes   66
Phonons or lattice vibrations   160
Photon diffusion   52
Physical singularity   241, 242, 270, 273
Physical time   241, 243, 267, 286, 297
Planetary nebulae
– central star   16, 17, 38, 39, 62, 72, 84
– formation   4, 39
– K1-16   72
– luminosity function   42, 84–87
– mass distribution of the central stars   40, 85
– NGC 1501   74
– Sand 3   74
Platelets   228, 230
Polar cap gap model   122, 124, 126, 127, 194
Polar cap or lighthouse model   112, 117
Pressure, in kinetic theory   22
Proton fluid   152
PSR 0540-69   132
PSR 0655+64   212
PSR 0826-B4   110
PSR 0833-45, Vela pulsar   112, 115, 128, 130, 132, 133, 159, 174, 182, 204, 219
PSR 1641-45   112
PSR 1855+09   201, 202
PSR 1913+16   156
PSR 1913+16, Hulse-Taylor pulsar   181–183, 185, 188, 189, 195, 197, 198, 211
PSR 1919   110, 111
PSR 1937+21   196, 198, 203
PSR 1953+29   197, 201
PSR 1957+20   198, 199, 203
PSR 2224+65   195
PSR1913+16, Hulse-Taylor pulsar   156
Pulsar
– γ-ray emission   128–133
– acceleration of particles   120
– age   126, 209
– amplitude of the pulse   110
– arrival times of the pulses   110
– binary   156, 181–195, 197, 198, 211, 212
– birthrate   181
– critical equilibrium period line   183
– current   171, 178, 191, 192
– death line   127, 171, 182
– dispersive delay   110, 112
– distance estimation   172, 178
– drifting subpulse phenomenon   113, 125
– electron-positron wind from   122
– evolution   127
– in the Crab Nebula   109, 128, 130–133, 161, 172, 174, 182, 193, 196, 204, 219
– lifetime   127, 171, 181, 202
– magnetic field   121, 122, 126, 171, 202
– magnetic-energy density   118
– magnetosphere   119, 123, 132, 134, 185, 187, 204
– millisecond   171, 196–205, 211, 212
– nulling   110
– number detected   111
– optical emission   128–133, 193
– particle energy density   118
– period   111, 122, 126, 174
– period derivatives   111, 126
– pulse polarization   110, 111, 113, 117, 121, 125, 206
– pulse structure   110, 125
– radio emission   110, 122, 128, 174, 205, 209, 210, 213
– recycled   189, 190, 197
– relativistic wind from   121
– rotational energy   128, 138
– spatial distribution   178–181, 201
– spin up line   183, 188, 211
– spindown rate   126, 138, 139
– velocity distribution   175, 180, 191–195
– X-ray emission   128–133, 187, 217

Quantum liquid state   30, 31
Quasars   311, 314, 320–322

R-region   248
Radiative diffusion coefficient   28
Radiative equilibrium   98
Radiative levitation   45, 46, 48
Radiative mass loss   48
Radiative opacity   29, 53
Radiative transport   28
Radiative zero solution   53, 54
Recycling scenario   189, 190, 197, 202
Reference frame
– chronometric or Killing   269, 272, 273, 278
– comoving   240, 251, 274, 289, 290, 292
– Eddington–Finkelstein   249, 252
– eulerian   272
– freely-falling   298

Index   339

- Lemaitre   246, 247, 249, 252
- Lorentz   271, 285
- of locally nonrotating observers or of zero angular momentum observers (ZAMO)   272–274, 286, 287, 289, 297
- Schwarzschild   241, 242, 244, 247, 258, 267

Roche Lobe   183
ROSAT   167
Ruderman-Sutherland model   124, 126

SAS-2   205, 206
Scaling factor   172
Schwarzschild radius   242
Schwarzschild sphere   242, 245, 247, 248, 250, 251, 254, 258
Semiclosed world   256, 257
Shear vibrations   229
Sonic radius   313
Specific heat   55, 57
- of the electrons   28
- of the ions   31, 53

Stars
- *pre*-white dwarf   15

Eri B   6–7, 26
- born-again AGB stars   40
- failed AGB   4
- G117-B15A   82
- G191-B2B   12
- GD 323   17
- GD 358   76, 89
- GD 378   17
- GD 50   26
- GW Vir   21, 72
- HL Tau 76   61
- PG 0112   12
- PG 1159   15, 16, 18–21, 47, 48, 62, 63, 72–78, 80–81
- PG 1707   74, 76
- PG 2131   74, 75
- Procyon   6
- RX J2117   72, 74, 76
- S 216   17
- Sirius B   6, 26
- van Maanen 2   7
- ZZ Ceti   21, 51, 61, 62, 68, 77, 78, 81–82, 89

Superconductivity   159–161, 207–210
Superfluid velocity   160
Superfluidity   108, 152, 159–163, 166, 207, 213, 221
Supernova   108, 159, 163, 185
Supernova rate   172, 173, 181

Supernova remnants   167, 172–174, 181
Superwind   39
Synchrotron radiation   130–132, 138, 139, 174, 175, 196, 204, 229, 313

T-*region*   248, 250–251, 253
Thermal bremsstrahlung   313
Thermodynamic equilibrium   21
Thompson scattering   188
Time Machines   311–312
Timescale
- for composition change   45
- for crossing the H-R diagram   39, 84
- free–fall time   61
- thermal   88, 89

Timing noise   128, 129
Tolman-Oppenheimer-Volkoff equation   150
Transport of energy
- in degenerate matter   28

Tycho's supernova remnant   176
Type I supernovae   169
Type Ia supernovae   173
Type Ib supernovae   173
Type II supernovae   173

URCA process   166
URCA reactions   164

V404 Cyg   319
Virgo Cluster   85

White dwarfs
- absolute visual magnitude   8
- accretion rate   47
- bolometric magnitude   9
- catalog   7
- chemical evolution   19, 50, 77
- color-magnitude diagram   7
- contraction   53
- convection   48
- cooling age   58, 83
- cooling curve   56, 57, 72, 80
- cooling timescale   2, 54–57, 78, 81
- helium–rich, DB/DO   43, 50–51, 62, 72, 76, 80, 87, 89
- hydrogen–rich, DA   43, 50–51, 80, 87, 89
- lack of helium-rich   12, 16, 51
- luminosity   53
- luminosity function   2, 9–10, 58–60, 82–84

- luminosity-core temperature relation 54
- mass determinations 2
- mass distribution 2, 4–5
- mass loss rate 47
- mass-radius relation 103, 105
- Mestel law 54–57, 82
- population 83
- progenitor evolution 32, 44, 50–51
- progenitor mass 1
- pulsation 20, 33, 60, 61, 71–82, 87–90
- rotation periods 20
- spectra 8, 10
- spectral classification 10–15
- surface composition 10, 43–51
- surface magnetic fields 19

white dwarfs
- Mass-radius relation 25

White holes 253, 257
Whole Earth Telescope 69, 71, 73, 82
Wormholes 311–312

X-ray binary 185, 200–202, 205, 211, 212, 214
X-ray emitting stars 155, 185, 187
X-ray sources 317–320

Zero boundary condition 53
ZZ Ceti instability strip 89–90